An Introductory Path to Quantum Theory

Stephen Bruce Sontz

An Introductory Path to Quantum Theory

Using Mathematics to Understand the Ideas of Physics

 Springer

Stephen Bruce Sontz
Matemáticas Básicas
Centro de Investigación en Matemáticas, A.C.
Guanajuato, Mexico

ISBN 978-3-030-40769-8   ISBN 978-3-030-40767-4   (eBook)
https://doi.org/10.1007/978-3-030-40767-4

Mathematics Subject Classification (2010): 81-01, 46C99, 46N50

This Springer imprint is published by the registered company Springer Nature Switzerland AG
The registered company address is: Gewerbestrasse 11, 6330 Cham, Switzerland

Otra vez a Lilia con mucho amor

Preface

> The greatest mathematicians,
> as Archimedes, Newton, and Gauss,
> always united theory and
> applications in equal measure.
> Felix Klein

This text book is based on my class notes for an introductory quantum theory course attended by advanced undergraduate mathematics students who had no or only very little prior knowledge of physics. However, it is aimed at those from the scientific community, such as advanced undergraduate and graduate students in chemistry, computer science, engineering, mathematics and quantum biology (to name some fields in alphabetic order), who can benefit provided that they arrive with the mathematical portfolio which is described in the following section on Prerequisites. It could serve some of the needs of undergraduate physics students as well. Scientists who have terminated their formal education can also use this book for independent study. And the occasional philosopher may find something of interest, too.

Despite the large number of texts on quantum theory, I could find none attuned to the needs of my students. On the one hand they were not physics students, and so many fine texts were beyond them. On the other hand their mathematics background was minimal, and so slugging through the rigors of functional analysis to do everything 'right' was not an option. Besides, quantum physics as presented here motivates the study of functional analysis and not *vice versa*. They needed clear explanations that explained. And that is how I got motivated to write this book.

The leitmotif of this book is that students in scientific fields who have enough mathematical knowledge can use that to learn quantum theory by seeing how certain mathematical structures are relevant to quantum physics. These structures include the Schrödinger equation, linear operators acting in a Hilbert space, commutation relations, and representations of Lie groups among others. These topics include most, but not all, of the basic material of an introductory physics quantum mechanics course together with some emphasis on the role that calculations play in providing insight. But even more emphasis is put on the underlying physics ideas and meanings and how the mathematics helps in understanding all that. An important point here is that many aspects of quantum theory are notoriously counter-intuitive, and so the mathematics is an anchor point which allows us to organize our thoughts about quantum theory. So there is an essential *unification* of the physics and its mathematics.

Yet make no mistake! This is primarily a physics book, although written in modern mathematical language, and so the reader's kit of mathematical tools will not be prerequisite enough without a healthy dose of interest in physics. Of course, the interest many people have in quantum theory comes from its 'weird' notions and their philosophical consequences. However, I feel strongly that these considerations can not be rationally addressed without first dominating at the very least the physics and mathematics as presented in this book. So another motivation for me was to write down here what I think is the 'theoretical minimum' for understanding quantum theory. But I could well have underestimated; even more may be required.

Also I think that it is important to state ever so clearly that this is not intended to be a text on the mathematical methods used in quantum theory. Of course, such methods can be studied quite independently of any interest in quantum theory itself. But this book is meant to show how the language and the logic of mathematics help us to understand the *ideas* in quantum theory. While at times this objective is not totally distinct from that of a study of mathematical methods, my primary aim is to get a better understanding of the physical ideas that make quantum theory what it is and how those ideas motivate the study of the underlying mathematics. For me quantum theory, such as that of atoms and molecules, motivates the study of the spectral theory of unbounded operators (whatever that is!) and not the other way around. So I feel that the physics should be the starting point, even for mathematicians. And, by the way, this was the historical process, the path traveled by the giants who founded quantum theory. For those interested in mathematical methods in and of themselves there are any number of excellent texts on that important topic.

The organization and emphasis of this book reflect in large part my own path to understanding quantum theory. Since certain topics do not help me or my intuition, I do not emphasize them, though I do try to explain why they are side issues at best for me. But sometimes they appear to me to be unnecessary distractions from the main ideas of quantum theory. This will brand me as a heretic in some eyes. The reader is free to analyze these topics and come to a different conclusion. This is welcome. After all, quantum theory is widely recognized as both a highly successful scientific enterprise as well as a work in progress.

My understanding is that the Schrödinger equation is the central structure in quantum theory in terms of both physics and mathematics. Actually, in the first paper of Schrödinger concerning his eponymous equation there is an idea that I have called *Schrödinger's razor*. His idea is that a partial differential equation *alone* tells us what its solutions are and that quantum theory should be based on such a partial differential equation, its solutions and nothing else. Here is an eminent physicist saying that a physical theory should be based strictly on a mathematical foundation.

The *uncertainty principle* on the other hand is not much needed nor much used in physics. It merits a chapter only because so much has been made of it that I wanted to show how marginal it is. It is an optional chapter, of course. *Complementarity*, including the famous *'wave/particle'* duality, also is rather a side issue at best and a distraction at worst. Therefore, it only deserves another optional chapter to support my opinion. No doubt, many will find this perspective of mine to be heretical. Or worse. The *correspondence principle* is so marginal, having only to do with a relation of quantum theory to classical theory, that I will only refer to it once more. Unfortunately, the principles mentioned in this paragraph, as well as *quantum fluctuations* and *decoherence*, are often used to provide buzz words instead of a valid scientific explanation. I try to avoid such errors and have even included an optional chapter on how to speak (correctly!) about quantum theory.

However, the interpretation of the solution of the Schrödinger equation is so important that it is given in three chapters, the first one for a mathematical interpretation and the second one for a physical interpretation. The latter then feeds back to motivate the use of the abstract mathematical structure of Hilbert spaces in quantum theory. So physics leads to this mathematics whose basics are presented *without proofs* in the first 15 pages of Chapter 9. Then there is a third chapter on the relation of measurement with the solution of the Schrödinger equation. This is part and parcel of what that solution means and so must be included. And this includes the famous collapse (or quantum jump) condition, which is used in the *theory* of quantum information and quantum computation and in *experiments* as a way to prepare known initial conditions. While some do object to one or more of these aspects of the interpretation of quantum theory, it is my intention to present the standard formulation of quantum theory as is. If others wish to change that theory, then they must know what they propose to change, since standard quantum theory has had so much success. Besides beginners are well advised to begin at the beginnings.

I present the harmonic oscillator and the hydrogen atom in as much careful detail as can be tolerated by a beginner. The point is to see Schrödinger's razor in action in concrete examples. The fact that these are *exactly solvable* systems is irrelevant, since all that matters is that everything boils down to understanding the solutions of the appropriate Schrödinger equation. If the only way such understanding can be achieved with currently available techniques is via approximate solutions, then so be it. But the purpose of this book is to aid the reader understand the basics of quantum theory, not to delve into those important technical aspects of approximation theory.

The role of symmetry in quantum theory is a key factor, too. This is first seen here with the hydrogen atom, where the spherical symmetry of the system plays a crucial role in its solution. The topic of symmetry then continues with chapters on angular momentum and its relation with the Lie group $SO(3)$ of the rotations of 3-dimensional space (leaving a point unmoved) and its universal covering Lie group $SU(2)$ (words which we will get to know better). This might give the reader enough motivation to study the general role in quantum theory of Lie groups and their associated Lie algebras, all of which can be found in other texts. One of my favorites is [38], which devotes over 650 pages to what the author refers to as an introduction to the role of groups and representations in quantum theory.

Other topics that are not usually included in an introductory text are presented, such as the Heisenberg picture, the quantum plane, and quantum probability. I do this since it seems to me that this is the right time in one's education to be exposed to these rather basic ideas. Important, yet rather more technical topics such as variational methods, the Hartree-Fock approximation, scattering theory and the Born-Oppenheimer approximation (among many others) can be learned in advanced courses or by independent reading. Another missing topic is a thorough discussion of the quantum theory of multi-particle systems, since the theory of tensor products is needed to do that cleanly. And I did not wish to limit my audience to those readers with a knowledge of that rather abstract mathematical subject. However, there is a brief chapter on bosons and fermions, just enough to get the flavor. Besides, this is something best understood in *quantum field theory (QFT)*, which is also beyond our limited terrain. Actually, this book mainly concerns the non-relativistic quantum theory of one or two body systems. Certainly QFT is an important part of quantum theory, but it is best left to serious study after learning the basics as presented in this and other books.

The role of time in quantum theory remains problematical. This physical observable does not fit well into current quantum theory since it has no corresponding self-adjoint operator, a crucial (though quite technical) detail that you will learn about later. The topic of time in itself merits its own book, but that is not going to be this book. Besides it is a subject open for further research and so is not appropriate for an introductory text. We will take the simple, everyday intuition of time as that irreversible flow embodied by the readings of a clock.

Another topic that will not be considered in any detail is the famous two-slit experiment. While this is an important experiment that defies both common sense and classical physics, it is not sufficient for justifying all of quantum theory. Think of justifying one's love for operatic arias by appealing only to Nessun Dorma. One over-performed aria is not able to demonstrate the value of an entire art form. Not even the greatest aria could do that. Think of the two-slit experiment as an over-invoked example of the success of quantum theory. And since there are a multitude of expositions of it, the reader is well advised to look for the best of them, keeping in mind that not all performances of Nessun Dorma are all so great. And remember that quantum theory has an enormous number of successes which explain things that classical physics does not. Only a paltry fraction of those successes is mentioned in this book. Think of this book as your first flight on an airplane. I want you to see how it flies and how things look from this new viewpoint. But it won't be a world tour, and so some standard important destinations will be missed.

There is a long chapter about the specific probability theory associated with quantum physics, known as *quantum probability*. I have various reasons for doing this. First, this non-deterministic aspect of quantum theory seems to me to be inescapable despite some famous strong objections. Second, I think that many, if not all, of the counter-intuitive aspects of quantum theory arise from its probabilistic interpretation. So it is best that we understand how this works. Actually, I think that this is the most important chapter in the book. Moreover, this was the first example historically speaking of a *non-commutative probability theory*, which has become an independent and quite active area of contemporary mathematical research.

I feel that an axiomatic approach is appropriate for quantum theory. But I describe that in the penultimate chapter, since the axioms are meant to express the most fundamental principles that have been introduced in the course of the book. So my exposition is not the axiomatic approach as that is usually understood in, say, Euclidean geometry. This seems to be the best way to do this, since the beginner has no preestablished intuitions about quantum theory in contrast to our innate spatial intuitions. Physicists tend to regard axioms as 'too mathematical' which is an understandable attitude. But strangely enough, quite a few mathematicians also dislike the axiomatic method when applied to physics. Nonetheless, D. Hilbert thought that the axiomatization of physics was important enough to be included in his famous list of *mathematical* problems for the 20th century. Since then opinion has diverged. Some think that the problem is not well stated and others that it is no longer scientifically relevant. Even others think the problem is both important and unsolved. Count me in the last group. Of course, antagonists to this approach can easily skip the chapter explicitly about axioms and remain unaware of what the list of the axioms is and just carry on. This is sometimes referred to as the philosophy of "shut up and compute". While this can keep one going for a long time, even for a lifetime (all of which is fine), I wish to advocate for "speak up and comprehend" as well.

The very last and shortest chapter concerns the enigmatic relation of quantum theory to modern gravitational theory as expounded in general relativity. It is one important indication that physics is an ongoing activity.

The book ends with Appendix A, a 14-page crash course on measure theory. This is done not only to make this presentation more self-contained, but also to convince the reader that the amount of material in measure theory that is essential for doing physics is quite small. And the only goal is simply to get a decent theory of integration. Actually, even this brief treatment of measure theory is probably more than twice what you might really need. However, I decided to put in the extra slack to give somehow a bit more context to the theory. Feel free to pick and choose. My only advice is that it is best to avoid the famous texts by the famous authors. They have a different agenda and so go into details of no interest for us.

Now a rather interesting aspect of the axiomatization of quantum theory is that it is almost certainly incomplete. Even though it is nearly a century old, the foundations of current quantum theory remain a work in progress. While writing this book, word of new experimental and theoretical research on the foundations of quantum theory reached me. It is quite hard to keep up with it all. I think the time has not yet arrived for a definitive monograph on quantum theory. This is a small part of the motivation for writing this book, namely that it is worthwhile to carefully write down what is widely accepted in order to see if it can indeed endure further scrutiny! No doubt some of the 'axioms' presented here will become as obsolete as the 'rule' prohibiting the splitting of an infinitive. Here are some of the puzzling, unanswered problems about quantum theory. And there are others. But let the reader be warned that all of these are topics for other books and for further research projects.

- Why do some observables, such as the mass of an electron, have only one value, and how can this fit into a probabilistic theory?

- How can observables of quantum systems be measured by 'classical' systems (which give us one unique reading), when in fact all systems, including the measuring system, are quantum systems (which can give many non-unique readings)?

- Can one even treat time as an observable in the context of quantum theory or some generalization of it? If so, how? If not, why not?

- Behind the probabilistic interpretation of quantum theory is there some other more intuitive, though not necessarily deterministic, theory which makes experimentally accessible predictions that distinguish it from standard quantum theory?

As a course textbook, Chapters 1 to 15 could be used for a one-semester course, either at the advanced undergraduate or graduate level, for students in any scientific field, provided they have the mathematical prerequisites described in the next section. If time permits some of the remaining chapters could be included too. I like to include some of the chapter on uncertainty, since the students already have that buzz word in their collective bonnet. The chapter on the Heisenberg picture is useful too, since it gets one to think outside of the box about what is 'really' moving in a quantum system.

It has become a cliché, but it bears repetition: Doing the exercises is an integral part of the learning process. The relevant verbs in the imperative in order of their importance are: understand, calculate, and prove. The first of these is behind all the exercises. Some problems will be too hard for you. They then serve as motivation for studying from other sources. Above all, do not be discouraged but learn to carry on. No illustrations are given. If the reader feels the need for them, then making them becomes an implicit exercise, too.

Students of all sorts—whether they happen to be registered in a course or not—are encouraged to use this book as a springboard for further studies. Be it in physics, mathematics, engineering, biology, chemistry, computer science, or philosophy, the literature on quantum theory is enormous. However, for those who wish to continue to the frontier of scientific research, please note that this book will not take you there, though it can be the start of a longer path to get you there. In that regard I especially like [20] for its wonderful mix of mathematical techniques and physical ideas. It can get you closer to that frontier. And it is chock full (in over 1000 pages!) with many topics not even mentioned here. But whatever your future interests may be, have fun!

My understanding of quantum theory comes from innumerable sources, including many colleagues whom I have known over many years as well as authors whom I have never met. If I were to make a list, it would surely be incomplete. Nonetheless I do greatly thank all those unnamed scientists in one fell swoop. But having named no one, it follows that I can not lay blame on anyone for my shortcomings, inconsistencies, minor mistakes, or major errors. I would dearly appreciate any and all comments on those flaws for which I, and I alone, bear the responsibility.

I do greatly thank and appreciate the support of Springer in bringing this work to the public. I especially thank my editor Donna Chernyk.

Some of my colleagues might disagree with some of my statements or even on the rational discourse I give to support them. We might disagree as to the meaning of 'rational discourse'. So be it. Please make your scientific arguments, and please let me make mine. I do try to base my assertions on the current state of the experimental science of physics, though I may fail in that regard. For one thing, experimental physics is an ongoing activity! And for that matter so is rational discourse.

Guanajuato, Mexico
November 2019

Stephen Bruce Sontz

Prerequisites

If I have seen further than others, it is by
standing upon the shoulders of giants
Isaac Newton

Since the 17th century scientific descriptions of physical phenomena have been encoded in equations, and so the reader must deal with the math behind those equations. Since then mathematics, and especially its role in physics, has been a growth industry. The reader will need at least the following prerequisites in mathematics to understand this book. Also you must think symbolically and logically with a gumption and dedication for learning.

- How to deal with the four arithmetic operations (addition, subtraction, multiplication, and division) of these number systems and how they relate to each other: positive and negative integers, real numbers $\mathbb{R}$ (that is, decimal numbers) and complex numbers $\mathbb{C}$, including $i = \sqrt{-1}$. Know what is the absolute value and the triangle inequality in each system. Know what is a square root, although not necessarily how to calculate it. Know how to extend $\mathbb{R}$ with the two 'infinite' symbols $+\infty$ and $-\infty$.

- How to manipulate algebraic expressions, including polynomials, using the four arithmetic operations. How to solve a quadratic equation.

- The concepts of set, subset and empty set as well as the notation used to describe sets. Cartesian product, union, intersection and complement. Finite set, countably infinite set. Functions between sets, the arrow notation used for functions and what it means for a function to be: one-to-one, onto, bijective. What is a relation, a partial order and, more specifically, an equivalence relation and the set of equivalence classes of an equivalence relation.

- Know what are the trigonometric functions sin and cos, the exponential function to the base e and the identity $e^{i\theta} = \cos\theta + \mathrm{i}\sin\theta$.

- Some analytic geometry including the relation of the plane Euclidean geometry with the Cartesian plane $\mathbb{R}^2$. The equation of a circle in $\mathbb{R}^2$. The relation of the 3-dimensional Euclidean space with the Cartesian space $\mathbb{R}^3$. The equation of a sphere in $\mathbb{R}^3$.

- The basic rules of calculus of functions of one real variable. How to take a derivative of elementary functions and how to evaluate some simple integrals, both definite and indefinite. Finite and infinite definite integrals. The Leibniz (or product) rule, the chain rule, integration by parts and change of variable formulas. What an ordinary differential equation (ODE) is and how to recognize a solution of one without necessarily knowing how to solve it. Know that linear ODEs have solutions and how initial conditions give a unique solution.

- Some basics of multi-variable calculus such as what a partial derivative is and what a multiple integral is. What a partial differential equation is and how to recognize a solution of one without necessarily knowing how to solve it. Some vector calculus such as grad and curl.

- Familiarity with limits, sequences, infinite series, convergence.

- Linear algebra with scalars in $\mathbb{R}$ or in $\mathbb{C}$. Vector space, linear map, linear functional, kernel, range, linear independence, basis. Relation between linear maps and matrices. Dimension of a vector space. Know these operations of matrices: matrix multiplication, determinant, trace. Invertible matrix and its inverse. Eigenvectors and eigenvalues. Know what is an eigenvalue problem and what its solution consists of. Know what isomorphism and isomorphic (denoted $\cong$) mean for vector spaces.

- Some words from topology such as open, closed, dense and boundary.

As for physics the only prerequisite is curiosity, probably having to do with a desire to understand quantum 'weirdness', plus some notion of what velocity is (such as that experienced in cars) and that mass is somehow related to weight. One should understand that energy is the coin of the realm in nature, that energy is what makes things move and that it comes in many forms that are interchangeable to some extent. You need not know how energy is represented mathematically. No prior physics study is required.

Contents

Notations and Abbreviations

Symbol	Meaning
$\mathbf{a}$ or a	Acceleration, the second time derivative of position as a function of time.
a_0	The Bohr radius.
$a \in A$	The element a is in the set A.
$a \notin A$	The element a is not in the set A.
$[a, b]$	The set of real numbers x satisfying $a \leq x \leq b$.
(a, b)	The set of real numbers x satisfying $a < x < b$.
$a \sim b$	The number a is approximately equal to b.
$\{a\}$	The set with exactly the one element a.
$A \backslash B$	The set of elements in the set A but not in the set B.
A^t	The transpose of the matrix A.
$[A, B]$	Commutator of the operators A and B, namely $AB - BA$
$A \cup B$	The union of the sets A and B.
$A \cap B$	The intersection of the sets A and B.
$B = (B_1, B_2, B_3)$	The magnetic field.
$\mathcal{B}(\mathbb{R})$	Borel subsets of $\mathbb{R}$.
°C	Degrees of temperature in Celsius.
C^2	Functions with continuous partial derivatives of all orders ≤ 2.
C^∞	Functions which have partial derivatives of all orders.
C_0^∞	C^∞ functions f identically zero in the complement of some compact set (that depends on f).
$\mathbb{C}$	The set of complex numbers.
$\mathbb{C}^n$	The complex Euclidean space of complex dimension n.
CCR	Canonical commutation relations.
curl	The curl (of a vector-valued function).
det	Determinant.
dim	Dimension.
$Dom(A)$	Domain of a densely defined operator A.
e	Electric charge on an electron.

e	Real number (about 2.818), which is the base for exponentials and (natural) logarithms.
e	The identity element of a group.
E	Energy.
E	Quantum event.
$E \wedge F$	The infimum of the quantum events E and F.
$E \vee F$	The supremum of the quantum events E and F.
$\mathcal{E}(\cdot)$	Expected value.
E_A	The pvm associated to the quantum observable A.
$\exp M = e^M$	The exponential of a matrix M.
$\mathbf{F}$	Force.
$f : S \to T$	f is a function that maps elements of the set S to elements of the set T.
$f^{-1}(t)$	$\{s \in S \mid f(s) = t\}$, where $f : S \to T$. This is called the *inverse image* of t by f.
f_+	The positive part of a real-valued function. N.B: $f_+ \geq 0$.
f_-	The negative part of a real-valued function. N.B: $f_- \geq 0$.
g	The g-factor in quantum magnetic theory.
G	Newton's gravitational constant.
G	A group.
$Gl(\cdot)$	The general linear group; the group of all invertible linear maps.
grad	The gradient (of a scalar function).
H	Hamiltonian.
$\mathcal{H}$	Hilbert space.
h	Planck's constant.
$\hbar$	Planck's (normalized) constant; $h/2\pi$.
i	The complex number $\sqrt{-1}$.
I	The identity operator, the identity matrix.
$\inf S$	The infimum or greatest lower bound of a set of real numbers S.
k or **k**	Wave number; wave number vector.
$\mathrm{Im}(\cdot)$	Imaginary part.
ker	Kernel of a linear operator.
l	Angular momentum (or azimuthal) quantum number.
$\mathbf{L}$	Angular momentum.
$L = (L_1, L_2, L_3)$	Angular momentum.
$[L]$	Dimensions of length.
$L^1(\cdot)$	Lebesgue space of integrable functions.
$L^2(\cdot)$	Lebesgue space of square integrable functions.
$\mathcal{L}(\mathcal{H})$	The space of bounded, linear operators acting in $\mathcal{H}$.
$L_j(x)$	The Laguerre polynomial of degree j.
$L_j^k(x)$	The associated Laguerre function for integer indices satisfying $0 \leq k \leq j$.

lim	The limit of a function or of a sequence.
m	Magnetic quantum number.
m	Mass.
$[M]$	Dimensions of mass.
m_e	Mass of an electron.
m_p	Mass of a proton.
$Mat(n \times n; \mathbb{C})$	The set of all $n \times n$ matrices with complex entries.
$Mat(n \times n; \mathbb{R})$	The set of all $n \times n$ matrices with real entries.
n	Principal quantum number for the hydrogen atom.
$\mathbb{N}$	The set of non-negative integers $n \geq 0$.
$O(3)$	The Lie group of all 3×3 orthogonal matrices.
$\mathbf{p}$ or p	Momentum.
P	Probability function, probability measure.
P_A	The pvm associated to the quantum observable A.
$P(E\mid\psi)$	The probability that the quantum event occurs given a system in the state ψ.
$P(A \in B\mid\psi)$	The probability that the quantum observable A when measured gives a value in the (Borel) subset B of $\mathbb{R}$ given a system in the state ψ.
$\mathrm{Pow}(S)$	The power set of the set S, i.e., the set whose elements are all the subsets of S.
$\mathrm{Proj}(\mathcal{H})$	The set of all projections in $\mathcal{L}(\mathcal{H})$.
$P_l(t)$	The Legendre polynomial of degree l.
$P_l^m(t)$	The associated Legendre function with integer indices l and m satisfying $\mid m\mid \leq l$.
pvm	projection valued measure.
$p \Rightarrow q$	p implies q, where p and q are propositions.
q_e	The electric charge of an electron.
qevm	Quantum event valued measure; a pvm.
Ran	Range of a linear operator or of a function.
$\mathrm{Re}(\cdot)$	Real part.
Ry	The Rydberg energy.
$\mathbb{R}$	The set of real numbers, the real line.
$\mathbb{R}^+$	The set of non-negative real numbers $r \geq 0$.
$\mathbb{R}^2$	Two-dimensional Euclidean space; the Euclidean plane.
$\mathbb{R}^3$	Three-dimensional Euclidean space.
$\mathbb{R}^n$	n-dimensional Euclidean space.
$\mathbb{S}^1$	The unit circle $\{z \in \mathbb{C}\mid \mid z\mid = 1\}$.
$S^\perp$	The set of all vectors orthogonal to the subset S of a Hilbert space. (Actually, a closed vector subspace.)
$s \cong t$	$\cong$ is an equivalence relation on a set S and the elements s and t in S are equivalent under $\cong$.
$S/\cong$	The quotient set of equivalence classes of the equivalence relation $\cong$ on the set.

$SO(3)$	The Lie group of all 3×3 orthogonal matrices with determinant 1.
$so(3)$	The Lie algebra associated to $SO(3)$.
$SU(2)$	The Lie group of all 2×2 unitary matrices with determinant 1.
$su(2)$	The Lie algebra associated to $SU(2)$.
span(S)	The span of the set S of vectors; the set of all finite linear combinations of elements of S.
Spec(A)	Spectrum of the operator A.
sup S	The supremum or least upper bound of a set of real numbers S.
t	Time.
T	Linear operator, or simply, operator.
T^*	The adjoint of the linear operator T.
$[T]$	Dimensions of time.
Tr	Trace.
U	Unitary operator.
$U(2)$	The Lie group of all 2×2 unitary matrices.
v or v	Velocity.
V	Potential energy.
Var	Variance, dispersion.
$V \perp W$	The vector subspaces V and W of a Hilbert space are orthogonal.
$V \oplus W$	The direct sum of the vector subspaces V and W of a Hilbert space.
$V \cong W$	The mathematical structures V and W (say, vector spaces) are isomorphic.
x or x	Position.
$Y_{l,m}$	The spherical harmonic function for integer indices satisfying $|m| \le l$.
z^*	The complex conjugate of $z \in \mathbb{C}$.
$\mathbb{Z}$	The set of all integers, including zero and the negative integers.
$\aleph_0$	The first infinite cardinal number.
$\delta_{i,j}$	Kronecker delta ($= 1$ if $i = j$ and $= 0$ if $i \ne j$).
∂	Partial derivative.
Δ	Laplace operator, Laplacian.
Δ	Variance, dispersion.
χ_B	Characteristic function of a (measurable) set B.
π	The ratio of the circumference of a circle to any diameter; the irrational number $3.14159\cdots$.
Π	Product over a finite index set.
$\vec{\mu} = (\mu_1, \mu_2, \mu_3)$	The magnetic moment.
μ_B	The Bohr magneton.

μ_L	Lebesgue measure, i.e., length, area, volume, etc., as the case may be.
ν	Frequency measured in cycles per second ($\equiv$ hertz).
Ψ	Wave function.
ψ, ϕ	Wave functions. Elements in a Hilbert space.
$\psi \perp \phi$	The elements ψ and ϕ of a Hilbert space are orthogonal.
ρ	Probability density function.
ρ	Representation.
$\rho(A)$	The resolvent set of the operator A.
σ	Standard deviation.
$\sigma_1, \sigma_2, \sigma_3$	The Pauli matrices.
Σ	Summation over an index set (finite or infinite).
ω	Angular frequency.
ω	An element in a sample space.
Ω	A sample space.
$\emptyset$	The empty set, the set with no elements in it.
∞	Infinity.
∇	The gradient (of a scalar function).
$\{a \mid P(a)\}$	The set of all elements a for which the statement $P(a)$ is true.
$\{a \in A \mid P(a)\}$	The set of all elements a in the set A for which the statement $P(a)$ is true.
$\langle \cdot \rangle$	Expected value.
$\langle \cdot, \cdot \rangle$	Inner product.
$\lvert \cdot \rvert$	Absolute value.
$\lVert \cdot \rVert$	Norm.
$\langle \cdot \rvert$	Dirac bra notation.
$\lvert \cdot \rangle$	Dirac ket notation.
$L := R$	The new left side L is defined to be the already known right side R.
$L =: R$	The new right side R is defined to be the already known left side L.

Chapter 1

Introduction to this Path

Si le chemin est beau, ne nous
demandons pas où il méne.
Anatole France

1.1 A New Physics

By the time of the start of the 20th century physics had settled down to what
we nowadays call *classical physics*, based on Newton's theories of motion and
gravitation as well as on the Maxwell equations concerning electricity and
magnetism. At this point the reader need not be concerned with what these
theories were. Rather what they were not is important. And they were not
adequate for explaining lots of properties of solids, liquids, and gases. Most
dramatically, the data coming from the new science of *spectroscopy*, which
we will describe later, seemed inconsistent with classical physics. And so it
was!

But then M. Planck in his celebrated paper of December 1900 on black
body radiation was lead to introduce the idea that the energy contained in
light sometimes comes in packets given by the famous equation $E = \hbar\omega$,
where E is the energy of a 'packet' of light that is represented by a 'wave' of
angular frequency ω, measured in the standard units of radians per second.
(The frequencies of radio stations are measured in Hertz, abbreviated Hz,
which are cycles per second. One cycle is 360 degrees or $2\pi \sim 6.28$ radians.)
Notice that in this simple equation we see a 'particle' property on the left
side being related to a 'wave' property on the right side, though Planck did

© Springer Nature Switzerland AG 2020
S. B. Sontz, *An Introductory Path to Quantum Theory*,
https://doi.org/10.1007/978-3-030-40767-4_1

not think of it that way. When the dust settles we will have to admit that we are dealing with entities that are neither particles nor waves, though they can have some properties of particles and some properties of waves.

It was already known that classical physics could not explain black body radiation. With his new hypothesis Planck was able to get the correct spectra (each of which depends *only* on the temperature of the black body) as well as to get a good approximation of $\hbar$, which does not depend on the temperature— nor on anything else. So, $\hbar$ was immediately recognized by Planck as a new constant of nature; it is nowadays called Planck's constant.

The challenge to the generation of physicists from 1901 to 1925 was to explain how Planck's hypothesis could be included in a new theory that would extend the classical theory, which had so many wonderful successes, and also accommodate the known, new measurements. It is important to realize that experimental physicists continued, as they always had, to measure physical quantities in laboratories and observatories. They did not stop measuring 'classical' energy in order to turn their attention to measuring 'quantum' energy, for example. Speaking of 'quantum' energy as a new distinct entity is the sort of doublespeak that I hope my readers will learn to avoid. And so it is for all the measured quantities: position, time, velocity, mass, and so on. These are neither 'classical' nor 'quantum' quantities. Rather they are *experimental* quantities that are to be explained by some sound theory.

This new theory had already been named as *quantum mechanics* before it was fully developed. This is not the best of names for this new theory, but we are stuck with it. I prefer to call it *quantum theory* as in the title of this book. Whatever. The first part of the process by which this new theory was produced out of existing classical ideas plus the new role of Planck's equation is known as *canonical quantization*, another unfortunate choice of wording.

This gives a broad overview of how quantum theory arose as a new theory of physical systems. But so far we have kept mathematics on the sideline, most particularly in its modern form (definitions + theorems + proofs). One of the great intellectual achievements of the 20th century was to put a lot of quantum theory on a sound, modern mathematical basis. But that is an abstract, technical development which can discourage many from studying quantum theory. Most mathematicians and other scientists lack an adequate background in physics to be able to appreciate the motivation for the math. This book presents the basic ideas of quantum theory in order to arrive at the accepted, precise definitions. The meaning of these definitions is then amplified in the statements of the theorems which themselves become steps to further ideas. Proofs are of secondary importance at this stage of learning, although they serve as a check that the original ideas have been correctly formulated. Otherwise, at a more advanced level, a feedback loop is initiated by modifying the ideas or the definitions.

One motivation that can be given is that the physicists need this abstract mathematics. But the physicists usually do not want it! The physicists are often happy to proceed using what they know about physical systems from experiment along with a good dose of 'physical intuition'. The latter is mainly mythical. How can one have intuition about quantum processes that defy our everyday experience? And even worse, physical intuition has been known to lead to incorrect explanations. Since there is only one universe (at least locally) it might make more sense if we all "get on the same page", as the saying goes. For example, we see the importance of this mathematical point of view in the rather new fields of *quantum information* and *quantum computing*. A strong interplay between physics and mathematics is nothing new; it is clearly evident already in the scientific works of Archimedes.

In this text we will present how that modern mathematics is relevant to quantum theory. One way to read this book is as a path for studying both physics and mathematics as united disciplines. Though some background in classical physics would be useful, it is not at all required since we will review what we will be needing. And most importantly, absolutely nothing about quantum theory is assumed to be known by the reader. Everything is organized for the most innocent of novices in that regard.

As far as topics go, I cover most of what is in an introductory one-semester course as taught at the college level in a physics department. Those topics are found in the Table of Contents and are described in the Preface.

1.2 Notes

While it is traditional to date the start of quantum theory with the 1900 paper by M. Planck, there were earlier experimental indications that things do consist of discrete entities. I do not refer to those ancient Greeks who advocated the existence of *atoms*, that is, indivisible structures composing all matter. Their idea was not supported by contemporary experimental evidence.

But on the other hand, J. Dalton at the start of the 19th century provided a way of understanding the measured weights of substances before and after chemical reactions in terms of *integer* multiples in the appropriate unit of each basic chemical (nowadays known as an *element*) involved in the reaction. The 'tricky bit' was finding the appropriate unit, which in modern science is called the *dalton*! This is known as the *Law of Multiple Proportions* and lead Dalton to propose the first version of the modern atomic theory, namely that chemicals consist of arrangements of discrete units, the latter leading to the modern idea of an 'atom' of matter, which we now know *can* be divided into pieces. It is interesting to note the comparison of the etymology of the word 'atom' with that of the word 'individual'. Atom derives from the ancient Greek whereas individual derives from Latin. Both originally meant something which can not be divided. In contemporary usage these words are applied to things which can be divided though the resulting parts are considered to be of a different type. For example, an individual frog can be dissected in order to study its anatomy, but none of the structures so identified in its body is considered to be a frog.

Subsequently, in 1869 D. Mendeleev introduced the first version of the *Periodic Table* which is a discrete sequence of the chemical elements, each one of which corresponds to a type of atom. But even the existence of atoms and molecules was a controversial issue that was not settled until the early 20th century. Again, the integers enter into this classification: there is a first element, a second, a third, and so forth. The role of the integers can be regarded with hindsight as essentially 'fingerprints' of quantum theory. However, as noted in Chapter 21, I regard Planck's constant as an essential feature of quantum theory. And Planck introduced his eponymous constant in 1900. The role of the integers as well as the periodicity itself of the Periodic Table are understood as consequences of quantum theory.

I once was talking with a rather clever person, who happened to be a bookseller. He asked me quite unexpectedly to describe what my work is about in one word! My immediate reaction was to tell him that he knew perfectly well that there is no such description. But before I could speak, it suddenly struck me that there was such a word. "Quantum," I said. "Oh," he replied, "I sell a lot of books with that word in the title." And that ended that part of the conversation. This book is intended to introduce the reader to what is the science behind the word 'quantum', which is so overused in popular culture. And which is so important in modern science.

Chapter 2

Viewpoint

> ... our wisdom is the point of view from which
> we come at last to regard the world.
> Marcel Proust

2.1 A Bit of Motivation

There are two famous sayings that describe our knowledge of quantum theory. The first says that in order to learn quantum theory one has to solve 500 problems. The other says that nobody understands quantum theory. There is no contradiction. The point of the first saying is that quantum theory consists of some rules of calculation that one can learn with enough experience. The point of the second saying is that nobody can explain the rules in and of themselves, except that they give a correct description of physical systems. The first says we can *build* some sort of intuition of quantum theory. The second says that this intuition is limited, a least for the current state of human knowledge.

Quantum theory, as any scientific theory, is an attempt to understand and explain, and not just to predict and control, natural events. Being a science, quantum physics is an intellectual discipline as well as a way of observing and experimenting. There have been various traditions for developing the intellectual side of physics, but nowadays there are just two established, living traditions in physics which are represented by two scientific communities centered on active professionals working in academic, industrial, government, and other institutions. These are called physicists and mathematicians. This division is artificial. And lopsided. All physicists work in physics, while only a minority of mathematicians work in what is called *mathematical physics*.

These two communities have so much in common that the general public can not distinguish them. But both communities are well aware that there

© Springer Nature Switzerland AG 2020
S. B. Sontz, *An Introductory Path to Quantum Theory*,
https://doi.org/10.1007/978-3-030-40767-4_2

are quite marked differences. The mathematical physics community tends to entirely ignore aspects of prediction and control and becomes, according to many in the physics community, "too mathematical". In the question period after a conference talk by a mathematical physicist, a (real!) physicist in the audience is apt to ask what are the experimental consequences of the material just presented. Too often the answer is: "I don't know".

On the other hand the physics community tends to produce explanations in an *ad hoc* manner that depends on everything from subjective intuitions to universal principles. Such explanations frustrate those in the mathematics community for their lack of a self-consistent, logical structure. But even worse, such "explanations" can even fail to pass the experimental test. In ancient times it was intuitive that certain objects move naturally in circles while others naturally remain at rest. This intuition is no longer accepted by physicists. "Nature abhors a vacuum" was once accepted as a universal principle by just about everybody who thought about it, but no more.

Each community has some good points to make. And each community could clean up its act. I hope to satisfy both communities for whom this book is intended. And the rest of the scientific world is invited to come along for what I hope is a fun ride. The physics community needs to be more aware that the mathematical, logical structure of quantum theory is an aid and not an obstacle. The mathematics organizes the rules, making some of them more basic while other rules are logical consequences of the basic rules. These basic rules are called, since the time of Euclid, the *axioms*. Some people, even some mathematicians, do not wish to consider the axioms of quantum theory. Why is this so? I suspect for mathematicians this is largely due to their lack of background in physics, so that the axioms ring hollow. So the mathematicians are invited to roll up their sleeves and learn some of the physics that motivates doing the math.

But one does not arrive at a Euclidean exposition without a lot of prior thought. The *Elements*, the textbook that Euclid wrote on geometry and other topics, was a result of a historical process of finding basic principles and deriving their consequences. Then Euclid wrapped it up in a very pretty package. So mathematicians think that what they need is such a pretty package of pure mathematics in order to learn quantum theory. But they are wrong. The mathematical sequence (axiom, definition, theorem, proof, corollary) lacks motivation. What the mathematicians need is the motivation to just start the mathematical sequence; otherwise they quickly lose their way along the 'mathematically correct' sequence. Without proper motivation they do not know what 'it' is all about. It seems that "there is no there there." For the topics in the *Elements* that motivation comes from common experiences in dealing with spatial and numerical relations.

But the motivation for quantum theory comes from attempts to explain some experiments by using rather new ideas. We will start with those ideas that survived and leave unmentioned those that fell by the wayside. We continue by showing how the language of mathematics expresses these ideas and how mathematics can relate and develop these ideas. We consider the ideas behind *first quantization* and how the *Schrödinger equation* fits into that setting, even though it is not a mathematical consequence of that setting. We follow two basic ideas of Schrödinger that we express in contemporary language as:

1. A *partial differential equation* (nowadays known as the Schrödinger equation), and nothing else, via all of its solutions gives us a complete description of a quantum system. I call this idea *Schrödinger's razor*.

2. *Eigenvalue problems* are at the heart of quantum theory.

This leads naturally to the delicate question of the interpretation of a solution, called a *wave function*, of the Schrödinger equation. We do this at two levels: mathematical and physical interpretations. The latter leads in turn to the introduction of a certain, specific *Hilbert space* and, consequently, to the mathematics of Hilbert spaces in general.

We also present the two fundamental physical systems of the *harmonic oscillator* and the *hydrogen atom*. We discuss the quantization of *angular momentum* (of classical physics) and how it leads to the introduction of a new type of quantum angular momentum, called *spin*.

Even though I regard the *Heisenberg uncertainty principle* to play a rather minor role in quantum theory, I discuss it in a lot of detail. There are two reasons for this. The first is that by long standing tradition it is presented in every introductory course on quantum theory. The second is that its rather straightforward probabilistic interpretation is too often confounded with a philosophical discussion about the limits of human knowledge of physical reality. Such a philosophical discussion belongs properly to a consideration of the probabilistic aspect in general of quantum theory, and not just to one small corner of quantum theory. Actually, I feel that it is such a small corner that this topic appears in an optional chapter near the end of the book.

And during all this, I try to maintain a reasonable, although by no means perfect, level of mathematically acceptable rigor, in part not to lose the mathematicians in a flurry of disorganized formulas and also in part to show the physicists that mathematical rigor is a useful razor for excluding wishful thinking—and more importantly for finding *all* of the solutions of a given Schrödinger equation. As mentioned earlier this is *Schrödinger's razor*.

And we will end, rather than start, this exposition with the axioms, which then give us the standard description of quantum theory. I will take the strict positivistic approach that anything that is not inferred from this axiomatic approach is not quantum theory. This is not because I think that all that strays from standard quantum theory is unpardonable heresy. Quite the contrary! I simply want all deviations from the straight and narrow road to be identified clearly as such, because these several different and intriguing possibilities must then be considered:

- We have actually arrived at an affirmation that is only apparently not part of quantum theory, but can be shown to be a logical consequence of the axioms. So our understanding of quantum theory has improved. An example of this is the possibility, as shown by J. von Neumann and E.P. Wigner in [33], of having a Schrödinger operator which has an eigenvalue embedded in its continuous spectrum.

- We have stated an affirmation that is demonstrably in contradiction with quantum theory. In this case it is important to identify exactly the experimental conditions which will give sufficient information for deciding whether quantum theory or the new affirmation is true. If experiment decides in favor of the new affirmation, then we have the intellectual challenge to find a new theory that replaces the currently accepted standard quantum theory with something that explains all previously explained experiments and is also consistent with the new affirmation. Bell's inequalities fall into this rubric.

- We have an affirmation that is logically independent from standard quantum theory, but makes clear experimental predictions. Such an affirmation then must be checked by said experiments; however, the experimental results will neither confirm nor reject quantum theory. Either way quantum theory will have to be modified to include an explanation of the new affirmation. It seems that *grosso modo* this was how spin came to be incorporated into quantum theory. A similar case was the proposal that parity violation could better explain what are now known as neutral kaon decays. Some physicists fiercely rejected this proposal at first as being in contradiction with quantum theory, which it is not. In this case some experiments came first, and they were followed up by other experiments which confirmed parity violation. This lead to an understanding of how the already established quantum theory could be amplified to allow for spin and for parity violation.

- We have an affirmation that is logically independent from quantum theory, but in no way changes quantum theory since it makes no new experimental predictions at all. Such explanations are often justified on the basis that they allow a person to better understand quantum theory. But such a subjective evaluation valid for one person may well be perceived subjectively by another person as being confusing and quite unnecessary. An example of this is the statement that the elementary particles possess the same sort of consciousness as humans have. Yes, I have heard this. Of course, Ockham's razor cuts away such statements.

Almost all statements about the relation between standard quantum theory and human consciousness seem to fall into this category. Such statements can not be said by any stretch of the imagination to be supported nor rejected by quantum theory. For the time being, they remain inaccessible to quantum theory. Human consciousness (as well as any other property of any biological organism, of course) is a part of the natural world, and quantum theory may well be the ultimate fundamental description of the natural world. If all this is true, then we may well expect an explanation of all biological phenomena, including human consciousness, in terms of quantum theory some day. But I guess that will not be very soon.

I hope the reader realizes the utility of analyses along these lines as an integral part of the development of scientific theory in general and of quantum theory in particular. Also, we all tend to think 'classically' in the sense that we believe that to change undesirable results of our actions we must change what we did to get those results. The quantum viewpoint denies this!

2.2 Notes

The two basic ideas that I ascribe to Schrödinger are implicit in his papers of 1926. I have not found them explicitly stated. What I call *Schrödinger's razor* is found in the opening words of the first of his 1926 papers, [27]. My rendition also has an implicit element to it. Explicitly, the solution must be in the Hilbert space associated with the partial differential equation. Also, the *elliptic regularity theorem*, may tell us that the solution must be in a certain subspace of that Hilbert space. The point that Schrödinger makes quite clearly is that integers are not put in at the start, but rather come out naturally by solving the partial differential equation. (It should be noted that during the period 1901–1925 there were proposed explanations that did put in the integers by hand. Schrödinger rejected that methodology.) However, Schrödinger puts conditions on the solution, which we now can understand as guaranteeing the square integrability (a Hilbert space condition) of the solution or as arising from regularity theorems. Of course, Schrödinger was not thinking in 1926 about Hilbert spaces or regularity theorems. All that came later.

The second basic idea comes from the very title of his 1926 papers. (See [27].) While Schrödinger equates quantum theory with eigenvalue problems, that is only strictly true for a part of the *spectrum* of the *Hamiltonian*. (The meaning of Hamiltonian and spectrum will be central topics for us.) However, due to a theorem of H. Weyl, every number in the spectrum of a (self-adjoint) Hamiltonian is an *approximate eigenvalue* as described in the optional Section 9.5. So this sort of generalized eigenvalue problem is indeed at the heart of quantum theory.

I can not emphasize strongly enough that Schrödinger, a physicist *par excellence*, proposed a strict mathematical formulation as the correct basis for quantum theory. However, we must not fall into a cult of personality. Schrödinger has it right not by some inherent birthright, but rather because experiment bears him out. Schrödinger's razor is a guiding principle behind almost all experimental and theoretic quantum physics. Because it almost works. It falls short with the collapse (or quantum jump) criterion as will be discussed in Chapter 10.

In science texts one emphasizes the ideas that have survived to the day and typically does not even mention those that have fallen to the wayside. So it is with Schrödinger's razor, which has been a success even though this particular name appears to be new. It is curious to point out that M. Born in 1924 had a similar idea that has not flourished. He proposed that a discrete, finite difference equation might be the correct way to describe the still incompletely understood quantum phenomena. There is a certain plausibility to the idea that the discrete transitions observed by spectroscopy could be described by discrete equations. But it was not to be.

My understanding is that the two sayings are due to R. Feynman, but that may be folklore.

The paper by Wigner and von Neumann has an error. The potential energy was incorrectly calculated. This error was recognized and corrected in the subsequent literature. Their basic idea still holds with the correction.

Bell's inequalities have generated an enormous literature, both on the experimental and theoretical sides. The interested reader is invited to look into this further.

Chapter 3

Neither Particle nor Wave

Neither fish nor fowl.
Folklore

The basic equation of quantum theory is the Schrödinger equation. It is the dynamical time evolution equation of the theory. It can no more be proved mathematically than the equally famous equation of motion $\mathbf{F} = m\mathbf{a}$ of classical mechanics, which is also called the Second Law of Motion of I. Newton. However, there are certain basic ideas—which become intuitions as time goes by—behind these equations. For example, behind Newton's Second Law is the idea that every material body has a mass (denoted by $m > 0$) which, when it is acted upon by a total force (denoted by $\mathbf{F}$), results in an acceleration (denoted by $\mathbf{a}$) of that material body. And moreover, that is all one has to understand in order to understand the motion of any thing. A material body at rest (so that its velocity is zero and therefore its acceleration $\mathbf{a}$ is also zero) is simply a particular case, which has an enormous importance in engineering. Aristotle's opinion was that motion and rest were quite different properties of a material object. This is not how Newton thought about the physics of motion and of rest.

After viewing the world though "Newtonian glasses" for a good while, one sees everything in motion and everything at rest as the result of forces and masses. Just forces and masses. And Newton's Second Law. And nothing else, except Newton's First Law and his Third Law. One can become quite comfortable with the idea that Newton's equation of motion is a universal law of nature. Indeed, it was considered as such by the scientific community for some 250 years.

© Springer Nature Switzerland AG 2020
S. B. Sontz, *An Introductory Path to Quantum Theory*,
https://doi.org/10.1007/978-3-030-40767-4_3

3.1   Basics of Quantum Theory

This chapter is devoted to discussing what some consider is the most central idea in quantum theory, namely, that all matter and interactions are made up out of basic constituents that are neither particles nor waves in the usual sense of those words. However, those basic constituents are described with some concepts that are associated with particles as well as with some other concepts that are associated with waves. Sometimes these constituents are called *wave/particles* (or *particle/waves*). I feel that it is misleading to say, as it is often done, that these constituents of matter are both particles and waves. I hope that saying that they are neither particles nor waves is a much clearer statement. After all, some bats can swim and fly, but that does not make them *fish/fowls*. Rather, those bats are a different sort of beast with some properties in common with both fish and fowl.

There is a lot of sound and fury used by those who say that the basic constituents of matter are both waves and particles. This is the infamous *wave/particle duality*. Do not feel intimidated by such rhetoric, even when it is encased in a verbal coating consisting of an undefined 'complementary principle.' None of this is actually used in quantum theory *a la* Schrödinger equation. If this sort of vagueness 'helps' some scientists to 'understand' quantum theory, all the more power to them. But I have no idea what they are talking about nor what 'intuition' they get from all that. However, when I say that the basic constituents of matter are neither waves nor particles, that is a statement with a meaning, even though it is a negative meaning. I could just as well have said that the basic constituents of matter are neither goose nor gander. But the way I have said it is intended to make a rhetorical point in opposition to the usual rhetorical clap-trap. I hope the reader will appreciate the spirit with which this *mot* has been presented.

3.2   Black Bodies

Behind all this is the idea that there are many, many more small things than large things, since every large thing is composed of many, many small things. Think of bacteria which are composed of molecules, each of which is composed of atoms, each of which is composed of electrons and nuclei. (Whether this is like a Russian *matryoshka* which eventually has a smallest doll or rather continues in infinite regress is an irrelevant question for this analogy, though most physicists lean toward the first possibility.) And the subsequent idea is that to be able to understand the large things, one has to understand the small things, at least to some degree. And, as it turns out, to understand the small things well enough, classical physics is sometimes not adequate. Although sometimes it is! The birth of quantum theory occurred with a seminal paper by M. Planck published in 1900. In it he was trying to understand a big thing, a type of physical system known as a *black body*. To achieve this he introduced for the first time in physics an inkling of the idea that something, in this case light, could have both a particle property as well as a wave property.

First off, a black body is not what you might expect it to be. A black body is some material object in *thermal equilibrium* with its environment. By 'material object' I mean some combination of chemical compounds, i.e., ordinary matter. The way to produce a black body in the laboratory is to heat a sealed, enclosed oven to some temperature and allow enough time for its interior and its contents to arrive at thermal equilibrium, that is to say, everything inside is at the same *temperature*. Then one quickly opens and closes a door in the oven in order to let out whatever electromagnetic energy (possibly light) might be inside. This is done so quickly that the thermal equilibrium inside the oven is essentially left unchanged. Then a *spectral analysis* is done on the emitted light. This means that the light is spread out, say by a prism, into the rainbow of its constituent colors (*frequencies*), and the intensity of the light for each color is measured. One next produces from this data the graph of the intensity as a function of the frequency. This type of analysis of light is called *spectroscopy.*

Then there is a mind-boggling result, namely that this graph has a shape that *only* depends on the interior temperature of the oven. Stated more strongly, the shape of the graph does not depend on the nature of the objects inside nor on the walls of the oven. You can not see any particular thing inside the oven because the emitted light coming out of the oven in any direction is identical to the emitted light coming out in any other direction. The resulting graph is called the *spectrum of black body radiation* at the given temperature.

There is a black body that is a common object of everyday experience. It is the Sun. This might seem absurd since the Sun emits enormous quantities of light, a small fraction of which in turn powers the weather and biological systems here on Earth. However, this emitted energy is a minute fraction of the energy in the *photosphere*, the visible surface of the Sun. So it is a trickle, that is, it does not effectively change the energy of the photosphere by much. (That lost energy is replenished by the energy coming up from the lower regions of the Sun.) And, sure enough, the measured spectrum of Sun light closely agrees with that of a black body with temperature of approximately 5500°C. This is how we can measure the surface temperature of the Sun—and of any star!

This is all well and good, but how do we explain this? The physicist W. Wien came up with a description of the black body spectra using the then known (so-called *classical*) physics. The problem was that it totally disagreed with the experimentally measured spectra. Then M. Planck published a watershed paper in December 1900 in which he could explain these spectra, provided he made an assumption in contradiction to classical physics. He assumed that the oven walls could absorb and emit electromagnetic energy (in the form of light) inside the oven only in discrete packets, each of which contained the quantity of energy E given by *Planck's equation,*

$$E = \hbar\omega = h\nu,$$

where ω is the angular frequency (measured in units of radians/second) of the transferred electromagnetic wave component. Similarly, ν is the (usual) frequency measured in cycles/second = Hertz. The constant h does not depend on the temperature of the black body and so is a new universal constant of nature. Strictly speaking, h is called Planck's constant. But the form $E = h\nu$ is rarely used anymore. Since 1 cycle is 2π radians, we see that $\omega = 2\pi\nu$. Consequently,

$$\hbar = \frac{h}{2\pi}.$$

Nowadays $\hbar$ rather than h is often called *Planck's constant*. As we shall see later when discussing angular momentum, it is more fundamental than h in some sense.

In 1905 Albert Einstein refined Planck's hypothesis by introducing the idea that $E = \hbar\omega$ is a universal formula for all electromagnetic radiation, including light as a special case. But now the formula is used to describe a *photon*, which is a discrete 'particle' packet of electromagnetic radiation of energy E associated with a 'wave' of angular frequency ω. This idea is really the first explicit reference to the composite 'wave/particle' nature of the constituents of physical systems, though Planck had the germ of this idea. With this photon hypothesis Einstein was able to explain the *photoelectric effect*. In 1921 Einstein was awarded the Nobel prize in physics for this work, and not for his theory of special relativity, also published in the same year 1905. Can you believe it?

After having introduced this new constant and having the experimental data of the spectra of black bodies at various temperatures, Planck was able to fit the data with a specific value of h. These fits were judged to be good. How that numerical analysis was done in 1900 is beyond my ken. But anyway Planck got a numerical value for h, which is not half-bad. Nowadays extremely accurate measurements of $\hbar$ have been made.

3.3   Dimensions and Units

Since $\hbar$ was a new quantity in physics, Planck knew that it was important to identify its dimensions of measurement. (Since 2π is dimension-less, the dimensions of h are the same as those of $\hbar$.) We can compute these dimensions directly from $E = \hbar\omega$ as follows:

$$\dim[E] = \dim[\hbar]\,\dim[\omega]$$

Recall that three basic dimensions are length, time, and mass, whose units in the *Système international d'unités (SI)* are meters, seconds, and kilograms, respectively. The notation for these dimensions are, respectively, $[L]$, $[T]$, and $[M]$. The dimensions for a dimension-less quantity (that is, a pure number) are denoted by $[\,]$. One can use any formula for an energy in order to compute $\dim[E]$. One such famous formula is $E = mc^2$, where m is a mass and c is the speed of light. Speed is a real number (or *scalar*), namely, the magnitude of a velocity vector, the derivative of a position vector with respect to time. So we see that dim $c = [L] / [T] = [L][T]^{-1}$ and therefore

$$\dim[E] = [M][L]^2[T]^{-2}.$$

This gives us an incredibly important result which we will see again:

$$\dim[\hbar] = \dim[E]\,(\dim[\omega])^{-1} = [M][L]^2[T]^{-2}[T] = [M][L]^2[T]^{-1}. \qquad (3.3.1)$$

But for now, we want to take note of what Planck had already noted in 1900. There were already identified two other fundamental constants of physics: the speed of light c as noted above and the constant G of Newton's law of gravitation. One can also compute the dimensions of G in terms of the three basic units $[L]$, $[T]$, and $[M]$.

Exercise 3.3.1 *Compute* dim G.
Hint: $G m_1 m_2/r^2$ *is the gravitational force between two objects of masses m_1 and m_2 with distance r between them. Also, force F is related to mass m and acceleration a by Newton's second law, $F = ma$. Finally, acceleration is the second derivative $x''(t)$ with respect to time t of the position $x(t)$ as a function of time.*

Next, Planck noted that certain combinations of the dimensions of c, G and $\hbar$ give the dimensions of $[L]$, $[T]$, and $[M]$, respectively.

Exercise 3.3.2 *Find real numbers α, β, γ such that* $\dim[c]^\alpha \dim[G]^\beta \dim[\hbar]^\gamma$ *is equal to i)* $[L]$, *ii)* $[T]$, *iii)* $[M]$.
   Then for each of these three cases compute the numerical value of $c^\alpha G^\beta \hbar^\gamma$ in the units of the SI.

The first case of this exercise gives us the *Planck length*, the second case gives the *Planck time* and the third case gives the *Planck mass*. This can be extended to other physical quantities. For example, there is *Planck energy, Planck frequency, Planck density* as well as Planck units for electromagnetism and thermodynamics. It is widely accepted that these Planck quantities have some sort of importance in fundamental physics, but no one knows what that might be exactly. So quantum theory starts with a still unsolved puzzle.

3.4 Notes

The prehistory of current quantum theory begins with Planck's paper in 1900 and ends with Heisenberg's in 1925. This time period is discussed in detail in many books, but here I have only used some of the basic ideas that have survived. The terminology *quantum mechanics* has also survived from that period and has become an immovable object. I do not like this terminology first since not all measured quantities come in discrete quanta and second since mechanics in classical physics is the science of motion. And quantum mechanics is not exactly a science of motion, but of another type of change. I prefer *quantum theory*, but that too has the overworked word 'quantum' in it. However, the pre-1925 theory did not disappear overnight. For example, I have heard that when Enrico Fermi arrived at the University of Chicago there still was a physics professor there teaching the 'old quantum mechanics' from before 1925.

The spectra of black bodies as well as of many heated materials were measured in the 19th century first by prisms and later in the same century more accurately due to the development of precision machined diffraction gratings. This new science of *spectroscopy* generated an enormous amount of data that classical physics could not explain. The structure of the visible light from the Sun and other stars was studied extensively, including the dark lines which were soon realized to be due to absorption in the stellar atmosphere. The history of spectroscopy and how it led to quantum theory is nicely summarized in Appendix F of [29].

The wonderful idea that every equation in physics can be analyzed in terms of physical dimensions seems to have been missed by many a physicist, starting at least as early as C. Huygens who seems to be the first to use equations in physics. I have heard any number of stories of a collaboration of a mathematician and a physicist, where the mathematician comes up with a formula which the physicist rejects at first sight. The mathematician then objects that his physicist colleague has not even read the long, rigorous proof. "I don't have to," responds the physicist. "The dimensions are wrong!"

Graphs of the classical (Wien) and quantum (Planck) spectra for black body radiation (and the measured spectra) are on the Internet.

Yes, surprisingly enough, some bats are rather good swimmers, and they are mammals, neither fish nor fowl. The ultimate irony of the 'wave/particle duality' is that the basic constituents of matter are nowadays usually called *particles*, though the meaning of this terminology is remote from the idea of very small dust particles, say. These *elementary particles*, which it seems are never called *elementary waves* nor anything else, are described by the *Particle Data Group* in periodic updates available on the Internet. The most commonly known elementary particles (in historical order of discovery) are the electron, the photon, the proton, the neutron, and the neutrino.

During my student days I once commented to a fellow student (but I forget who) that I did not understand what 'wave/particle duality' really means. The response, after a reflective pause, was that a quantum system is described by a wave function, a concept which we will be getting to shortly. That's a good answer, but we are going to spend three chapters interpreting the wave function!

The idea that the fundamental constituents of matter are neither particles nor waves is not new at all. There is a discussion of this in 1963 in chapter 37 of the first volume [12] of the Feynman Lectures on Physics.

Chapter 4

Schrödinger's Equation

> Nature is the realization
> of the simplest conceivable
> mathematical ideas.
> Albert Einstein

Some physicists will tell you the Heisenberg uncertainty principle is the most basic aspect of quantum theory. This is clearly wrong. The uncertainty principle as Heisenberg originally proposed is not even an exact quantitative statement; it basically says some sorts of situations can not happen. Later it was cast as an *inequality*. But no inequality can be used to predict what will occur in nature, although it can predict what can not occur.

The basic principle of quantum theory simply has to be a time evolution *equation*. And that is exactly what Schrödinger's equation is, and what the Heisenberg uncertainty principle is not. And just as Newton's equation of motion starts by describing a massive point particle, Schrödinger's equation starts by describing a massive 'wave/particle'.

We start with Planck's equation $E = \hbar\omega$, which we now use as a general relation for a 'wave/particle' such as a photon or an electron. Since energy zero states seem to be undetectable, we take $E \neq 0$ or equivalently $\omega \neq 0$.

But photons also have *(linear) momentum* as has been measured. We write $\mathbf{p} = (p_1, p_2, p_3) \in \mathbb{R}^3$ in general for the momentum vector. For classical massive point particles with mass $m > 0$, this is defined by $\mathbf{p} = m\mathbf{v}$, where $\mathbf{v} = (v_1, v_2, v_3) \in \mathbb{R}^3$ is the velocity vector of the particle. But the momentum $\mathbf{p}$ of a photon can not be given by $m\mathbf{v}$. Firstly $m = 0$ for a photon and secondly $||\mathbf{v}|| = c$, the speed of light, is a universal constant.

In analogy with $E = \hbar\omega$, we are lead (following an idea of L. de Broglie) to write for photons and for massive particles the formula

$$\mathbf{p} = \hbar\mathbf{k},$$

© Springer Nature Switzerland AG 2020
S. B. Sontz, *An Introductory Path to Quantum Theory*,
https://doi.org/10.1007/978-3-030-40767-4_4

where the vector $\mathbf{k} \in \mathbb{R}^3$ needs to be better understood. In analogy with energy, it seems that momentum zero states are also undetectable, and so $\mathbf{p} \neq 0$ or equivalently $\mathbf{k} \neq 0$.

(By the way, from the theory of special relativity, we know that $(E, \mathbf{p}) = (E, p_1, p_2, p_3)$ is a 4-vector if we choose our units so that $c = 1$. Even if you do not understand special relativity, the only point here is that energy is naturally associated with linear momentum and not with velocity. Explicit reference to the mass, whether zero or non-zero, is avoided this way.)

What are the dimensions of $\mathbf{k}$? Clearly,

$$\dim[\mathbf{k}] = \dim[\mathbf{p}] \dim[\hbar]^{-1}.$$

We recall from (3.3.1) that

$$\dim[\hbar] = \dim[E] \dim[\omega]^{-1} = \left([M][L]^2[T]^{-2}\right)[T] = [M][L]^2[T]^{-1}.$$

Since $\hbar$ is a fundamental constant of nature, this result is rather important. For example, one question about this is what quantities in physics have the dimensions $[M][L]^2[T]^{-1}$. We will come back to this curious consideration. But for now we continue as follows:

$$\dim[\mathbf{k}] = \dim[\mathbf{p}] \dim[\hbar]^{-1} = \left([M][L][T]^{-1}\right)\left([M]^{-1}[L]^{-2}[T]\right) = [L]^{-1}.$$

So, $\mathbf{k}$ has dimensions of inverse length in close analogy with ω, which has dimensions of inverse time. (In special relativity with the speed of light $c = 1$, we have that $(t, \mathbf{x}) = (t, x_1, x_2, x_3)$ is also a 4-vector, where t is time and $\mathbf{x} = (x_1, x_2, x_3) \in \mathbb{R}^3$ is an ordinary spacial vector.)

In fact the cosine wave, $\cos(\omega t)$ as a function of t, where t has dimensions of time makes sense, since then $\dim(\omega t) = [\,]$, i.e, ωt is dimension-less. Recall that is our units, ωt has the units of radians. So we can use the rules of calculus for $\cos \theta$, where θ is a dimension-less number whose units are radians. By the way, the radian is a unit in the SI. We remind the reader that $\cos (Q)$ does not have a sensible meaning if Q has non-trivial dimensions.

Of course, we can also associate to ω the sine wave function, $\sin(\omega t)$. So the parameter ω in Planck's equation $E = \hbar \omega$ can be associated to either of the waves $\cos(\omega t)$ or $\sin(\omega t)$. But for many reasons (simpler formulas, relation to Fourier analysis) we prefer to use the complex wave where $i = \sqrt{-1}$:

$$e^{-i\omega t} = \cos(\omega t) - i \sin(\omega t).$$

(The reader might wonder why the minus sign is used. Well, the minus sign is just a convention, though a very standard one. If I had used the plus sign, would you wonder why I used that equally valid convention?) Notice that this wave has a *periodicity* in time. Its *period* is $T_p = 2\pi/\omega$, that is,

$$e^{-i\omega(t+T_p)} = e^{-i\omega t}$$

and no other positive number strictly smaller that T_p has this property. Recall that $\omega \neq 0$.

A similar, 3-dimensional analysis works for $\mathbf{k} = (k_1, k_2, k_3) \in \mathbb{R}^3$, to which we associate the complex wave $e^{i\mathbf{k}\cdot\mathbf{x}}$ as a function of $\mathbf{x} = (x_1, x_2, x_3) \in \mathbb{R}^3$. Here $\mathbf{k} \cdot \mathbf{x} = k_1 x_1 + k_2 x_2 + k_3 x_3$, the inner product of the vectors $\mathbf{k}$ and $\mathbf{x}$. Then this wave has a periodicity in space. Its period is $\mathbf{x}_p = (2\pi/k^2)\,\mathbf{k}$, namely

$$e^{i\mathbf{k}\cdot(\mathbf{x}+\mathbf{x}_p)} = e^{i\mathbf{k}\cdot\mathbf{x}},$$

where $k^2 = ||\mathbf{k}||^2 = \mathbf{k} \cdot \mathbf{k}$, the usual Euclidean norm squared, is a standard notation. We call $\mathbf{k}$ the *wave vector*. Its norm $||\mathbf{k}||$ is called the *de Broglie wavelength*. The period $\mathbf{x}_p$ is *not* unique since $\mathbf{x}_p + \mathbf{y}$ has the same periodicity property provided that $\mathbf{k} \cdot \mathbf{y} = 0$. So we can chose $\mathbf{y}$ to be any vector in the two-dimensional subspace orthogonal to $\mathbf{k}$. However, $\mathbf{x}_p$ is the vector with smallest length as well as being parallel to $\mathbf{k}$ with the periodicity property. Recall that a vector $\mathbf{w}$ is *parallel* to $\mathbf{k}$ if $\mathbf{w} = r\mathbf{k}$ for some real number $r > 0$.

So $Ae^{i\mathbf{k}\cdot\mathbf{x}}$ is a 'wave' representing a 'particle' with momentum $\hbar\mathbf{k}$, provided that A does not depend on $\mathbf{x}$. And $Be^{-i\omega t}$ is a 'wave' representing a 'particle' with energy $\hbar\omega$, provided that B does not depend on ω.

Finally, by taking a product, we see that $e^{i\mathbf{k}\cdot\mathbf{x}} e^{-i\omega t} = e^{i(\mathbf{k}\cdot\mathbf{x}-\omega t)}$ is a 'wave' representing a 'particle' with momentum $\mathbf{p} = \hbar\mathbf{k}$ and energy $E = \hbar\omega$. Yet another way to write this fundamental 'wave' is

$$e^{i(\mathbf{p}\cdot\mathbf{x}-Et)/\hbar} = e^{-i(Et-\mathbf{p}\cdot\mathbf{x})/\hbar}.$$

(The expression $Et - \mathbf{p}\cdot\mathbf{x}$ arises naturally in special relativity, while $Et + \mathbf{p}\cdot\mathbf{x}$ does not.) The 'wave/particle' in all of these statements could be any massive particle, such as an electron, a proton, a neutron, etc., as well as the mass zero photon.

4.1 Some Classical Physics

Now we need an interlude in classical mechanics. Suppose that the *curve* $\gamma : (-b, b) \to \mathbb{R}^3$ for some $b > 0$ solves Newton's equation of motion, namely $\mathbf{F} = m\mathbf{a}$ for some mass $m > 0$ and acceleration $\mathbf{a} = \mathbf{x}''(t)$, the second derivative with respect to t of the position $\mathbf{x}(t)$. Here the variable $t \in (-b, b)$ represents *time*. By the way, any such solution γ is called a *trajectory*. We also assume that $\mathbf{F} : \mathbb{R}^3 \to \mathbb{R}^3$. We say that $\mathbf{F}$ is a *force field*, which itself is a special case of a *vector field*. This says that the force at any point $\mathbf{x}$ in space depends only on that point $\mathbf{x}$, nothing else. This is a big, big assumption. It means that the force does not depend on the velocity of the classical particle, its temperature, the ambient pressure, the time, etc. Now we make an even more restrictive assumption. We assume that there exists a scalar valued function $f : \mathbb{R}^3 \to \mathbb{R}$ such that $\mathbf{grad}\, f = \nabla f = \mathbf{F}$. Recall from calculus that the gradient of f is defined by

$$\mathbf{grad}\, f = \nabla f := \left(\frac{\partial f}{\partial x_1}, \frac{\partial f}{\partial x_2}, \frac{\partial f}{\partial x_3} \right).$$

If such a function f exists, we say in mathematics that $\mathbf{F}$ is *exact*. (Or more precisely, that the 1-form associated to the vector field $\mathbf{F}$ is exact.) If such a function f exists, we say in physics that the force field $\mathbf{F}$ is *conservative*. A necessary condition for $\mathbf{F}$ to be *conservative* is that $\mathbf{curl}\,\mathbf{F} = \nabla \times \mathbf{F} = 0$. If this is so, we say that $\mathbf{F}$ is *closed*. This is so, since we have the calculus identity

$$\nabla \times (\nabla f) = 0,$$

provided that f is a C^2 function. Here we are using:

Definition 4.1.1 *A function which has continuous partial derivatives of all orders ≤ 2 is said to be a C^2 function.*

In general, $\nabla \times \mathbf{F} = 0$ is not a sufficient condition for $\mathbf{F}$ to be conservative.

Exercise 4.1.1 *Without looking in a book, find a closed vector field $\mathbf{F}$ which is not exact.* **Hint:** *The domain of $\mathbf{F}$ must be some proper open subset of $\mathbb{R}^3$.*

Now when the force field $\mathbf{F}$ is exact, we write

$$\mathbf{F} = -\nabla V,$$

where $V = -f : \mathbb{R}^3 \to \mathbb{R}$. The reason for changing the sign will soon become apparent. Whenever we introduce a new quantity into physics, we should identify what its dimensions are. We leave it to the reader to check that $\dim[V] = \dim[Energy]$. So we call V the *potential energy* associated to the force field $\mathbf{F}$. Sometimes we simply say that V is the *potential*. This is a misuse of the definite article 'the' in English, since $V + c$ is also *a* potential energy, provided that c is a constant, i.e., a real number.

There is another energy that is important in classical mechanics. If a particle with mass m has velocity $\mathbf{v} = (v_1, v_2, v_3) \in \mathbb{R}^3$, then we say that

$$E_K := \frac{1}{2}mv^2$$

is the *kinetic energy*. (As usual, $v^2 = \mathbf{v} \cdot \mathbf{v} = ||\mathbf{v}||^2$.)

Finally, a particle of mass m and velocity $\mathbf{v}$ moving in a conservative force field with potential $V(\mathbf{x})$ has *total (mechanical) energy* defined by

$$E_T(\mathbf{x}, \mathbf{v}) := E_K + V = \frac{1}{2}mv^2 + V(\mathbf{x}).$$

Now why is all this important? Well, let's consider how the total energy changes with time as our particle moves according to Newton's equation of motion. So we consider $E_T(\gamma(t), \gamma'(t))$, where $\gamma'(t) = d\gamma/dt$ is the usual derivative with respect to the time variable t. So, using the Leibniz rule and the chain rule of multi-variable calculus we see that

$$\frac{d}{dt}\left(E_T(\gamma(t), \gamma'(t))\right) = \frac{d}{dt}\left(\frac{1}{2}m\,\gamma'(t) \cdot \gamma'(t) + V(\gamma(t))\right)$$

$$= \frac{2}{2}m\,\gamma''(t) \cdot \gamma'(t) + \nabla V(\gamma(t)) \cdot \gamma'(t)$$

$$= \left(m\,\gamma''(t) + \nabla V(\gamma(t))\right) \cdot \gamma'(t)$$

$$= \left(m\,\gamma''(t) - \mathbf{F}(\gamma(t))\right) \cdot \gamma'(t)$$

$$= 0.$$

The last equality is true simply because $\gamma(t)$ is a trajectory, that is, solves $\mathbf{F} = m\mathbf{a}$. Since $(-b, b) \ni t \mapsto E_T(\gamma(t), \gamma'(t))$ is a function defined on an interval and has derivative equal to zero in that interval, it follows that it is a constant function. So, there exists a constant, say E_0, such that $E_T(\gamma(t), \gamma'(t)) = E_0$ for all $t \in (-b, b)$. Of course, E_0 has dimensions of energy. The moral of this story is that if we measure the total energy of our particle at any time, then that will be (or was) its total energy at all other times. The kinetic energy may change, and the potential energy may change. But any increase in one of these energies corresponds *exactly* to a decrease of the other. This is called the *conservation of total mechanical energy* for a particle acted on by a conservative force. And now you see how the terminology *conservative force* arose. It has nothing to do with politics.

We summarize this discussion in words as follows:

Proposition 4.1.1 *Suppose that a classical massive particle moves in $\mathbb{R}^3$ according to Newton's equation of motion in a conservative force field. Then the total mechanical energy of the particle is constant as it moves along its trajectory.*

Another way to think of this is to consider the so-called *tangent space* $T(U) := U \times \mathbb{R}^3 = \{(\mathbf{x}, \mathbf{v}) \mid \mathbf{x} \in U, \mathbf{v} \in \mathbb{R}^3\}$, where U is an open set in $\mathbb{R}^3$. Then $T(U)$ is the domain of definition of the force field and its potential, and the *total energy* functional

$$E_T : T(U) \to \mathbb{R}$$

is defined by $E_T(\mathbf{x}, \mathbf{v}) := \frac{1}{2}mv^2 + V(\mathbf{x})$ for all $\mathbf{x} \in U$ and $\mathbf{v} \in \mathbb{R}^3$. Then for a given initial energy E_0 of our classical particle, its trajectory must lie for all times (past as well as future) on the hypersurface

$$S(E_0) := \{(\mathbf{x}, \mathbf{v}) \in T(U) \mid E_T(\mathbf{x}, \mathbf{v}) = E_0\}.$$

Notice that even though we are studying a classical motion in 3 dimensions, we are lead to considering the six-dimensional space $T(U)$ and a family of 5-dimensional hypersurfaces $S(E_0)$, where $E_0 \in \mathbb{R}$.

Exercise 4.1.2 *Actually, there are potentials such that for some values of E_0 the set $S(E_0)$ is empty. Understand how this can happen.*

We end here this very brief interlude in classical mechanics, which continues
to this day as a rich and active area of research.

4.2   Introducing Schrödinger's Equation

Next, let's try some intuitive reasoning, namely that if a massive 'particle'
such as an electron also is a 'wave', then it should be described as the solution
of a partial differential equation that looks like (though need not be exactly)
the usual wave equation. We start from the equation for total energy

$$E_T = \frac{1}{2}mv^2 + V(\mathbf{x}). \tag{4.2.1}$$

Let's consider the kinetic energy first. Since we do not have the velocity $\mathbf{v}$ in
the function $e^{i(\mathbf{k}\cdot\mathbf{x}-\omega t)}$ (recall $i = \sqrt{-1}$), we rewrite the kinetic energy as

$$E_K = \frac{1}{2}mv^2 = \frac{1}{2m}(mv)^2 = \frac{1}{2m}p^2 = \frac{1}{2m}\hbar^2 k^2.$$

But $k^2 = k_1^2 + k_2^2 + k_3^2$ also is not seen in the expression $e^{i(\mathbf{k}\cdot\mathbf{x}-\omega t)}$. However,
if we take a partial derivative with respect to x_1 we get

$$\frac{\partial}{\partial x_1}e^{i(\mathbf{k}\cdot\mathbf{x}-\omega t)} = ik_1\, e^{i(\mathbf{k}\cdot\mathbf{x}-\omega t)} \tag{4.2.2}$$

But this still is not what we want. So we take the second partial and get

$$\frac{\partial^2}{\partial x_1^2}e^{i(\mathbf{k}\cdot\mathbf{x}-\omega t)} = -k_1^2\, e^{i(\mathbf{k}\cdot\mathbf{x}-\omega t)}$$

This is closer to what we are looking for. Then passing on to the *Laplacian
operator*

$$\Delta := \frac{\partial^2}{\partial x_1^2} + \frac{\partial^2}{\partial x_2^2} + \frac{\partial^2}{\partial x_3^2}$$

we get

$$\Delta\left(e^{i(\mathbf{k}\cdot\mathbf{x}-\omega t)}\right) = (-k_1^2 - k_2^2 - k_3^2)e^{i(\mathbf{k}\cdot\mathbf{x}-\omega t)} = -k^2\, e^{i(\mathbf{k}\cdot\mathbf{x}-\omega t)}.$$

(An operator is just a way of changing a function into another function. This
is what the Laplacian does; it changes a function f into Δf. We will be seeing
a lot of operators!) Multiplying by a constant gives us

$$-\frac{\hbar^2}{2m}\Delta\, e^{i(\mathbf{k}\cdot\mathbf{x}-\omega t)} = \frac{\hbar^2}{2m}k^2\, e^{i(\mathbf{k}\cdot\mathbf{x}-\omega t)} = E_K\, e^{i(\mathbf{k}\cdot\mathbf{x}-\omega t)}.$$

Similarly, for the total energy we take a partial derivative with respect to t
as follows:

$$\frac{\partial}{\partial t}e^{i(\mathbf{k}\cdot\mathbf{x}-\omega t)} = -i\omega\, e^{i(\mathbf{k}\cdot\mathbf{x}-\omega t)}.$$

Identifying E_T with $\hbar\omega$ we see that

$$i\hbar\frac{\partial}{\partial t}e^{i(\mathbf{k}\cdot\mathbf{x}-\omega t)} = \hbar\omega\, e^{i(\mathbf{k}\cdot\mathbf{x}-\omega t)} = E_T\, e^{i(\mathbf{k}\cdot\mathbf{x}-\omega t)}.$$

We go back to the fundamental relation (4.2.1) from classical physics, now written as

$$\hbar\omega = \frac{1}{2m}\hbar^2 k^2 + V(\mathbf{x}). \tag{4.2.3}$$

We multiply through by $e^{i(\mathbf{k}\cdot\mathbf{x}-\omega t)}$ and get

$$\hbar\omega e^{i(\mathbf{k}\cdot\mathbf{x}-\omega t)} = \frac{1}{2m}\hbar^2 k^2 e^{i(\mathbf{k}\cdot\mathbf{x}-\omega t)} + V(\mathbf{x})e^{i(\mathbf{k}\cdot\mathbf{x}-\omega t)}.$$

Next, we identify two of these terms as derivatives of $e^{i(\mathbf{k}\cdot\mathbf{x}-\omega t)}$ as noted above, thereby getting:

$$i\hbar\frac{\partial}{\partial t}e^{i(\mathbf{k}\cdot\mathbf{x}-\omega t)} = -\frac{\hbar^2}{2m}\Delta e^{i(\mathbf{k}\cdot\mathbf{x}-\omega t)} + V(\mathbf{x})e^{i(\mathbf{k}\cdot\mathbf{x}-\omega t)}. \tag{4.2.4}$$

Here the expression $\Delta e^{i(\mathbf{k}\cdot\mathbf{x}-\omega t)}$ must be interpreted as $\Delta\left(e^{i\mathbf{k}\cdot\mathbf{x}}\right)e^{-i\omega t}$, since ω depends on $\mathbf{x}$ according to equation (4.2.3), which itself is called a *dispersion relation*. In this case the dispersion relation assigns an angular frequency (or equivalently, an energy) to every point $(\mathbf{x}, \hbar\mathbf{k}) - (\mathbf{x}, \mathbf{p})$ in the *phase space* of positions and momenta.

Next comes an enormous leap. You could even call it a *non sequitur*. In analogy to the prototype equation (4.2.4) we write down the following *Schrödinger equation*:

$$i\hbar\frac{\partial}{\partial t}\psi(t,\mathbf{x}) = -\frac{\hbar^2}{2m}\Delta\psi(t,\mathbf{x}) + V(\mathbf{x})\psi(t,\mathbf{x}) \tag{4.2.5}$$

and say that this is the fundamental partial differential equation of quantum theory for a massive 'particle' of mass $m > 0$ that is subject to a potential energy (scalar) field $V(\mathbf{x})$. In one form or another (4.2.5) or equations much like it are the fundamental equations of quantum theory. These equations always include two important constants: Planck's constant $\hbar$ and $i = \sqrt{-1}$. This is the first fundamental equation in physics that involves these constants. In fact, it may be that just the one constant $i\hbar$ is what is physically important. It is worthwhile to remind the reader that $\hbar$ is a real, positive number.

Notice that $\psi(t,\mathbf{x})$ is the unknown in (4.2.5). In 1926 in one of his four seminal papers on quantization Schrödinger wrote down this very equation, but he justified it differently, though he did not (and could not) prove it. This equation (and others similar to it) is called the *time dependent Schrödinger equation*. For now we simply call it the *Schrödinger equation*. (The *time independent Schrödinger equation* will be discussed shortly.)

Many, many comments are in order. First of all, the above discussion is not a mathematical proof. The only way to justify the correctness of the Schrödinger equation is by comparing its predictions with experimental results. Approximately 10^5 scientific papers have been written that refer, one way or another, to the Schrödinger equation. Despite criticisms in some of these papers, the bulk of them do support the Schrödinger equation. In 1933 Erwin Schrödinger shared the Nobel prize in physics with Paul Dirac. Other Nobel laureates have been decried as being undeserving of their prize. But not these guys!

The Schrödinger equation is a time evolution equation due to the partial derivative with respect to time. Thus it is the quantum analogue of Newton's equation of motion. Newton's equation of motion is, in general, non-linear in the unknown trajectory, while Schrödinger's equation is linear in its unknown $\psi(t, \mathbf{x})$. Despite many attempts, to date no instance of non-linearity has been observed experimentally to contradict this. In particular the constant function $\psi(t, \mathbf{x}) \equiv 0$ always solves Schrödinger's equation, but as we will see it has no physical meaning. In fact, it is not clear at all at this point what physical meaning a solution $\psi(t, \mathbf{x})$ might have.

One extremely important fact here is that the potential energy $V(\mathbf{x})$ in the Schrödinger equation (4.2.5) is the same potential energy that gives the conservative force $\mathbf{F}(\mathbf{x}) = -\nabla V(\mathbf{x})$ in classical mechanics.

Some other important points are that the Schrödinger equation is *one* partial differential equation while Newton's equation is a *system* of ordinary differential equations, and this system is typically coupled. Moreover, the Schrödinger equation is linear. Despite many experimental attempts, to date no instance of non-linearity has been observed to contradict this. However, Newton's equation is typically non-linear. Also, Newton's equation is second order in time, while Schrödinger's equation is first order in time.

But there is an equivalent way (known as *Hamiltonian mechanics* in physics) of formulating the Newtonian theory for *systems with conservative forces* so that the equation of motion is first order in time. This equivalent way is a well-known technique in the theory of ordinary differential equations, though it might not have a name. Anyway, in part it consists in replacing the derivatives of the unknown with new variables, which are then taken to be new unknowns. However, Hamiltonian mechanics is only a special case of Newtonian mechanics, which is the theory that allows one to also study non-conservative forces such as friction forces. If you think friction forces are not important, I recommend that you think about how automobile brakes work and why ships don't have brakes.

Any partial differential equation that is first order in time leads one to a *Cauchy problem*, which asks whether a unique solution exists for that partial differential equation combined with any 'reasonable' *initial condition* for the solution. This holds for the Schrödinger equation provided that the initial condition is taken to be an element in an appropriate *Hilbert space*.

Sound, light, and other wave phenomena had already been well studied by the time Schrödinger's seminal paper appeared in 1926. They are described by *wave equations*, the simplest of which is

$$\frac{1}{v^2}\frac{\partial^2 u}{\partial t^2} = \Delta u,$$

where v is a constant with dimensions of $[L]\,[T]^{-1}$, that is, v is a speed. Wave equations are second order in the time variable. So, Schrödinger's equation is *not* a wave equation. Nonetheless, by a standard abuse of language, everyone (mathematicians as well as physicists) calls a non-zero solution $\psi(t, \mathbf{x})$ of (4.2.5) a *wave function*.

4.3   The Eigenvalue Problem

Now Schrödinger's equation is like Newton's equation of motion in that it is meant to be a universal law of nature. Even though it describes a multitude of physical systems, there are some general facts about Schrödinger's equation. Here is one of those facts. The idea is to identify those solutions that have the form

$$\psi(t, \mathbf{x}) = \psi_1(t)\psi_2(\mathbf{x}). \tag{4.3.1}$$

We do not assert that every solution has this form, since that is not true as we shall see shortly. But we simply wish to identify as best we can those solutions that have this particular form as a product of a function of t alone and a function of $\mathbf{x}$ alone. So we substitute the above expression into (4.2.5) to get:

$$i\hbar\frac{\partial}{\partial t}\big(\psi_1(t)\psi_2(\mathbf{x})\big) = -\frac{\hbar^2}{2m}\Delta\big(\psi_1(t)\psi_2(\mathbf{x})\big) + V(\mathbf{x})\psi_1(t)\psi_2(\mathbf{x})$$

Equivalently,

$$i\hbar\,\psi_1'(t)\psi_2(\mathbf{x}) = \Big(-\frac{\hbar^2}{2m}\Delta\psi_2(\mathbf{x}) + V(\mathbf{x})\psi_2(\mathbf{x})\Big)\psi_1(t)$$

Next, dividing both sides by $\psi_1(t)\psi_2(\mathbf{x})$ we see that

$$i\hbar\frac{\psi_1'(t)}{\psi_1(t)} = \frac{1}{\psi_2(\mathbf{x})}\Big(-\frac{\hbar^2}{2m}\Delta\psi_2(\mathbf{x}) + V(\mathbf{x})\psi_2(\mathbf{x})\Big)$$

The point of all this is that the left side does not depend on $\mathbf{x}$ while the right side does not depend on t. But the two sides are equal! So, each side does not depend on $\mathbf{x}$ and does not depend on t. And there are no other variables in play. So the two (equal) expressions in (4.3.2) must be equal to a constant complex number, which we denote by E. Why E? Because it has the dimensions of an energy. In the general theory of partial differential equations one calls E the *separation constant*. It is not clear at this point that E must be a *real* number, though a physicist would say that it obviously must be real since energies are measured quantities, and hence real. So we actually have that

$$E = i\hbar \frac{\psi_1'(t)}{\psi_1(t)} = \frac{1}{\psi_2(\mathbf{x})}\Big(-\frac{\hbar^2}{2m}\Delta\psi_2(\mathbf{x}) + V(\mathbf{x})\psi_2(\mathbf{x})\Big), \qquad (4.3.2)$$

where E is a real number.

Exercise 4.3.1 *Verify that the dimensions of E are those of energy.*

From the first two expressions in (4.3.2) we conclude that

$$i\hbar\,\psi_1'(t) = E\psi_1(t).$$

The unknown is the function $\psi_1(t)$. And the general solution of this first order ordinary differential equation is known from calculus to be $\psi_1(t) = C_1 e^{-itE/\hbar}$, where C_1 is a *constant of integration*. Since the constant C_1 can be absorbed into the factor $\psi_2(\mathbf{x})$ we simplify things by putting $C_1 = 1$. So we find that

$$\psi_1(t) = e^{-itE/\hbar}. \qquad (4.3.3)$$

From the first and third expressions in (4.3.2) we conclude that

$$-\frac{\hbar^2}{2m}\Delta\psi_2(\mathbf{x}) + V(\mathbf{x})\psi_2(\mathbf{x}) = E\psi_2(\mathbf{x}). \qquad (4.3.4)$$

A non-zero solution $\psi_2(\mathbf{x})$ of this is also called a *'wave' function* and never a *'particle' function*. Another name for a non-zero solution of (4.3.4) is a *stationary state*. Here 'stationary' means that $\psi_2(\mathbf{x})$ does not depend on the time t. The justification for saying 'state' will come later with the physics interpretation of $\psi_2(\mathbf{x})$.

It is very important to emphasize that the energy E here in this equation is the same separation constant energy that appears in the equation for $\psi_1(t)$. We call equation (4.3.4) the *time independent Schrödinger equation*, since the time variable t does not appear in it. It is a linear second order partial differential equation whose unknowns are $\psi_2(\mathbf{x})$ and the energy E. Being linear, the constant function $\psi_2(\mathbf{x}) \equiv 0$ is always a solution, no matter what value the energy E has. But this leads to the physically irrelevant solution $\psi(t,\mathbf{x}) \equiv 0$. So we are only interested in non-zero solutions $\psi_2(\mathbf{x})$. But the question of whether such solutions exist will depend on the value of E. In general, for some values of E we will have no non-zero solution. But for other values of E we will have a non-zero vector space all of whose non-zero elements will be non-trivial solutions of (4.3.4).

Exercise 4.3.2 *Let $E \in \mathbb{C}$ be given. Define*

$$V_E := \{\psi_2 \mid -\frac{\hbar^2}{2m}\Delta\psi_2(\mathbf{x}) + V(\mathbf{x})\psi_2(\mathbf{x}) = E\psi_2(\mathbf{x})\}.$$

Prove that V_E is a complex vector space. (We understood that $\psi_2 : \mathbb{R}^3 \to \mathbb{C}$ and that ψ_2 is of class C^2. Recall Definition 4.1.1.)

Definition 4.3.1 *If V_E is non-zero, we say that E is an* eigenvalue *for the* linear operator

$$H := -\frac{\hbar^2}{2m}\Delta + V(\mathbf{x})$$

and that the non-zero elements in V_E are eigenfunctions *or* eigenvectors *of that linear operator. We say that H is a* Schrödinger operator, *and also that H is the* (quantum) Hamiltonian *of the physical quantum system. The time independent Schrödinger equation $H\psi = E\psi$ is called an* eigenvalue problem.

Remarks: A *linear operator A* (usually just called an *operator*) is a mapping of functions ψ to other functions $A\psi$ that is *linear* in the same sense as in linear algebra: $A(\psi_1 + \psi_2) = A\psi_1 + A\psi_2$ and $A(\lambda\psi) = \lambda A\psi$ for all $\lambda \in \mathbb{C}$. Notice that parentheses are only used when absolutely necessary.

The solution of an eigenvalue problem consists in finding those eigenvalues (for us the energy E) and the associated non-zero eigenfunction (for us the function ψ). However, if $\psi \in V_E$ is non-zero and $\lambda \in \mathbb{C}$ is non-zero, then the non-zero function $\lambda\psi$ also is an eigenfunction associated with the eigenvalue E. For this reason, the 100% correct way to solve the eigenvalue problem $H\psi = E\psi$ is to find those numbers E such that $V_E \neq 0$. Then the solution consists of those numbers E together with the non-zero elements in V_E. In practice however, one usually specifies E and a corresponding eigenfunction. If the dimension $\dim_{\mathbb{C}} V_E \geq 2$, then one usually specifies a basis of V_E, which will not be unique, of course. The reader should compare this with the situation in linear algebra where one has an $n \times n$ matrix M and the eigenvalue equation is $Mv = \lambda v$ whose the solution consists of those scalars λ that have a corresponding non-zero vector v.

Another point must be noted here. The separation constant E can be a non-real complex number as far as we can say mathematically as of now. But we know that physically it represents an energy, and so it should be a real number. We shall see later on that the Schrödinger operator H must satisfy a mathematical property which will guarantee that its eigenvalues E are real numbers.

The study of Schrödinger operators is just one of many active areas of contemporary research in mathematical physics. See [2], [5], [10], [15], [30].

If E is an eigenvalue, then for every non-zero $\psi_2 \in V_E$, we have that

$$\psi(t, \mathbf{x}) = e^{-iEt/\hbar}\psi_2(\mathbf{x}) \tag{4.3.5}$$

is a non-zero solution of the time dependent Schrödinger equation. Finite linear combinations of functions of the form (4.3.5) for various values of the energy eigenvalue will again be solutions of the time dependent Schrödinger equation, since it is a linear equation. However, such solutions will not have the product form $\psi(t, \mathbf{x}) = \psi_1(t)\psi_2(\mathbf{x})$. We can also hope that some infinite sums (or integrals) of functions of the form (4.3.5) will be solutions of the time dependent Schrödinger equation. But that would require further analysis to make sure that things do indeed work out correctly.

The problem of finding the eigenvalues of a given linear operator is called an *eigenvalue problem*. As Schrödinger says in the very title of his seminal paper, quantum theory is viewed as an eigenvalue problem. It actually is a spectral theory problem in general. This is so, since the spectrum (which we have not yet defined!) of an operator always contains the eigenvalues, but sometimes includes other types of spectrum. The complete analysis of the relevant operators in quantum theory requires an understanding of all the spectrum, not just the eigenvalues, but as a beginner you will do just fine by thinking of the spectrum as the eigenvalues plus other real numbers that are 'almost' eigenvalues. But Schrödinger pointed the scientific community in the right direction, and research along these lines in mathematical physics continues to this day.

At this point in the exposition the mathematicians among the readers will feel uneasy because the operator H was not given a precise domain of definition of functions ψ. This is a technicality which we will have to face. The physicists among the readers will feel uneasy because, even though the eigenvalues of H have been given a physical interpretation as energies, the eigenfunctions of H so far do not have a physical interpretation. This is not a technicality, but rather a fundamental problem at the very heart of quantum theory. In fact, only after resolving this problem will we be able to see why $\psi(t, \mathbf{x}) \equiv 0$ does not have a sound physical interpretation and so must be discarded from the theory.

4.4   Notes

In 1926 physicists were familiar with partial differential equations, but not with matrices. In those days linear algebra was a technical mathematical topic that only very few physicists learned. More generally, non-commutative multiplication, such as that of the quaternions, was considered in the physics community as a strange abstract concept of interest only to mathematicians. When Heisenberg discovered his version of quantum theory, he was amazed to find a physically natural non-commutative multiplication law. He was told this multiplication was exactly that of matrices. And so Heisenberg's theory was dubbed *matrix mechanics*. But it was Schrödinger's quantum theory that swept through the scientific world like a wild fire, because it is expressed in a more familiar language. We will follow the path that Schrödinger opened up. Its relation to Heisenberg's theory will be taken up later in Chapter 17.

For those who are interested, the precise definition of the spectrum of a linear operator will be given in the optional Section 9.5. But if you prefer to think intuitively of the spectrum as the set of real numbers which are almost eigenvalues of a Schrödinger operator, then no harm will be done.

Chapter 5

Operators and Canonical Quantization

> One can measure the importance of
> a scientific work by the number of
> earlier publications rendered
> superfluous by it.
> David Hilbert

One essential feature of quantum theory should have become apparent by now. And that is the central role played by *linear operators*, which we shall merely call *operators*. We assume that the reader is familiar with *vector spaces* and *linear transformations* from linear algebra of finite dimensional spaces. In this book the vector space *can* be infinite dimensional. And we say 'operator' instead of 'transformation.' The field of scalars will usually be the complex numbers, in which case we are then considering *complex* vector spaces. First as a quick refresher here are some basic definitions.

Definition 5.0.1 *Suppose that V and W are vector spaces and $T : V \to W$ is a function from V to W. We say that T is a* (linear) *operator if*

$$T(v_1 + v_2) = Tv_1 + Tv_2 \quad \text{for all } v_1, v_2 \in V,$$
$$T(\lambda v) = \lambda Tv \quad \text{for all } v \in V, \lambda \in \mathbb{C}.$$

Notice that parentheses are omitted from the notation if no ambiguity arises. Then for an operator T the kernel *(or* null space) *of T is defined by*

$$\ker T := \{v \in V \mid Tv = 0\},$$

© Springer Nature Switzerland AG 2020
S. B. Sontz, *An Introductory Path to Quantum Theory*,
https://doi.org/10.1007/978-3-030-40767-4_5

while the range *of T is defined by*

$$\text{Ran}\, T := \{ w \in W \,|\, \text{there exists } v \in V \text{ with } Tv = w \}.$$

Also we advise the reader to review the case when $T : V \to V$ is a linear operator and V has finite dimension. In this case from an introductory linear algebra course, there are two important functions of T, the trace $Tr(T)$ and the determinant $\det(T)$, which also could be briefly reviewed before reading Chapter 13.

5.1 First Quantization

For a first look at this topic in the setting of infinite dimensional spaces we shall leave to a side questions of what are the domain and co-domain spaces of these operators. For now they just take certain functions as input and produce other functions as output. In the time independent Schrödinger equation we saw that the kinetic energy $p^2/2m$ in classical mechanics was replaced with the operator for the kinetic energy

$$K := -\frac{\hbar^2}{2m}\, \Delta,$$

that is, this is the operator that sends the function ψ to the function

$$\mathbb{R}^3 \ni \mathbf{x} \mapsto K\psi(\mathbf{x}) = -\frac{\hbar^2}{2m}\, \Delta \psi(\mathbf{x}).$$

So, the expression $p^2 = p_1^2 + p_2^2 + p_3^2$ was replaced by the operator

$$-\hbar^2 \Delta = -\hbar^2 \left(\frac{\partial^2}{\partial x_1^2} + \frac{\partial^2}{\partial x_2^2} + \frac{\partial^2}{\partial x_3^2} \right).$$

This is a *second order differential operator*.

On the other hand, the potential energy $V(\mathbf{x})$ in classical mechanics was replaced by the operator that sends the function ψ to the function

$$\mathbb{R}^3 \ni \mathbf{x} = (x_1, x_2, x_3) \mapsto V(\mathbf{x})\psi(\mathbf{x}),$$

which is called a *multiplication operator*. Three examples of such a classical potential are x_1, x_2, and x_3. These then correspond to the three multiplication operators given by x_1, x_2, and x_3. This leads one to wonder what are the operators that correspond to each component of $\mathbf{p} = (p_1, p_2, p_3)$. The answer is that each p_j is given by the *first order differential operator*

$$p_j = \frac{\hbar}{i} \frac{\partial}{\partial x_j} \qquad \text{for } j = 1, 2, 3.$$

(Recall $i = \sqrt{-1}$.) This can also be seen in the eigenvalue equation (4.2.2), which is equivalent to saying that $p_1 = \hbar k_1$ is an eigenvalue of the operator

$$\frac{\hbar}{i}\frac{\partial}{\partial x_1}.$$

This almost algorithmic way of replacing functions of $(\mathbf{x}, \mathbf{p}) \in \mathbb{R}^3 \times \mathbb{R}^3$, the *phase space*, with operators is called the *canonical quantization* of classical mechanics or *first quantization*. (At some later date you might learn about *second quantization*, though it seems that there are no third or higher quantizations.) And, in fact, this is an algorithm for all functions defined on phase space having the form $f(\mathbf{x}, \mathbf{p}) = f_1(\mathbf{x}) + f_2(\mathbf{p})$, provided that f_1 and f_2 satisfy some mild condition.

But the method does not work so easily for functions on phase space such as $f(\mathbf{x}, \mathbf{p}) = x_1 p_1 = p_1 x_1$, since the operators associated to the factors x_1 and p_1 do not commute. (We will come back to this point in a moment.) This leads to the so-called *ordering problem* of quantization. Should we use the order $x_1 p_1$ or $p_1 x_1$ or something else as the 'correct' expression for quantizing? This ordering problem has been solved, and the upshot is that there are many non-equivalent solutions. Put another way, we can say that quantization is not unique.

Physicists use canonical quantization without giving it a second thought. They find the formula for the total energy of a classical system in terms of positions and momentum. This is the *classical Hamiltonian*. Then they apply the canonical quantization rules to produce a linear operator, the *quantum Hamiltonian*, which is then declared to be appropriate for the corresponding quantum system. Persons who do not understand classical physics think that physicists are some sort of magicians.

To study the lack of commutativity that often occurs in quantum theory, we introduce the extremely important notation of the *commutator* of two operators A and B acting in a vector space:

$$[A, B] := AB - BA. \tag{5.1.1}$$

Therefore, A and B commute (that is, $AB = BA$) if and only if $[A, B] = 0$. Also, notice that $[A, B] = -[B, A]$ and, in particular, $[A, A] = 0$.

We wish now to study all possible commutators among the six operators $x_1, x_2, x_3, p_1, p_2, p_3$. First, we note that

$$[x_j, x_k] = 0 \tag{5.1.2}$$

for all $1 \le j, k \le 3$, since multiplication by x_k first and then by x_j gives the same result as in the opposite order. Next, we note that

$$[p_j, p_k] = 0 \tag{5.1.3}$$

for all $1 \le j, k \le 3$, since differentiation with respect to x_k first and then with respect to x_j gives the same result as in the opposite order. (For $j \ne k$ this is the theorem about *mixed partial derivatives* from multi-variable calculus, provided that one is differentiating relatively 'nice' functions, namely those of class C^2.)

The remaining case is for the commutator $[p_j, x_k]$ with $1 \leq j, k \leq 3$. We compute this explicitly by letting these operators act on an appropriate function $\psi = \psi(x_1, x_2, x_3)$ as follows:

$$\begin{aligned}
[p_j, x_k]\psi &= (p_j x_k - x_j p_j)\psi \\
&= \frac{\hbar}{i}\frac{\partial}{\partial x_j}\left(x_k \psi\right) - x_k\left(\frac{\hbar}{i}\frac{\partial}{\partial x_j}\psi\right) \\
&= \frac{\hbar}{i}\frac{\partial x_k}{\partial x_j}\psi + x_k\frac{\hbar}{i}\frac{\partial \psi}{\partial x_j} - x_k\frac{\hbar}{i}\frac{\partial \psi}{\partial x_j} \\
&= \frac{\hbar}{i}\delta_{jk}\psi,
\end{aligned}$$

where we used nothing other than the product (or Leibniz) rule of calculus and the *Kronecker delta* defined by $\delta_{jk} = 1$ if $j = k$ and $\delta_{jk} = 0$ if $j \neq k$. This is usually rewritten as

$$i[p_j, x_k] = \hbar\,\delta_{jk}\,I, \tag{5.1.4}$$

where I denotes the *identity operator*, which is defined by $I\psi = \psi$ for all admissible functions ψ. (We won't worry here about what those admissible functions might be except to say that there are a lot of them.) One says that (5.1.4) is the *canonical commutation relation* or briefly the *CCR*. For some authors the commutation relations (5.1.2) and (5.1.3) are included in the CCR, but usually not.

In particular, as noted above, $[p_1, x_1] \neq 0$, since $\hbar \neq 0$. For those who understand a bit of classical mechanics we remark that in the classical theory the position and momentum observables are represented by functions (that is, not by operators as here) and any pair of functions whatsoever commute. In short, the lack of commutativity of p_j and x_j for each $j = 1, 2, 3$ is a purely quantum property. However, it is a curious fact that the quantum commutation relations (5.1.2), (5.1.3), and (5.1.4) for $j \neq k$ are the same as the corresponding classical commutation relations.

5.2   The Quantum Plane

A standard procedure used in the mathematics motivated by physics is to take the relevant physical variables or observables and consider all possible ways of adding and multiplying them together. If we also have a scalar multiplication (by real or complex numbers), this will typically give us a vector space which also has a product for every pair of its elements; such a structure is called an *algebra*. For now this is the idea that counts, and we leave the technical, rigorous definition to a side. In the case at hand the variables are going to be position x and momentum p for a simple (i.e., spin zero, single particle) physical system constrained to the one-dimensional real line $\mathbb{R}$. (See Chapter 14 for *spin zero*.) Using the last section as motivation we posit that x and p satisfy the commutation relation

$$px - xp = [p, x] = \frac{\hbar}{i} 1 \tag{5.2.1}$$

and no other relations. Here we could take $\hbar$ to be another (formal) variable, but for now we consider $\hbar$ to be a positive real number. The algebra we get this way is denoted as $\mathcal{Q}_\hbar$. Here 1 denotes the *identity element* of the algebra $\mathcal{Q}_\hbar$, which satisfies $1 \cdot a = a \cdot 1 = a$ for all $a \in \mathcal{Q}_\hbar$ as well as $x^0 = 1$ and $p^0 = 1$. The algebra $\mathcal{Q}_\hbar$ is what I prefer to call the *(Weyl-Heisenberg) quantum plane*, even though it seems that M. Born was the physicist who discovered the commutation relations (5.2.1). The standard name for this algebra is the *Weyl-Heisenberg algebra*, which conceals its geometric structure.

I now ask the reader to accept the idea (and definition!) that a *point of an algebra* $\mathcal{A}$ with an identity element $1_\mathcal{A}$ is a linear function $l : \mathcal{A} \to \mathbb{C}$ which is also multiplicative (that is, $l(ab) = l(a)l(b)$ for all $a, b \in \mathcal{A}$) and satisfies $l(1_\mathcal{A}) = 1$. Then it is an easy exercise to show that the quantum plane has no points, i.e., there does not exist any such linear function l for the algebra $\mathcal{Q}_\hbar$ defined above. (Hint: If such a point l existed, apply it to equation (5.2.1) to get a contradiction.) This is a fundamental example of a *non-commutative space*, which in turn one studies in *non-commutative geometry*. The quantum plane $\mathcal{Q}_\hbar$ is a non-commutative space with no points that arises directly from the most basic aspects of quantum theory.

The next exercise motivates this strange definition of point of an algebra.

Exercise 5.2.1 *Let* $\mathbb{R}[x, y]$ *be the (commutative!) algebra of all polynomials in the variables* x *and* y *with real coefficients. For each point* (r, s) *in the Euclidean plane* $\mathbb{R}^2$ *(that is, a pair of real numbers) define* $l_{r,s} : \mathbb{R}[x, y] \to \mathbb{R}$ *by* $l_{r,s}(p) := p(r, s)$ *for every polynomial* $p(x, y)$. *(So,* $l_{r,s}$ *is evaluation of the polynomial at the point* (r, s) *in the Euclidean plane* $\mathbb{R}^2$.*) Prove that* $l_{r,s}$ *is a point of the algebra* $\mathbb{R}[x, y]$.

It turns out that any point l *of the algebra* $\mathbb{R}[x, y]$ *is equal to* $l_{r,s}$ *for a unique point* (r, s) *in the Euclidean plane. But this is more difficult to prove. Anyway, the points of the algebra* $\mathbb{R}[x, y]$ *correspond to the points (in the usual sense) of the Euclidean plane* $\mathbb{R}^2$.

The next exercise identifies the symmetries of the quantum plane and thereby serves to justify why we call it a *plane*. But first we need a definition of a fundamental mathematical object that is used to describe symmetries.

Definition 5.2.1 *Suppose that* G *be a set that has a binary operation, that is, a function* $G \times G \to G$ *which is denoted as* $(g, h) \mapsto gh$ *for* $g, h \in G$. *Then we say that* G *(together with this operation, which is called* multiplication*) is a* group *if these properties hold:*

- *(Associativity)* $(gh)k = g(hk)$ *for all* $g, h, k \in G$.

- (*Identity element*) *There exists a unique element* $e \in G$ *such that* $eg = g$ *and* $ge = g$ *for all* $g \in G$. *In certain contexts the identity element* e *is denoted as* I *or as* 1. *Or even as* 0 *when ones uses an additive notation.*

- (*Inverses*) *For each* $g \in G$ *there exists a unique element denoted as* g^{-1} *such that* $g\,g^{-1} = e$ *and* $g^{-1}g = e$. *We say that* g^{-1} *is the* inverse element *of* g.

If $gh = hg$ *for all* $g, h \in G$, *then we say the group* G *is* abelian. *Otherwise, we say that* G *is* non-abelian. *It turns out that non-abelian groups play important roles in physics.*

We say that a group is represented (as symmetries) *of a set* S *or* G *acts on* S *if there is a function* $G \times S \to S$, *denoted as* $(g, s) \mapsto g \cdot s$, *satisfying*

- $e \cdot s = s$ *for all* $s \in S$,

- $(gh) \cdot s = g \cdot (h \cdot s)$ *for all* $g, h \in G$ *and all* $s \in S$.

In this case we also say that there is an action *of* G *on* S. *For* $s, t \in S$ *we write* $s \equiv_G t$ *if there exists some* $g \in G$ *such that* $g \cdot s = t$.

A morphism *of groups is a function* $\alpha : G_1 \to G_2$, *where* G_1 *and* G_2 *are groups, such that* $\alpha(gh) = \alpha(g)\alpha(h)$ *and* $\alpha(g^{-1}) = (\alpha(g))^{-1}$ *for all* $g, h \in G_1$ *and* $\alpha(e_1) = e_2$, *where* e_j *is the identity element of* G_j *for* $j = 1, 2$. (*This is also called a* homomorphism.)

Exercise 5.2.2 *Prove that* $\equiv_G$ *is an equivalence relation on* S. (*Compare with Exercise 9.3.11. The set of its equivalence classes is denoted* S/G.)

Exercise 5.2.3 *Suppose that* p *and* x *satisfy* (5.2.1). *Find necessary and sufficient conditions on the six complex numbers* a, b, c, d, e, f *such that the pair* (p', x') *satisfies* $i[p', x'] = \hbar 1$, *where*

$$p' = ap + bx + c \quad and \quad x' = dp + ex + f.$$

Identify the set of all such 'change of variables' as a group and represent it as a group of symmetries of the complex vector space $\mathbb{C}^2$, *the complex 'plane'.*

Identify the dimensions of a, b, c, d, e, *and* f *assuming that* p *and* p' *have dimensions of momentum and that* x *and* x' *have dimensions of length.*

5.3   Notes

The role of quantization in quantum theory is problematic at best. It is not clear why the merely approximate classical theory of a system should be the point of departure for arriving at the correct, exact quantum theory of that same system. Some scientists maintain that quantization should not only be excluded from quantum theory, but also that it has no role whatsoever in science. Unfortunately, this is not how physicists actually arrive at a suitable quantum theory for a system of interest; canonical quantization is used in practice by practicing physicists. But the mathematical fact is that the way of converting functions on phase space to operators in Hilbert space is not unique, at least given our current understanding. This is encapsulated in the first part of a saying due to E. Nelson: "First quantization is a mystery." The second part goes: "Second quantization is a functor." But second quantization is for another day, for another book.

It is my opinion that the problem of quantization properly understood has nothing to do in any way with classical physics, because in principle quantization simply deals with finding the correct operators, including the (quantum, of course) Hamiltonian, for any given quantum system. And nothing else. The touchstone for knowing if this has been done correctly is experiment and not, as far as I can see, either mathematics or classical physics. For this reason the discussion of quantization given here, which is also usually given in other texts, is rather problematic. The concepts—based on experiment—that have been central to the development of the theory of classical physics have been taken over into quantum theory. For example, we should understand energy in the first instance as a property of physical nature. Energy is energy, and that's that. There is no 'classical energy' nor 'quantum energy'. Moreover, experiments are neither classical nor quantum, despite the claims of some famous physicists. It may well be that quantum theory is fundamental, but even if it is not, we have now arrived at the clear understanding that classical physics is surely not fundamental. Thus, it makes more sense logically to introduce quantum theory with absolutely no reference to classical physics. But I have followed the historical, rather than the logical, path.

It is unfortunate that many texts introduce the Hamiltonian formulation of classical physics as a motivation for the quantum Hamiltonian. This is because an audience of physics students is being addressed, not the audience of this book. As I am trying to make clear, it should be the other way around, namely that quantum theory should motivate classical theory. So, kind reader, do not fret about your lack of knowledge of classical physics!

However, it is widely accepted that classical theory should arise as an *emergent* theory from quantum theory. There are several ways this might be done, though the *semi-classical limit* of quantum theory, that is, the limit as $\hbar \to 0$, is the most common. The *correspondence principle* is another. These relations of quantum theory to classical theory will not be discussed in this book, since I consider classical physics to be a side issue for us.

Chapter 6

The Harmonic Oscillator

Music of the spheres.
Ancient philosophy

Having his equation well in hand, Schrödinger proceeded to solving it for various important physical systems. We now will consider the first such system, the harmonic oscillator moving along a straight line.

6.1   The Classical Case

First, we study the harmonic oscillator in classical mechanics. This physical system consists of a massive point particle constrained to move on a straight line, which we model by the real numbers $\mathbb{R}$. There is also a force given as a function of position $x \in \mathbb{R}$ by $F = -kx$, where $k > 0$ is a constant. Of course, k has dimensions which we now compute:

$$\dim[k] = \dim[F](\dim[x])^{-1} = [M][L][T]^{-2}[L]^{-1} = [M][T]^{-2}$$

We can think of the harmonic oscillator as a mathematical model of a physical system consisting of massive block, thought of as a point particle, sliding on a friction-less horizontal plane subject to the force applied to it by a massless spring attached to the block at one end and attached to a fixed wall at its other end. By a standard and useful convention the position where the spring applies zero force is taken to be the origin $0 \in \mathbb{R}$. Because of this system, we say that k is the *spring constant*. The condition $k > 0$ means that the force is in the opposite direction to any non-zero displacement of the particle from position $x = 0$. We say that F is a *restoring force*. Without any further ado we write down Newton's equation of motion for the particle

© Springer Nature Switzerland AG 2020
S. B. Sontz, *An Introductory Path to Quantum Theory*,
https://doi.org/10.1007/978-3-030-40767-4_6

with mass $m > 0$:

$$m\frac{d^2x}{dt^2} = -kx \qquad \text{or} \qquad \frac{d^2x}{dt^2} = -\frac{k}{m}x.$$

Next we note that $\dim[k/m] = [T]^{-2}$ and that $k/m > 0$. So we introduce $\omega > 0$ by

$$\omega := +\sqrt{\frac{k}{m}}.$$

Then $\dim[\omega] = [T]^{-1}$. These formulas show that ω is an angular frequency associated to the physical system by this explicit function of its physical parameters k and m. So Newton's equation now has this form:

$$\frac{d^2x}{dt^2} = -\omega^2 x.$$

This is a linear ordinary differential equation of order 2 with constant coefficients. As such it is explicitly solvable and has a 2-dimensional vector space of solutions. By just thinking a little about calculus we see that

$$x_1(t) = \cos(\omega t) \qquad \text{and} \qquad x_2(t) = \sin(\omega t)$$

are two linearly independent solutions. So the general solution is

$$x(t) = \alpha\cos(\omega t) + \beta\sin(\omega t),$$

where the real numbers α and β are determined by the initial position and the initial velocity of the particle. Another common way to write this solution is

$$x(t) = A\cos(\omega t + \phi),$$

where A is called the *amplitude* and ϕ is called the *phase*.

Exercise 6.1.1 *Find formulas that express A and ϕ in terms of α and β.*

It turns out that

$$F = -kx = -\frac{dV}{dx},$$

where $V(x) = \frac{1}{2}kx^2$. So F is a conservative force whose potential energy is $V(x)$. So the total energy of the classical harmonic oscillator for $(x, p) \in \mathbb{R}^2$ is

$$E_T(x, p) = \frac{1}{2m}p^2 + \frac{1}{2}kx^2$$

in terms of the parameters k and m. One often uses the relation $k = m\omega^2$ to write

$$E_T(x, p) = \frac{1}{2m}p^2 + \frac{1}{2}m\omega^2 x^2$$

in terms of the parameters ω and m.

Recall that for a function $f : X \to Y$ between sets X and Y we define the range of f by $\operatorname{Ran} f := \{y \in Y \,|\, \text{there exists } x \in X \text{ with } f(x) = y\}$.

Exercise 6.1.2 *Prove $\operatorname{Ran} E_T = [0, \infty)$. This means that the total energy of the classical harmonic oscillator can have any non-negative value.*

6.2   The Quantum Case

Putting $\hbar = 1$, $m = 1$ and $\omega = 1$, dimension-less quantities, the quantum harmonic oscillator Hamiltonian is given by the canonical quantization as

$$H_{osc} = \frac{1}{2}\left(p^2 + x^2\right). \qquad (6.2.1)$$

Also, both x and $p = -i\,d/dx$ are now dimension-less. We want to solve the time independent Schrödinger equation $H_{osc}\psi = E\psi$. Now Dirac had an uncanny knack for seeing the basic structure behind a problem. So we are going to see how he solved this problem.

First off, we have the identity $\alpha^2 + \beta^2 = (\alpha - i\beta)(\alpha + i\beta)$, where α and β are real numbers. And p and x are real quantities, though not real numbers. Anyway, Dirac wanted to factor the oscillator Hamiltonian more or less the same way. He then considered the factorization

$$(x - ip)(x + ip) = x^2 + p^2 - i(px - xp) = x^2 + p^2 - i[p, x]$$

and realized that he was dealing with a non-commutative multiplication and so he had to evaluate the commutator $[p, x]$. But we have already seen this essentially in (5.1.4), which in this 1-dimensional setting becomes $i[p, x] = I$, since $\hbar = 1$. So, we have

$$(x - ip)(x + ip) = x^2 + p^2 - I.$$

Dividing by 2 then gives

$$\frac{1}{2}(x - ip)(x + ip) = H_{osc} - \frac{1}{2}I.$$

We rewrite this as

$$H_{osc} = a^* a + \frac{1}{2}I,$$

where $a = (x + ip)/\sqrt{2}$ and $a^* = (x - ip)/\sqrt{2}$. Since $p = \frac{1}{i}\frac{d}{dx}$, it follows that a and a^* are first order linear ordinary differential operators.

As the cartoon says: Now a miracle occurs. How? For suppose that a non-zero ψ solves the first order ordinary differential equation (ODE) $a\psi = 0$. Then ψ also solves the second order ODE

$$H_{osc}\psi = \frac{1}{2}\psi$$

and, since ψ is non-zero, this means that $1/2$ is an eigenvalue of H_{osc}. But $a\psi = 0$ is equivalent to $(x + d/dx)\psi = 0$ or $\psi'(x) = -x\psi$. The general complex valued solution of this ODE is

$$\psi(x) = Ce^{-x^2/2}, \qquad (6.2.2)$$

where C is any complex number. (That this solves the ODE we leave to the reader as an exercise. Then the theory of ODE's says that this gives the *general* solution.) Picking any $C \neq 0$ we get a non-zero function $\psi(x)$. Hence, $1/2$ is an eigenvalue of H_{osc}. For the time being we will simply put $C = 1$, but later we will see that there is a 'better' choice. But that does not matter now. So, we define $\psi_0(x) = e^{-x^2/2}$, which is called a *Gaussian function* or simply a *Gaussian*. Then ψ_0 is an eigenfunction whose eigenvalue is $E_0 = 1/2$. This also proves that ker $a = \{z\psi_0 \,|\, z \in \mathbb{C}\}$.

Exercise 6.2.1 *Due to choosing dimension-less quantities $m = \omega = \hbar = 1$, the eigenvalue E_0 does not look like an energy. But it is! Show by reinstating the standard dimensions that we have*

$$E_0 = \hbar\omega/2,$$

where we recall that ω is the angular frequency of the associated classical harmonic oscillator. Also, show that (6.2.1) changes back to its original form with the standard dimensions.

We will see later on that E_0 is the smallest eigenvalue for this physical system. For reasons that also come later on, the eigenfunction ψ_0 represents a *state* of the system. Since its associated energy is the lowest possible, we say that ψ_0 is the *ground state* of the harmonic oscillator and that E_0 is the *ground state energy* of the harmonic oscillator.

This is quite different from the classical harmonic oscillator which can have any energy $E \geq 0$ as you have already proved in Exercise 6.1.2. But the quantum harmonic oscillator has no energy eigenvalue E with

$$0 \leq E < E_0 = \hbar\omega/2.$$

The interval $[0, E_0)$ is called a *gap*. For some physical systems it is a major result that the system even has a ground state (Theorem I) with a gap (Theorem II). An entire, highly technical research paper is often required to prove such a result. In fact, one of the Millennium Problems of the Clay Institute is essentially of this type. The quantum harmonic oscillator is a relatively trivial system in this regard.

Next we see how to get the rest of the eigenvalues and eigenfunctions of the quantum harmonic oscillator. To achieve this we will use more commutation relations, but unlike the CCR proved above, these are only valid for this particular physical system. For convenience we write $H = H_{osc}$ for the rest of this chapter. First, we commute $[H, a^*]$ as follows:

$$[H, a^*] = [a^*a + (1/2)I, a^*] = [a^*a, a^*]$$
$$= a^*aa^* - a^*a^*a = a^*(aa^* - a^*a)$$
$$= a^*[a, a^*]. \tag{6.2.3}$$

This leads us to evaluate another famous *canonical commutation relation* which is also abbreviated as CCR:

$$[a, a^*] = aa^* - a^*a = \frac{1}{2}[x + ip, x - ip]$$

$$= \frac{1}{2}\Big([x, x] - i[x, p] + i[p, x] + [p, p]\Big)$$

$$= \frac{1}{2}\Big(i[p, x] + i[p, x]\Big) = i[p, x]$$

$$= I. \tag{6.2.4}$$

Here we used two basic identities for commutators, namely, $[A, A] = 0$ and $[A, B] = -[B, A]$, where A and B are any operators.

Substituting (6.2.4) back into (6.2.3) we obtain

$$[H, a^*] = a^*. \tag{6.2.5}$$

While this is a very, very nice formula in itself, we now will see how it lets us construct new eigenfunctions. We suppose that $\psi \neq 0$ is an eigenfunction of H with eigenvalue $E \in \mathbb{C}$. So, $H\psi = E\psi$.

Next, we consider the function $a^*\psi$ and ask whether it is an eigenfunction of H as well. So, using (6.2.5) in the first equality, we compute:

$$Ha^*\psi - a^*H\psi + a^*\psi - a^*E\psi + a^*\psi = (E + 1)a^*\psi.$$

One has to be very careful at this step, since there are two mutually exclusive possibilities:

- $a^*\psi = 0$.

- $a^*\psi \neq 0$ and so $E + 1$ is also an eigenvalue of H.

At this point we can see why we call a^* a *raising operator*. After all, it raises eigenvalues by $+1$. We will soon see that these eigenvalues are real.

Now $a^*\psi = 0$ is a first order, linear ordinary differential equation which we can easily solve, since it is equivalent to

$$\psi'(x) = x\psi(x).$$

The general solution of this is easily seen to be $\psi(x) = Ce^{x^2/2}$, where C is any complex number. Of course, this is *not* a Gaussian. We summarize this by noting that

$$\ker a^* = \{\psi \mid a^*\psi = 0\} = \{Ce^{x^2/2} \mid C \in \mathbb{C}\}.$$

Now we start with the eigenfunction $\psi_0(x) = e^{-x^2/2}$ of H with eigenvalue $1/2$ and construct an infinite sequence of eigenfunctions and eigenvalues of H. First, $\psi_1 := a^*\psi_0$ is non-zero (since $\psi_0 \notin \ker a^*$) and so has eigenvalue $(1/2) + 1 = 3/2$.

We can evaluate $a^*\psi_0$ to see that $\psi_1(x) = C_1 x e^{-x^2/2}$ for some constant complex number $C_1 \neq 0$. Consequently, $\psi_2 := a^*\psi_1 \neq 0$ is an eigenfunction of H (since $\psi_1 \notin \ker a^*$) whose eigenvalue is $(3/2) + 1 = 5/2$. Again, we can get more specific information, namely that $\psi_2(x) = H_2(x)e^{-x^2/2}$, where $H_2(x)$ is a polynomial of degree 2. And so $\psi_2 \notin \ker a^*$.

This proceeds (by mathematical induction), and we see for each integer $k \geq 0$ that

$$\psi_k(x) := (a^*)^k \psi_0(x) = H_k(x)e^{-x^2/2} \tag{6.2.6}$$

is an eigenfunction of H with eigenvalue $k + (1/2)$, where $H_k(x)$ is a specific polynomial of degree k which we can compute. Up to constant factors which will depend on k these are the famous *Hermite polynomials*. Again, up to conventional constant factors, ψ_k is called a *Hermite function*.

At this point in the argument we have shown that the harmonic oscillator has an infinite sequence of eigenvalues. And we have (more or less) identified the corresponding eigenfunctions. Now comes the 'tricky bit'. We want to show that there are no more eigenvalues! This is an amazing result, since it says that the quantum harmonic oscillator can only be in stationary states for the energies $(k + 1/2)\hbar\omega$ in the standard notation. This contrasts with the classical harmonic oscillator which can have any energy $E \geq 0$.

How will we do this? Well, if a^* is a raising operator, then what is the operator a? Again we compute its commutation relation with H. So we get

$$\begin{aligned}
[H, a] &= [a^*a + (1/2)I, a] = [a^*a, a] \\
&= a^*aa - aa^*a = (a^*a - aa^*)a \\
&= [a^*, a]a \\
&= -a, \tag{6.2.7}
\end{aligned}$$

where we used (6.2.4). This differs by a very important algebraic sign from the commutation relation (6.2.5) for H and a^*.

Again we ask what happens to an eigenfunction $\psi \neq 0$ of H when we act on it with the operator a. So, we have $H\psi = E\psi$ for a complex number E. Using (6.2.7), we then compute:

$$Ha\psi = aH\psi - a\psi = aE\psi - a\psi = (E - 1)a\psi.$$

One has to be even more careful at this step than with the argument for a^*. Again, there are two mutually exclusive possibilities:

- $a\psi = 0$ if and only if $\psi \in \ker a$ if and only if $\psi = C\psi_0$ for some $C \in \mathbb{C}$.

- $a\psi \neq 0$ and so $E - 1$ is also an eigenvalue of H.

At this point we can see why we say that a is a *lowering operator*. After all, it lowers eigenvalues by $+1$. It is time to understand why these eigenvalues of H are real—and a bit more.

Proposition 6.2.1 *Suppose that* $H\psi = E\psi$ *for some non-zero* ψ *and some complex number* E. *Then* E *is a real number and* $E \geq 0$.

Remark: As commented on earlier, we even have that $E \geq (1/2)\hbar\omega > 0$. But for now we will be happy just knowing that $E \geq 0$.

Sketch of Proof: The first part of the conclusion holds for any H that satisfies the *symmetry condition*

$$\int_{\mathbb{R}} dx\, \varphi_1(x)^* H\varphi_2(x) = \int_{\mathbb{R}} dx\, (H\varphi_1(x))^* \varphi_2(x), \qquad (6.2.8)$$

where z^* denotes the complex conjugate of $z \in \mathbb{C}$. This condition can be written in terms of the *inner product* of functions defined as

$$\langle f_1, f_2 \rangle := \int_{\mathbb{R}} dx\, f_1^*(x) f_2(x), \qquad (6.2.9)$$

where $f_1, f_2 : \mathbb{R} \to \mathbb{C}$ are functions such that this integral is defined. For now this is just convenient notation, but later will be seen as an essential structure for Hilbert spaces. In the notation of (6.2.9) we write (6.2.8) as

$$\langle \varphi_1, H\varphi_2 \rangle = \langle H\varphi_1, \varphi_2 \rangle, \qquad (6.2.10)$$

and we say that H is a *symmetric operator*. For the eigenfunction ψ we have

$$\langle \psi, H\psi \rangle = \langle \psi, E\psi \rangle = E\langle \psi, \psi \rangle.$$

On the other hand we have that

$$\langle H\psi, \psi \rangle = \langle E\psi, \psi \rangle = E^*\langle \psi, \psi \rangle.$$

By the symmetry condition (6.2.10) for H we see that

$$E\langle \psi, \psi \rangle = E^*\langle \psi, \psi \rangle.$$

Next, the condition that ψ is non-zero really means that $\langle \psi, \psi \rangle > 0$. And consequently, $E = E^*$, that is, E is real.

Actually, the operators x and p are symmetric. Hence, in the case of x we have

$$\langle \varphi, x^2\varphi \rangle = \langle x\varphi, x\varphi \rangle \geq 0.$$

Similarly, $\langle \varphi, p^2\varphi \rangle \geq 0$ by integration by parts. These two facts imply that $\langle \varphi, H\varphi \rangle = (1/2)\langle \varphi, p^2\varphi \rangle + (1/2)\langle \varphi, x^2\varphi \rangle \geq 0$. Therefore,

$$0 \leq \langle \psi, H\psi \rangle = \langle \psi, E\psi \rangle = E\langle \psi, \psi \rangle$$

But again we use $\langle \psi, \psi \rangle > 0$ to conclude that $E \geq 0$. ∎

The above proof has gaps in it. The integrals must be shown to exist and the partial integration must be justified. This can be done by specifying exactly which functions we are using in that proof. This will eventually lead us to study *Hilbert spaces*, which have a nice inner product. But then linear operators such as x and p give us headaches since $x\psi$ and $p\psi$ will not necessarily be elements of the Hilbert space even though ψ is an element in the Hilbert space. This will eventually lead us to consider *densely defined (linear) operators*, which are unfortunately often called *unbounded operators*. We only mention this poorly chosen latter expression since so many texts use it. The point is that some naturally arising, densely defined operators are bounded and so not unbounded, if my gentle reader will permit me a double negative. More on densely defined operators can be found in the optional Section 9.6.

Now we suppose that we have $H\psi = E\psi$ for some non-zero ψ and some $E \in \mathbb{R} \setminus \{1/2, 3/2, 5/2, \dots\}$. We want to show this leads to a contradiction. What do we do? We keep lowering ψ by repeatedly applying the lowering operator a. Since we assumed $E \neq 1/2$, we know that ψ is not a constant multiple of ψ_0. This gives us an eigenfunction $a\psi \neq 0$ with the eigenvalue $E - 1$. If $E - 1 < 0$ we have arrived at a contradiction.

If $E - 1 \geq 0$, we note that $E - 1 \neq 1/2$ since $E \neq 3/2$ was assumed. So $a\psi$ is not a constant multiple of ψ_0. Then we lower again to get $a^2\psi \neq 0$. Then $E - 2$ is an eigenvalue. If $E - 2 < 0$ we have a contradiction. Otherwise, we continue lowering. Eventually, we have that $E - k < 0$ for some $k \geq 1$ is an eigenvalue. And this is a contradiction. So the set $\{1/2, 3/2, 5/2, \dots\}$ contains all the eigenvalues of H. We would like to understand the dimension of the *eigenspaces* $V_k := \{\psi \mid H\psi = (k + 1/2)\psi\}$.

Exercise 6.2.2 *Prove that* $\dim_{\mathbb{C}} V_k = 2$.
Hint: *Use the theory of linear ordinary differential equations.*

Schrödinger's razor is going to tell us that not all of the solutions of the Schrödinger equation are physically relevant, but only those that are square integrable. We will see why this is so when we discuss the physical interpretation of the 'wave' function. This gives context for the next exercise.

Exercise 6.2.3 *Recall the definition of the Hermite functions* ψ_k *in* (6.2.6). *Show that* $\psi_k \in V_k$ *is square integrable, that is*

$$\int_{\mathbb{R}} dx\, |\psi_k(x)|^2 \text{ is finite which we write as :} \quad \int_{\mathbb{R}} dx\, |\psi_k(x)|^2 < \infty.$$

Show that any $\phi \in V_k \setminus \mathbb{C}\psi_k$ *is not square integrable.*
Hint: *For the second part prove this for one such* ϕ *first and then prove that this implies the result for all such* ϕ. *Be warned that this problem will require some knowledge of ODE's and integrals that you might not yet have. In that case be patient and use this problem as motivation for learning more!*

We would also like to say that H has only eigenvalues in its spectrum, and not other types of spectrum. But we do not yet have the definition of spectrum, so we can not prove this for now. However, with more tools from functional analysis this can be proved. Patience, dear reader.

Exercise 6.2.4 *Solve the Schrödinger equation $H\psi = E\psi$ for $\psi : \mathbb{R} \to \mathbb{C}$ where $E \in \mathbb{R}$ and $H = -\frac{\hbar^2}{2m}\frac{d^2}{dx^2} + V$ in these cases:*

- $V(x) = 0$ *for all $x \in \mathbb{R}$.*
  (This is always called the free particle *and never the* free wave.*)*

- *The* finite square well:

$$V(x) = \left\{ \begin{array}{ll} 0 & \text{if } x \in [0,2], \\ E_0 > 0 & \text{if } x \notin [0,2]. \end{array} \right.$$

We say that the energy E_0, a real number, is the depth *of the well.*

- *The* infinite square well:

$$V(x) = \left\{ \begin{array}{ll} 0 & \text{if } x \in [0,2], \\ +\infty & \text{if } x \notin [0,2]. \end{array} \right.$$

In this case take $\psi(x) = 0$ for $x \notin [0,2]$ and use Dirichlet boundary conditions, namely, $\psi(0) = \psi(2) = 0$. Or, if you prefer, take the limit as $E_0 \to +\infty$ of the solution of the previous part of this exercise.

- *A* potential barrier:

$$V(x) = \left\{ \begin{array}{ll} 0 & \text{if } x < 0, \\ E_0 > 0 & \text{if } x \geq 0. \end{array} \right.$$

We say that the energy E_0, a real number, is the height *of the barrier. In this case we seek solutions ψ such that $\psi(x), \psi'(x)$ are continuous at $x = 0$ and $\lim_{x \to +\infty} \psi(x) = 0$.*

Warning! A solution of the eigenvalue equation $H\psi = E\psi$ consists in finding two unknowns: the eigenvalue E and the eigenvector $\psi \neq 0$.

6.3 Notes

The cartoon is by S. Harris and is available on a T-shirt from the American Mathematical Society. It seems that Harris is not a scientist, but that he has an uncanny understanding of how we think.

The seven Millennium Problems of the Clay Institute can be found on the Internet. I referred to the one on the existence of a quantum gauge field theory. It is generally considered to be a formidable problem.

Operators with the same CCR as the raising operator a^* and the lowering operator a arise in more advanced topics such as quantum field theory. In those contexts the raising operators are given a new name: *creation operators*.

And the lowering operators are then called the *annihilation operators*. The reason for this colorful terminology comes from the new context.

The standard physics texts for quantum theory include the harmonic oscillator, the hydrogen atom (which we will discuss in Chapter 11) and a slew of other examples of solutions, some approximate, of Schrödinger's equation, such as square wells, double wells and potential barriers. While this is great practice and slowly but surely builds up some sort of 'intuition' about quantum theory, I leave most of that to those fine standard texts. Here I am aiming for another sort of 'intuition' which arises by taking the path indicated by the mathematical structures, which in turn arise naturally from the physics. So, I continue with the first of three chapters dedicated to a detailed interpretation of the 'wave' function ψ.

Chapter 7

Interpreting ψ: Mathematics

> The limits of my language
> are the limits of my world.
> Ludwig Wittgenstein

Before entering into the physics interpretation of the wave function, we give its mathematical interpretation, which will be our starting point. This also introduces *state* into the mathematical language.

7.1 A Cauchy Problem

Schrödinger's 1926 paper introduced the wave function ψ as the solution of what is now called the Schrödinger equation. But there was no physical interpretation of ψ in that paper. That seems to be an important missing part of the theory. Actually, we introduced two related versions of the Schrödinger equation, the first dependent on time and the second independent of time. We have yet to give a physical interpretation of the solutions of these equations. Before addressing this fundamental and controversial problem in physics, we will first discuss the problem of the mathematical interpretation. This will aid us in understanding the physical interpretation as well as motivating the role that functional analysis plays in quantum theory.

Let's denote by $\Psi(t, \mathbf{x})$ the unknown of the time dependent equation and by $\psi(\mathbf{x})$ the unknown of the time dependent equation for the energy E. The time dependent equation is first order in the time derivative. In fact it has the form

$$i\hbar\frac{\partial\Psi}{\partial t} = H\Psi, \tag{7.1.1}$$

where H is a special operator, the quantum Hamiltonian. So we expect (mathematically as well as physically!) that this is a good Cauchy problem,

© Springer Nature Switzerland AG 2020
S. B. Sontz, *An Introductory Path to Quantum Theory*,
https://doi.org/10.1007/978-3-030-40767-4_7

which means that for each initial value $\Psi_0(\mathbf{x})$ there should exist a unique solution $\Psi(t, \mathbf{x})$ of (7.1.1) such that

$$\Psi(0, \mathbf{x}) = \Psi_0(\mathbf{x}).$$

This fits well with an idea from classical physics, known as *determinism*, that complete knowledge of a physical system at some time determines the complete knowledge of that system at all future times. The seemingly obvious concept of knowledge, let alone complete knowledge, remains problematical to this day in the interpretation of quantum theory and is one of the reasons that some philosophers are actively interested in quantum theory.

One of Dirac's clever ways of speaking about quantum theory was to use the expressions *c-number* and *q-number*. While this terminology is not much used any more, it has a certain charm. Let's see why. First, c-number just means a complex number. And that is clear enough. But q-number means a quantum number, whatever that means! Well, Dirac tells us what that means; it is a linear operator. But this seemingly strange terminology comes from an equally strange intuition, namely, that the q-numbers are pretty much like the c-numbers with the notable exception that one q-number need not commute with another q-number. We consider the time dependent Schrödinger equation (7.1.1) in this light, and then we ask ourselves what the solution of (7.1.1) must be. Well, if H were a c-number, then the solution would be

$$\Psi(t, \mathbf{x}) = e^{-itH/\hbar}\Psi_0(\mathbf{x}),$$

where $\Psi_0(\mathbf{x})$ is the initial value at time $t = 0$, just as above. Of course, in this case the expression $e^{-itH/\hbar}$ will be a c-number, too.

So we make the quantum leap (sorry about that!) and suppose that when H is a q-number the same formula holds for the solution, except that now the expression $e^{-itH/\hbar}$ is a q-number. This means we would like to define an operator $e^{-itH/\hbar}$ (given that H is a q-number) in such a way that all this really makes sense. And this can be done, although the full details require the spectral theorem for a self-adjoint densely defined operator. We leave that full story to any good functional analysis text. The conclusion of that story is that the time dependent Schrödinger can be made, as we anticipated earlier, into a *Cauchy problem*, namely, for every initial condition (in some 'nice' space) there is a unique solution. We will say more about this in Chapter 9 where we introduce the appropriate 'nice' space, which will be a Hilbert space.

For now here are some partial details. Suppose that the initial condition Ψ_0 lies in the eigenspace $V_E = \{\psi \mid H\psi = E\psi\}$ where E is an eigenvalue (or *eigenenergy*) of H. Then H when restricted to this subspace is the same as the operator of multiplication by E, that is

$$H|_{V_E} = EI,$$

where I is the identity operator on V_E. On this subspace V_E the q-number $e^{-itH/\hbar}$ becomes (the operator of multiplication by) the c-number $e^{-itE/\hbar}$

and the solution on that subspace is indeed

$$\Psi(t, \mathbf{x}) = e^{-itH/\hbar}\Psi_0(\mathbf{x}) = e^{-itE/\hbar}\Psi_0(\mathbf{x}).$$

We have seen this all before, of course. Only in our earlier presentation we wrote ψ_2 instead of Ψ_0 for the solution of the time independent Schrödinger equation with eigenvalue E. (See (4.3.1) and (4.3.3).) Now here are two elementary comments. First, we can do this for every possible eigenvalue of H. Second, suppose that we can write the initial condition as a sum of functions, each of which lies in some V_E, say

$$\Psi_0 = \sum_{j \in J} \lambda_j \psi_j, \tag{7.1.2}$$

where $\lambda_j \in \mathbb{C}$ and $\psi_j \in V_{E_j}$. Then by the linearity of the time independent Schrödinger equation we would expect that

$$\Psi(t, \mathbf{x}) = \sum_{j \in J} \lambda_j \, e^{-itE_j/\hbar} \, \psi_j(\mathbf{x})$$

is a solution of it. And this works out, even when the index set J is infinite, provided we use the appropriate sense of the convergence of the infinite series. Recall that the quantum harmonic oscillator has infinitely many eigenvalues. So the presence of infinite sums is not avoidable in this discussion.

This exposition motivates an *ad hoc* definition, namely that

$$e^{-itH/\hbar}\Psi_0 = \sum_{j \in J} \lambda_j \, e^{-itE_j/\hbar} \, \psi_j,$$

provided that the expansion (7.1.2) holds. Ultimately, we want a definition of the exponential for a large class of operators. Actually, we want a definition of any 'reasonable' function of an operator for a large class of operators. The *functional calculus*, which is one form of the *spectral theorem*, does that. And these topics are part of a mathematical subject called *functional analysis*.

Another item on our wish list is that every possible initial condition Ψ_0 can be expanded out as in (7.1.2) in terms of eigenfunctions. Here 'possible' means that Ψ_0 lies in the appropriate Hilbert space. This is a wish upon a sometimes star. Because sometimes we can do this. But sometimes we can not. For example, we can do this for the harmonic oscillator. And that is an important theorem about the harmonic oscillator. But we can not do this for the free particle, which is described by the Schrödinger operator with $V \equiv 0$. Obviously, we can not avoid discussing this rather simple example in quantum theory. So, in that case the quantum Hamiltonian is

$$H = -\frac{\hbar^2}{2m}\Delta. \tag{7.1.3}$$

What goes wrong? Again, the appropriate Hilbert space comes into the discussion, because there are no eigenfunctions in that Hilbert space for this operator. So, there are no eigenvalues. But the operator does have non-trivial spectrum, and the spectral theorem does apply to it. And using (7.1.3),

$$e^{-itH/\hbar} = e^{it\hbar^2\Delta/(2m\hbar)} = e^{(it\hbar/2m)\Delta}$$

is an operator in that Hilbert space for every real number t. There is a formula for this operator as a *convolution integral operator*. In this way we have an explicit solution for any initial condition for the free particle. The mathematical details require familiarity with the *Fourier transform*. This is worked out very carefully in various books, such as [11].

There is here an underlying unexpressed idea, which says that a quantum physical system is *completely* described by the solution of its associated time dependent Schrödinger equation. Again, this is Schrödinger's razor. Also, we see now that the time dependent Schrödinger equation is *deterministic*, that is to say, given an initial condition there exists a unique solution of it satisfying that initial condition. Also, that solution as a function of $\mathbf{x}$ will remain in the appropriate Hilbert space as t changes if the initial condition also lies in that Hilbert space. So, the function that completely describes the system at time $t = 0$ evolves to the unique function that completely describes the system at time $t = T$ at any later (or earlier!) time T. And then that function for time $t = T$ can be used as the initial condition for the same time dependent Schrödinger equation. In other words, 'initial condition' and 'final condition' are just two cases of the general concept of 'condition'. For this reason we say that any non-zero solution $\Psi(t, \mathbf{x})$ is the *state* of the physical system at time t. (Already we can see two problems with the identically zero solution, $\Psi(t, \mathbf{x}) \equiv 0$. First, it has trivial time dependence. And second, it is a solution for *every* quantum system.) In the next chapter this notion of state will be modified somewhat due to a physics criterion, but this basic idea will remain. That criterion will also give another reason for excluding the solution $\Psi(t, \mathbf{x}) \equiv 0$ from further consideration.

As an aid in understanding the idea of the state of a physical system, we can refer to the classical theory. (For those without previous familiarity with this topic, you can skip to the next paragraph.) For example, Newton's equation of motion for a single particle is second order in time, and so we need initial conditions for both position and velocity. That solution is usually taken to be the trajectory $\mathbb{R} \ni t \mapsto \gamma(t) \in \mathbb{R}^3$ of the particle, that is, its position as a function of time t. But the position alone at a future time does not describe the physical state of the system; we also need the velocity. So the state of this physical system is $(\gamma(t), \gamma'(t)) \in \mathbb{R}^3 \times \mathbb{R}^3$. In this way, the state of the system at any time can be used as the initial condition for the motion of the particle. An alternative and equivalent way of describing the state of this physical system is to use the *phase space* of position and momentum variables. But I digress.

If we have the case $\Psi(t, \mathbf{x}) = e^{-itE/\hbar} \psi(\mathbf{x})$, where ψ is an eigenfunction with eigenvalue E for the time independent Schrödinger equation, we note that the time evolution of $\Psi(t, \mathbf{x})$ is rotation in the one-dimensional subspace

$\mathbb{C}\psi$ of V_E. However, we shall see later on that this phase factor $e^{-itE/\hbar}$ does not really change the state "as time goes by". (A *phase factor* is a complex number whose absolute value is 1.) So we say that $\Psi(t, \mathbf{x})$ is a *stationary state* with energy E. We also say that $\psi(\mathbf{x})$ is an *eigenstate* with *eigenenergy* E. Earlier we said that $\psi(\mathbf{x})$ is also called a stationary state. One just learns to live with this mild sort of ambiguity.

The concept of a state depends very much on context. It survives in the theory of C^*-algebras, for example. That in turn feeds back into probability theory, where a probability measure is recognized as a special type of state. But the logical, though maybe not historical, origin of the word 'state' comes from physics and is then incorporated into mathematics.

7.2 Notes

The material in this chapter is basically mathematical, though some of the statements require a more precise presentation of the mathematics of Hilbert space. However, the outline given here is correct. The only statement that goes beyond mathematics *per se* is that the time dependent Schrödinger equation gives a complete description of a quantum system, that is to say, Schrödinger's razor. And that is a controversial statement because of (among many other reasons) the concept of collapse (or quantum jump) of the wave function as we shall discuss in Chapter 10. But, as a matter of practice, there is only one fundamental quantum time evolution equation, though possibly in a disguised form, used by physicists. And that's the Schrödinger equation.

Determinism as a philosophical principle has a long history, going back at least to some ideas of the Pre-Socratics, who seem to be the first to try to explain nature using reason. Other major adherents in the Western tradition include Aristotle, Leibniz, and Spinoza. The Cauchy problem is a more recent development in the study of differential equations and clearly has intellectual connections with deterministic doctrines.

For whatever reason determinism is often defended by citing the example of an object falling due to the gravity of the Earth. So, suppose you have a ball in your hand and you release it. And it falls. And using classical physics you can quite well describe—even predict—how it will fall and when it will hit. Even taking into account air resistance. You have defended determinism. But wait! You only described—and predicted—the motion of the ball for a few seconds. The deterministic viewpoint says that one can do this indefinitely into the future. And that future is vast, much more than a matter of seconds. The current age of the (known) universe is around 10^{17} seconds, and we expect it to last at least that long into the future, though probably much longer. Let's say that you dropped your ball from the top of a very high skyscraper. The time until it hits ground might be around 10 seconds. The argument in favor of determinism requires an extrapolation of some 16 orders of magnitude. Yet it is extremely difficult to say where the ball will be even 10 seconds after it hits ground. Will it bounce? How? Will it break into pieces? How many? Where will it be in just one year, let alone 10 years?

Chapter 8

Interpreting ψ: Physics

Sir, I have found you an argument;
but I am not obliged to find you
an understanding.

Samuel Johnson

A physics theory must be provided with ways for explaining empirical data, whether experimental or observational. Otherwise, there is no method for testing that theory. This methodology is called its interpretation.

8.1 The Entrance of Probability

While the comments in the previous chapter give a lot more meaning to the wave functions $\Psi(t, \mathbf{x})$ and $\psi(\mathbf{x})$, they still lack a more direct physical interpretation. That standard interpretation was given by M. Born who used an unpublished suggestion provided by A. Einstein. This is one of those "bolts from the blue" as the saying goes. In other words, do not expect any motivation for this!

The idea is that one wants a solution $\Psi(t, \mathbf{x}) = \Psi_t(\mathbf{x})$ of the Schrödinger equation that also satisfies the *normalization condition*

$$\int_{\mathbb{R}^3} d\mathbf{x} \, |\Psi(t, \mathbf{x})|^2 = 1 \qquad (8.1.1)$$

for every time t. In the case of a stationary state $\Psi(t, \mathbf{x}) = e^{-itE/\hbar} \psi(\mathbf{x})$ with $E \in \mathbb{R}$ this normalization condition becomes $\int_{\mathbb{R}^3} d\mathbf{x} \, |\psi(\mathbf{x})|^2 = 1$. Notice that this normalization condition is a non-linear, inhomogeneous condition on the solution of the linear, homogeneous Schrödinger equation. We next define

$$\rho_t(\mathbf{x}) := |\Psi(t, \mathbf{x})|^2 \geq 0 \qquad \text{and} \qquad \rho(\mathbf{x}) := |\psi(\mathbf{x})|^2 \geq 0. \qquad (8.1.2)$$

© Springer Nature Switzerland AG 2020
S. B. Sontz, *An Introductory Path to Quantum Theory*,
https://doi.org/10.1007/978-3-030-40767-4_8

Then $\rho_t(\mathbf{x})$ is a *probability density function on* $\mathbb{R}^3$ for every time t, and $\rho(\mathbf{x})$ is a *probability density function on* $\mathbb{R}^3$. This terminology comes from classical probability theory. Here is the definition:

Definition 8.1.1 *We say that a (measurable) function* $f : \mathbb{R}^n \to \mathbb{R}$ *is a* probability density function on $\mathbb{R}^n$ *provided that*

$$f(\mathbf{x}) \geq 0 \text{ for all } x \in \mathbb{R}^n \quad \text{and} \quad \int_{\mathbb{R}^n} d^n \mathbf{x}\, f(\mathbf{x}) = 1.$$

This definition could lead to a detailed, boring discussion about measure theory, measurable functions, and Lebesgue integrals. But in practice the probability density function f in quantum theory will be continuous and Riemann integration theory from an elementary calculus course works just fine. If you wish to know a bit more (including all the technical terms in parentheses), see the crash course on measure theory in Appendix A.

The 'wave' function $\Psi_t(\mathbf{x})$ is interpreted as saying that the 'particle' which it describes has the probability of being in a ('nice') subset B of $\mathbb{R}^3$ at time t given by

$$\text{Prob}(B \mid \Psi_t) := \int_B d\mathbf{x}\, \rho_t(\mathbf{x}) = \int_{\mathbb{R}^3} d\mathbf{x}\, \chi_B(\mathbf{x}) \, |\Psi_t(\mathbf{x})|^2 = \langle \Psi_t, \chi_B \, \Psi_t \rangle,$$

where we used the inner product notation (6.2.9) in the last expression. Here χ_B is the *characteristic function of the subset* B, defined as $\chi_B(x) = 1$ if $x \in B$ and $\chi_B(x) = 0$ if $x \notin B$. Let's note that $\text{Prob}(B \mid \Psi_t)$ is indeed a real number between 0 and 1. This is so since

$$0 \leq \int_B d\mathbf{x}\, \rho_t(\mathbf{x}) \leq \int_{\mathbb{R}^3} d\mathbf{x}\, \rho_t(\mathbf{x}) = 1.$$

This first inequality follows because the integral of a non-negative function is a non-negative number and the second inequality because integrating a non-negative function over a bigger set than B gives us a bigger number.

For those who are savvy about such things in measure theory, B should be a *Borel set*. And for those who are not so blessed, do not worry since you do not have the sophisticated mathematical tools needed even to consider or construct a set which is not Borel. Lucky you! In practice, B will be a closed, solid ball, or a rectangular box or some quite other simple geometrical objects, all of which are Borel sets. And the integrals will be Riemann integrals. If these words are not enough for you, then have a look at Appendix A.

If we consider a system described by a stationary state ψ, then

$$\text{Prob}(B \mid \psi) := \int_B d\mathbf{x}\, \rho(\mathbf{x}) = \int_{\mathbb{R}^3} \chi_B(\mathbf{x}) |\psi(\mathbf{x})|^2 = \langle \psi, \chi_B \, \psi \rangle$$

is interpreted as the probability that the associated 'particle' is in the (Borel) subset B for all times t. We note one very immediate consequence that we already mentioned earlier. And that is that $\psi \equiv 0$ is excluded from consideration by this physical interpretation. Also, $\Psi_t \equiv 0$ is excluded for each value of t.

An implication of this interpretation is that one can experimentally test if a theoretically given 'wave' function is correct by doing the experiment to measure the location of the 'particle' repeatedly (and identically in some sense) many times. At the atomic scale and below such experiments for measuring position are non-trivial and for many systems have never been done nor are expected to be feasible.

Another implication of this is that $e^{i\theta} \Psi(t, \mathbf{x})$ describes the same physical system as does $\Psi(t, \mathbf{x})$ provided that θ is real. Similarly, we can multiply a stationary state ψ by any such *phase* $e^{i\theta}$ and get a state describing the same physical system. A consequence of this probabilistic interpretation is that we want to define differently what is the state of a physical system. We will come back to this when we discuss Hilbert spaces.

Yet another important consequence of this interpretation is that we do not look for a curve $t \mapsto \gamma(t) \in \mathbb{R}^3$ for t in some interval in $\mathbb{R}$ in order to describe, even partially, the time evolution of a quantum physical system. Instead the time dependent probability density function ρ_t in (8.1.2) gives the time evolution of the location of the 'particle'. In a stationary state, the time independent probability density function ρ in (8.1.2) describes the location of the 'particle' for all times t, that is, the probability density remains the same for all time and the 'particle' can not be thought of as moving on some complicated trajectory that somehow produces the probability density ρ.

This is infamously difficult to understand. Without going now into more details, which will be discussed later in Chapter 16, *quantum probability theory* says that *identically* prepared systems which undergo *identically* performed measurements will give *different* results in general (though with some rather important exceptions). Moreover, it says that there are no further hidden properties that distinguish among the seemingly identical systems or experiments. This fundamental lack of determinism is sometimes referred to as *quantum fluctuations*. If one is thinking of this expression as a way of encoding the essential probabilistic aspect of quantum theory, then that is fine. But if one thinks that the quantum fluctuations are varying physical properties that distinguish between 'identical' systems, then one is thinking classically. By the way, such variations are often considered to be actual changes in time of a stationary (i.e., time independent) quantum state, quite a contradiction! Quantum probability excludes any such hidden physical property. Of course, one must not be too dogmatic about this, just a little dogmatic. I think it is clearly true that quantum probability *says* that such hidden properties are excluded. However, it is a question for experiment to verify or falsify that assertion of quantum probability. And experiment, nothing else, always has the final say.

In general, the time dependent 'wave' function $\Psi_t(\mathbf{x})$ is complex valued, while its associated probability density function $\rho_t(\mathbf{x}) := |\Psi(t, \mathbf{x})|^2$ is real valued and non-negative. Consequently, another common name for $\Psi_t(\mathbf{x})$ is the *probability amplitude*. As expected, the time independent 'wave' function $\psi(\mathbf{x})$ is also called a probability amplitude. In classical probability, there are very many probability density functions, but there are no complex valued probability amplitudes. While this comment is elementary, it indicates a profound, fundamental difference between classical and quantum probability.

8.2   Expected Value

The probability density function $\rho_t(\mathbf{x})$, as introduced in (8.1.2), is the sort of object studied in classical probability theory, even though what it comes from, namely $\Psi(t, \mathbf{x})$, is not. As such we can use all the tools of classical probability to study $\rho_t(\mathbf{x})$ once we have calculated it using quantum theory. For example, for every (integrable) function $f : \mathbb{R}^3 \to \mathbb{C}$ we have its integral with respect to this probability density function, namely

$$\mathcal{E}(f) = \langle f \rangle = \int_{\mathbb{R}} f(\mathbf{x}) \, \rho_t(\mathbf{x}) \, d^3\mathbf{x},$$

which in classical probability theory is called the *expected value of f* although in some other contexts it also is referred to as the *average value of f*. Here we are introducing some standard notations for the well known integral. The notation $\mathcal{E}(f)$ comes from classical probability theory, while $\langle f \rangle$ comes from physics. These notations are ambiguous since they omit reference to the probability density function $\rho_t(\mathbf{x})$ or to its underlying probability amplitude $\Psi(t, \mathbf{x})$. In order to remove that ambiguity sometimes these notations are embellished with the appropriate symbols as subscripts. But no matter how you write it, the integral is the integral is the integral. Or in other words, old wine in new bottles.

We also have that $\mathcal{E}(f) = \langle \Psi_t, f \, \Psi_t \rangle$, which shows how we calculate the classical expected value in terms of the 'wave' function of quantum theory.

One of the most common aspects of classical probability and statistics is the study of the *moments* of a probability density function, which in this case are given for each integer $i \geq 1$ and each coordinate x_k of $\mathbf{x} = (x_1, x_2, x_3)$ as

$$\langle x_k^i \rangle := \int_{\mathbb{R}} x_k^i \, \rho_t(\mathbf{x}) \, d^3\mathbf{x}$$

provided that this integral converges absolutely. Of course, this is simply the expected value of x_k^i. Historically, the first moments (that is, the case when the exponent $i = 1$) have received much attention. It is convenient to put them into an *expected (or average) position vector*

$$\langle \mathbf{x} \rangle := (\langle x_1 \rangle, \langle x_2 \rangle, \langle x_3 \rangle) \in \mathbb{R}^3,$$

again provided that the integrals defining these three expected values are absolutely convergent. It is critically important to realize that the expected position vector does *not* tell us where we will measure the position in any one particular experiment, but rather gives us the expected (i.e., average) position that comes from a lot of identically prepared experiments. Moreover, the basic object that experiment should address is the probability density function itself which always exists even if its first moments do not. Again, this must be done by using a large ensemble of identically prepared experiments. This is done using classical *statistics*, a science which was invented precisely to deal with this sort of problem. Due to quantum theory we might want to measure the probability amplitude, a fundamentally quantum object, as well. We will come back later to this important problem.

8.3   Copenhagen and Philosophy

> Pour digérer le savior, il faut
> l'avoir avalé avec appétit.
>
> Anatole France

A much fuller development of Born's interpretation was given by N. Bohr, a Danish physicist who worked in Copenhagen. Without fleshing out that further for now, let me note that this combination of the works of Born and Bohr (and others) has come to be known as the *Copenhagen interpretation*.

Some scientists hold that some of these consequences are ridiculous, and so the Copenhagen interpretation must be wrong. Basically, this is because the consequences fly in the face of the standard ways of considering physical phenomena in classical physics. Two of Aristotle's four *causes* (which I prefer to call four types of *explanation*) already fell to the wayside with the advent of the modern scientific period in (approximately) the early 17th century. Those that fell were the *final cause* and the *formal cause*. The arguments for discarding these two types of explanations, or even what they are, need not concern us. The success of the subsequent scientific achievements over the next four centuries serves as a partial justification *a posteriori* for their elimination.

The two causes that survived were the *material cause* and the *efficient cause*. This was a major contribution by F. Bacon, who curiously enough is not remembered for his scientific activities. In Newtonian physics this corresponds to the statement that everything (with the exception of light) consists of masses (the material causes) subject to nothing other than forces (the efficient causes). This is still enough to have a deterministic theory. But quantum theory in the Copenhagen interpretation admits no determinism as had been previously understood, due to the probabilistic interpretation of the wave function. Consequently with the advent of quantum theory, these two remaining causes have also come under reconsideration. Maybe they should go, too! But some physicists want to preserve determinism and, at least in some form, the material and efficient causes. This leads to non-trivial questions about what 'material' means and what properties account for the 'reality' of 'material'. The concept of 'reality' itself has even been rejected by a vocal community! I don't want to get deeper into these considerations, let alone what one might mean by 'reality'.

It is interesting to note that light has played the central role *twice* in revising, one can even say in overturning, the Newtonian world view. The second time (in 1905) was the special theory of relativity as introduced by A. Einstein. We leave that discussion to other texts. The first time was the quantum theory as already foreseen by M. Planck in 1900. The whole confused argument over whether light either is only a 'wave' or is only a 'particle' can be seen as an attempt to see light as material (i.e., particles) which interacts with other materials (lenses, mirrors, retinas, etc.) by some unspecified efficient (i.e., intermediating) cause or as a wave (riding on some background, the unspecified material *ether*) which is nothing other than the efficient cause mediating interactions between one material object and another material object. The identification in the middle of the 19th century via *Maxwell's equations* of light as a time dependent *electromagnetic field* only added fuel to the fire, since it is not clear which side of the scale this weighs on, even though it was considered by most physicists at the time as definitive evidence for the wave interpretation, and therefore indirect evidence for the existence of a material ether. The current view is that there is no ether whose physical modes of vibration are light.

But is the electromagnetic field some sort of physical material in its own right or an efficient cause acting between material objects? My guess is that most physicists nowadays would find this question, even on the level of *classical electromagnetic theory*, to be annoyingly besides the point and, if some answer were insisted open, they would say that the electromagnetic field has physical meaning as a material (though zero mass) object in itself *and* as an interaction between other materials. The subsequent quantization of this *classical* 'wave/particle duality' of light does not help us better understand the nature of things, as far as I can make out. Rather I argue that it helps us realize that the distinction into two distinct types of causation (material and efficient) imposes a certain duality on nature, a duality which apparently is not there. But what does this have to do with what we do in physics? We have equations; we solve them (whether exactly or approximately is quite unimportant); we compare with appropriate experiments. And the whole idea of deterministic causes becomes a forgotten issue. What remains is the task of formulating new general ways for explaining quantum phenomena.

The wide success of the Baconian approach should be chalked up its being adequate for the study of classical theories, such as almost all of biology, but not having absolute universal applicability. Similarly, the classical Newtonian theory also has had enormous success in very many fields, but it also is not a universal theory. I shall comment later on about non-deterministic explanations ('causes') for quantum phenomena. These will be probabilistic explanations, but in terms of the appropriate concept of probability called *quantum probability*, which is explained in detail in Chapter 16.

There have been numerous intents to modify or replace the Copenhagen interpretation. We will not concern ourselves with those. Rather we remark that this *physical* interpretation feeds back to the *mathematical* underpinning of quantum theory. It tells us that we want to work with solutions of the Schrödinger equation that lie in a specific Hilbert space: $L^2(\mathbb{R}^3)$. This aspect of the theory deserves its own chapter, which will be Chapter 9.

8.4 Trajectories (Optional)

This brief section is optional, since it concerns the contrast of quantum theory to classical theory, a topic outside the scope of this book. A *trajectory* is the one-dimensional, continuous path along which a classical point particle moves according to Newton's Second Law of Motion. This is a fundamental concept in classical theory. Trajectories of quantum 'particles' do exist; however, they are not fundamental but rather secondary concepts in quantum theory. The solutions of the time dependent Schrödinger equation are fundamental in quantum theory.

For example, the observation of an electron in a *bubble chamber* should be understood within the context of quantum theory. The electron is observed at a finite number of positions, namely, where the bubbles are formed. Now we think of a classical point particle as moving continuously along its trajectory, despite the fact that even the highest speed movie or video only measures that particle at a finite number of places at a corresponding finite number of times. This is much like the observation of an electron in a bubble chamber. However, in classical theory the underlying theoretical concept is a function $\mathbf{x}(t) \in \mathbb{R}^3$ of time t; this function is what we think of as the *motion* of the classical particle. On the other hand, in quantum theory the time evolution of a quantum 'particle' is a function $\psi(t)$ in a Hilbert space, as will be explained in the next chapter. (Here again, t denotes the time.) In this sense there is no motion in quantum theory, at least in the classical sense of continuous motion along a theoretical trajectory in three-dimensional space.

Unfortunately, many scientists fall into the fallacy of thinking that there is an underlying classical, continuous, one-dimensional trajectory for a quantum 'particle'. However, I do not wish to be overly critical, since it does make sense to think of a discrete version of a trajectory in quantum theory as is the case with bubble chamber observations.

8.5 Notes

Aristotle was not the first philosopher of science, nor of knowledge in general. But for all his flaws, he was a person who took these issues seriously. Probably his errors are in large measure due to the historical context; there were no serious contemporaries to challenge him. He was surrounded by students and was no doubt an excellent expositor of his ideas as evidenced in his books, which appear to be what we would call lecture notes. But there does not seem to have been a strong tradition that valued constructive criticism of the teacher. Of course, in those days it was nigh well impossible to think of the motions of celestial objects in the same terms as those of terrestrial objects. So some of Aristotle's thinking was essentially doomed to be wrong from the start. But his 'intuitive' idea that a heavy object gravitates to Earth faster than a light one is an unforced error.

The problem for contemporary philosophy of science is that quantum theory throws a monkey wrench into whatever 'intuitive' approach with which *Darwinian natural selection* has provided our brains. We have evolved to interact with a world of 'classical reality' since that has had survival value. In this regard Aristotle is one of us. So we had best be careful lest we be doomed also to be wrong from the start.

Entirely too much emphasis is put on measure theory in the standard mathematics curriculum. The general picture is presented in Appendix A in about 14 pages. Though not everything is proved there, the basic facts are all clearly stated. Unless one plans on doing research in measure theory or in fields where measures are important in and of themselves, there is little reason to delve much deeper. The real motivation for knowing a bit of measure theory is that it allows one to calculate real life integrals, not just integrals of characteristic functions. And as a physicist once remarked to me: "In physics we evaluate the integral." So one must get used to rolling up one's sleeves and doing the integral to develop some sort of 'intuition' for quantum theory. In this regard almost all the famous measure theory texts by the famous mathematicians are utterly useless. But I do admire the authors of [7] for their enthusiastic approach and abundance of examples. Also [14], [19] and [23] are some other handy references for doing integrals.

Chapter 9

The Language of Hilbert Space

Da wäre noch eine Sache.
Was ist ein Hilbertraum?
David Hilbert

The basic mathematical language of quantum physics is the theory of Hilbert spaces. The basic facts are presented without giving proofs. The first three sections form a mini-course of some 14 pages and can be read as need arises in order to understand concepts and notation. The next four sections are explicitly optional. This chapter overlaps the material in many fine *functional analysis* texts, such as [34], where proofs can be found.

The amazing idea which has motivated so much of modern mathematical analysis is to view a function not only as a way of sending points in one set to points in another set, but also as itself a point in an infinite dimensional space with its own mathematical structure. It is in these *function spaces*, such as Hilbert spaces, where we search for the solutions of differential equations.

9.1 Facts and Definitions, No Proofs

Let's understand what the Hilbert space $L^2(\mathbb{R}^3)$ is and how it dovetails with Born's interpretation. Unfortunately, this leads us into the world of technical details of measure theory. (See Appendix A for those details.) Anyway, as a first approximation we oversimplify by defining $L^2(\mathbb{R}^3)$ to be all functions $\phi : \mathbb{R}^3 \to \mathbb{C}$ such that the integral $\int_{\mathbb{R}^3} d\mathbf{x} \, |\phi(\mathbf{x})|^2$ exists and is finite. (Actually, we should also assume that ϕ is *measurable*, for those who speak the language of measure theory.) Such a function ϕ is said to be *square integrable*. The first important mathematical consequence is this:

© Springer Nature Switzerland AG 2020
S. B. Sontz, *An Introductory Path to Quantum Theory*,
https://doi.org/10.1007/978-3-030-40767-4_9

Proposition 9.1.1 *If $\phi_1, \phi_2 \in L^2(\mathbb{R}^3)$, then the integral*

$$\int_{\mathbb{R}^3} d\mathbf{x}\, \phi_1(\mathbf{x})^* \phi_2(\mathbf{x})$$

exists. (Here for any complex number $\lambda = a + ib \in \mathbb{C}$ with $a, b \in \mathbb{R}$ we define its complex conjugate by $\lambda^* := a - ib \in \mathbb{C}$. Of course, $i = \sqrt{-1}$.)

What this means is that this integral makes sense within the context of measure theory, and in particular is equal to a complex number. The proof of this proposition essentially is an application of the Cauchy-Schwarz inequality, though it uses a 'tricky bit' from measure theory (called *absolute integrability*) as well. This proposition invites the next definition.

Definition 9.1.1 *If $\phi_1, \phi_2 \in L^2(\mathbb{R}^3)$, then we define their* inner product *by*

$$\langle \phi_1, \phi_2 \rangle := \int_{\mathbb{R}^3} d\mathbf{x}\, \phi_1(\mathbf{x})^* \phi_2(\mathbf{x})$$

So, $\langle \phi_1, \phi_2 \rangle \in \mathbb{C}$. This inner product satisfies the usual properties of the Hermitian inner product defined on the finite dimensional complex vector space $\mathbb{C}^n$, where $n \geq 1$ is an integer. That inner product is defined by

$$\langle w, z \rangle := w_1^* z_1 + \cdots + w_n^* z_n, \tag{9.1.1}$$

where $w = (w_1, \ldots, w_n) \in \mathbb{C}^n$ and $z = (z_1, \ldots, z_n) \in \mathbb{C}^n$.

This turns out to be the natural generalization to complex vector spaces of the more familiar inner product defined on the real vector space $\mathbb{R}^n$ by

$$\langle x, y \rangle := x_1 y_1 + \cdots + x_n y_n \tag{9.1.2}$$

for $x, y \in \mathbb{R}^n$, where $x = (x_1, \ldots, x_n)$ and $y = (y_1, \ldots, y_n)$.

Here are the properties of these inner products on complex vector spaces expressed in the notation of $L^2(\mathbb{R}^3)$. We take $\psi, \phi, \phi_1, \phi_2 \in L^2(\mathbb{R}^3)$ and $\lambda \in \mathbb{C}$.

1. $\langle \psi, \phi_1 + \phi_2 \rangle = \langle \psi, \phi_1 \rangle + \langle \psi, \phi_2 \rangle$ and $\langle \phi_1 + \phi_2, \psi \rangle = \langle \phi_1, \psi \rangle + \langle \phi_2, \psi \rangle$.

2. $\langle \psi, \lambda \phi \rangle = \lambda \langle \psi, \phi \rangle$ and $\langle \lambda \psi, \phi \rangle = \lambda^* \langle \psi, \phi \rangle$.

3. $\langle \phi, \phi \rangle \geq 0$.

4. Moreover, $\langle \phi, \phi \rangle = 0$ implies that $\phi = 0$, except on a set of *measure zero*, another concept from measure theory.

5. $\langle \phi, \psi \rangle = \langle \psi, \phi \rangle^*$ (Complex Symmetry)

Property 1 is called *bi-additivity*. Combining this with Property 2 we have *sesqui-linearity*, which is not to be confused with bi-linearity where there would be no complex conjugation in the second equation. Property 3 together with $\langle \phi, \phi \rangle = 0 \implies \phi = 0$ is called *positive definiteness*.

The complex vector space $L^2(\mathbb{R}^3)$ with this inner product turns out to be a particular case of what is known as a *Hilbert space* (Well, almost. Item 4 on the above list is a thorn in our side. In a Hilbert space we have that the inner product is positive definite. So, we are sweeping under the rug this technical detail.) And this Hilbert space is the 'nice' space referred to several times earlier. J. von Neumann recognized that this is the space for doing the spectral analysis of quantum Hamiltonian operators. That theory includes the eminently important *spectral theorem* and its associated *functional calculus*, which can be found in any good functional analysis text. Here is the definition:

Definition 9.1.2 *We say that a vector space $\mathcal{H}$ over the field of complex numbers $\mathbb{C}$ is a* (complex) *Hilbert space if there is a bi-additive, sesqui-linear, positive definite, complex symmetric inner product* $\langle \cdot, \cdot \rangle$ *mapping $\mathcal{H} \times \mathcal{H} \to \mathbb{C}$ with the* completeness property:

- *Define $||\phi|| := +\sqrt{\langle \phi, \phi \rangle}$ and $d(\phi_1, \phi_2) := ||\phi_1 - \phi_2||$, which are a norm and a metric, respectively. (Look up definitions for these concepts if need be.) Then we require that $\mathcal{H}$ is a* complete *metric space, that is, every Cauchy sequence of elements in $\mathcal{H}$ converges to a (necessarily unique) element in $\mathcal{H}$.*
  Recall from analysis that a sequence $\{\phi_j \,|\, j \in \mathbb{N}\}$ is said to be Cauchy if for every $\epsilon > 0$ there exists an integer $M > 0$ such that $||\phi_j - \phi_k|| < \epsilon$ whenever $j, k \geq M$.

Then the *Riesz-Fischer theorem* from functional analysis says precisely that $L^2(\mathbb{R}^3)$ is a complete metric space. The other defining conditions for a Hilbert space are easily checked. So we have that $L^2(\mathbb{R}^3)$ is a Hilbert space.

9.2 Unitary Operators

Whenever we introduce new objects into mathematics, we also really must introduce the ideas of isomorphism and isomorphic, (which are concepts in *category theory*). This is because someone could be studying a mathematical object, say a group G_1, and someone else (or even the same person!) could be studying the group G_2 and they could come to realize that every group property of G_1 was also a group property of G_2. And vice versa. This could lead to the hunch that the groups G_1 and G_2 are actually the 'same' group. Specifically, they may be *isomorphic* groups, which means that there exists an isomorphism $f : G_1 \to G_2$ of groups. And an *isomorphism* of groups is a *bijective* function that preserves all the group operations. And if that turns out to be true, we can conclude that every property of G_1 as a group is a property of G_2 as a group, and vice versa. Put more bluntly, we do not want to waste our time studying isomorphic groups G_1 and G_2 individually. We would rather prove that they are isomorphic groups and be done with it.

So we are led to ask what is the appropriate definition of an isomorphism of Hilbert spaces $\mathcal{H}_1$ and $\mathcal{H}_2$. In analogy with other situations (such as an isomorphism of groups) we want to have a bijective function $\mathcal{H}_1 \to \mathcal{H}_2$ which preserves all the operations of a Hilbert space. And these operations are those of a vector space and the inner product. The functions preserving the vector space operations, namely the vector sum and scalar multiplication, are the *linear transformations,* also known as *(linear) operators.* See Definition 5.0.1. So here is the definition:

Definition 9.2.1 *Suppose that $\mathcal{H}_1$ and $\mathcal{H}_2$ are Hilbert spaces. We say that $U : \mathcal{H}_1 \to \mathcal{H}_2$ is a* unitary operator *if U is a linear bijection that satisfies*

$$\langle U\varphi_1, U\varphi_2 \rangle_{\mathcal{H}_2} = \langle \varphi_1, \varphi_2 \rangle_{\mathcal{H}_1}$$

for all $\varphi_1, \varphi_2 \in \mathcal{H}_1$. (In particular, U is onto.)

We say that $\mathcal{H}_1$ is isomorphic to $\mathcal{H}_2$ if there exists a unitary operator $U : \mathcal{H}_1 \to \mathcal{H}_2$.

For historical reasons we still usually say to this day 'unitary operator' instead of the completely correct 'isomorphism of Hilbert spaces'. But the latter expression shows exactly why unitary operators are so important.

Exercise 9.2.1 *In this problem $\mathcal{H}$, $\mathcal{H}_1$, $\mathcal{H}_2$, and $\mathcal{H}_3$ denote Hilbert spaces. Here are some basic properties of unitary operators:*

1. *The identity map $I : \mathcal{H} \to \mathcal{H}$ is unitary.*

2. *If $U : \mathcal{H}_1 \to \mathcal{H}_2$ is unitary, then its inverse $U^{-1} : \mathcal{H}_2 \to \mathcal{H}_1$ is unitary.*

3. *If $U : \mathcal{H}_1 \to \mathcal{H}_2$ and $V : \mathcal{H}_2 \to \mathcal{H}_3$ are unitary, then the composition $VU : \mathcal{H}_1 \to \mathcal{H}_3$ is also unitary. Composition is also denoted by $V \circ U$.*

4. *Prove that*
$$\mathcal{U}(\mathcal{H}) := \{U : \mathcal{H} \to \mathcal{H} \,|\, U \text{ is unitary}\}$$
   is a group where the multiplication is composition of operators. This is called the unitary group *of the Hilbert space $\mathcal{H}$.*

5. *Any unitary operator $U : \mathcal{H}_1 \to \mathcal{H}_2$ is norm preserving, that is to say, $\|U\varphi\| = \|\varphi\|$ for all $\varphi \in \mathcal{H}_1$. (Here $\| \cdot \|$ denotes the norm in the appropriate Hilbert space.)*

Whenever we study mathematical objects together with isomorphisms between certain pairs of objects, we always have a *classification problem,* namely to identify all such objects up to isomorphism. As an example, the classification problem for finite groups is a monstrously difficult open problem. However, the classification problem for finite simple groups has been claimed to have been solved, though only a handful of experts may have read the entire proof. What about the classification problem for Hilbert spaces? Here is the relevant result.

Theorem 9.2.1 *For every Hilbert space* $\mathcal{H}$ *there is an associated cardinal number* dim $\mathcal{H}$ *called its* dimension, *which will be defined in the next section.*

The Hilbert spaces $\mathcal{H}_1$ *and* $\mathcal{H}_2$ *are isomorphic (as Hilbert spaces) if and only if* dim $\mathcal{H}_1 = $ dim $\mathcal{H}_2$.

For example, dim $\mathbb{C}^n = n$ for every finite integer (finite cardinal) n. Also, most of the commonly used, infinite dimensional Hilbert spaces such as $L^2(\mathbb{R}^n)$, the Sobolev spaces, the Bergmann spaces, etc., have dimension $\aleph_0$, the first infinite cardinal. So why study so many seemingly different infinite dimensional Hilbert spaces given that they are all the same? The answer is we are interested in operators acting in these spaces, and often an operator can have a manageable analytic formula in one of these Hilbert spaces, but nothing easy to deal with in an arbitrary isomorphic Hilbert space.

We have already seen examples of unitary operators, though that was not mentioned when they were introduced because we did not have enough available concepts to say much. But anyway here is a sketch of what happens. If H is any densely defined self-adjoint operator acting on the elements in a dense subspace of the Hilbert space $\mathcal{H}$, then the operator $U_t := e^{-itH/\hbar}$ for every $t \in \mathbb{R}$ is a unitary operator acting on *every* element in $\mathcal{H}$. Moreover, these unitary operators satisfy $U_0 = I_{\mathcal{H}}$, the identity map, and $U_s U_t = U_{s+t}$ for all $s, t \subset \mathbb{R}$. This means that the map $t \mapsto U_t$ is a group morphism from the additive group $\mathbb{R}$ to the unitary group $\mathcal{U}(\mathcal{H})$. The moral of this story is that the solution of the time dependent Schrödinger equation with any initial state $\varphi \in \mathcal{H}$ is given by its *unitary flow* $U_t \varphi$, which is also a state for every time t since U_t is norm preserving. Everything in this paragraph is discussed thoroughly and proved in any good functional analysis text.

9.3 More Facts without Proofs

In this section we collect some more facts about Hilbert spaces which will be needed later. For the sake of brevity proofs are not given, though some proofs are left as optional exercises. In any case, the reader should understand what each of these facts asserts. So read the exercises carefully, too.

The metric on a Hilbert space $\mathcal{H}$ gives it a corresponding *topology*, which roughly means a way to define convergence (and limits) of sequences. For us this is a language more than anything else. We let S denote a subset of $\mathcal{H}$ in the rest of this paragraph. We say that S is *open* if S is the union of *open balls* $B_r(\phi) := \{\psi \in \mathcal{H} \mid ||\psi - \phi|| < r\}$, where $\phi \in \mathcal{H}$ and $r > 0$. On the other hand, a subset S is *closed* if its complement $\mathcal{H} \setminus S$ is open or, equivalently, if the limit of any convergent sequence of points in S is itself in S. And S is *(sequentially) compact* if every sequence in S has a convergent sub-sequence. If $\mathcal{H}$ has finite dimension, then every subspace is closed; this is not so for infinite dimension. The *closure* of S, denoted $\overline{S}$, is the smallest closed set containing S or, equivalently, S together with all of its limit points. We say that S is dense if $\overline{S} = \mathcal{H}$. For an infinite dimensional Hilbert space there are (infinitely many!) subspaces that are not closed, and there are (infinitely many!) subspaces that are dense.

Let $\mathcal{H}$ be a Hilbert space. Suppose that $\Phi = \{\phi_j \in \mathcal{H} \mid j \in J\}$ is given, where J is an index set. In most of quantum theory J will be finite or countably infinite, but for the time being it can be arbitrary. We say that the set Φ is *orthonormal* provided that $\|\phi_j\| = 1$ for each $j \in J$ and that $\langle \phi_j, \phi_k \rangle = 0$ for all $j, k \in J$ with $j \neq k$.

We say that Φ is an *orthonormal basis* of $\mathcal{H}$ if it is orthonormal and *spanning*. Spanning means that the subspace of all finite linear combinations of elements in Φ is dense in $\mathcal{H}$. Equivalently, Φ is an orthonormal basis if and only if for every element $\phi \in \mathcal{H}$ there are unique complex numbers c_j for $j \in J$ such that $\sum_j |c_j|^2 < \infty$ and $\phi = \sum_j c_j \phi_j$. The latter (possibly infinite) sum is defined with respect to the convergence using the metric, as defined above, associated to the norm of the Hilbert space. Every Hilbert space $\mathcal{H} \neq 0$ has a (non-unique!) orthonormal basis Φ, whose cardinality does not depend on the choice of Φ. This leads to the definition $\dim \mathcal{H} := \operatorname{card}(\Phi)$, the *dimension* of $\mathcal{H}$.

We say that a Hilbert space $\mathcal{H}$ is *separable* if it has a dense, countable subset. It turns out that $\mathcal{H}$ is separable if and only if $\dim \mathcal{H} \leq \aleph_0$.

One says $\phi, \psi \in \mathcal{H}$ are *orthogonal* if $\langle \phi, \psi \rangle = 0$ with the notation: $\phi \perp \psi$. If S_1 and S_2 are subsets such that $\psi_1 \in S_1$ and $\psi_2 \in S_2$ imply that $\psi_1 \perp \psi_2$, then we say that S_1 and S_2 are *orthogonal* and write this as: $S_1 \perp S_2$.

If S is any subset of $\mathcal{H}$, we define $S^\perp$ (read: S perp), to be the set of all elements orthogonal to all the elements in S, namely

$$S^\perp := \{\psi \in \mathcal{H} \mid \langle \psi, \phi \rangle = 0 \text{ for all } \phi \in S\}. \tag{9.3.1}$$

Then $S^\perp$ is a closed subspace of $\mathcal{H}$.

Definition 9.3.1 *Let $\mathcal{H}$ be a Hilbert space and $T : \mathcal{H} \to \mathcal{H}$ be a (linear) operator. (See Definition 5.0.1.) Then we say that the operator T is* bounded *if there exists some real number $M \geq 0$ such that for all $\phi \in \mathcal{H}$ we have that $\|T\phi\| \leq M\|\phi\|$.*

Given a subset of the *real* numbers, it is critically important to understand what its supremum and infimum are.

Definition 9.3.2 *Let S be any subset of $\mathbb{R}$. The* supremum *(or least upper bound) of S is the smallest number u such that*

- *u is an upper bound of S, that is, $s \leq u$ for all $s \in S$,*

- *u is the least such upper bound, that is, if $s \leq v$ for all $s \in S$, then $u \leq v$.*

If S has some upper bound, then an important property of the real numbers is that a unique supremum exists. Notation: $u = \sup S$. If the nonempty set S has no upper bound, then we put $\sup S = +\infty$.

We also put $\sup \emptyset = -\infty$, where $\emptyset$ denotes the empty set.

This definition does not apply to the complex numbers, for which we do not define a relation $\leq$ for arbitrary pairs of complex numbers.

If $\sup S \notin S$, then S has no maximal (or largest) element.

Exercise 9.3.1 *The quickest definition of the* infimum *(or greatest lower bound) of a subset S of $\mathbb{R}$ is $\inf S := -\sup(-S)$, where $-S := \{-s \mid s \in S\}$. Alternately, one can rewrite Definition 9.3.2 by reversing inequalities so that $\inf S$ is defined as the lower bound that is larger than any other lower bound. First, carefully write out this alternative definition. Second, prove that it is equivalent to the quickest definition.*

Definition 9.3.3 *If $T : \mathcal{H} \to \mathcal{H}$ is a bounded operator, then we define its* operator norm *by*

$$\|T\| := \sup\{\|T\phi\| \mid \phi \in \mathcal{H}, \|\phi\| \leq 1\}. \tag{9.3.2}$$

Exercise 9.3.2 *Prove $\|T\| = \inf\{M \geq 0 \mid \|T\phi\| \leq M\|\phi\|$ for all $\phi \in \mathcal{H}\}$. This is often given as the definition of $\|T\|$ instead of (9.3.2).*

A central problem in Hilbert space theory is to prove that a given linear operator is bounded and then to calculate its operator norm. The latter is a *non-linear optimization problem*. What do these words mean? Well, define the unit sphere in a non-zero Hilbert space $\mathcal{H}$ as $S := \{\phi \in \mathcal{H}, \|\phi\| = 1\}$. Then the operator norm of T is the 'maximal' (or 'optimal') value of the non-linear function $S \to [0, \infty)$ given by $\phi \mapsto \|T\phi\|$. If the dimension of $\mathcal{H}$ is finite, then S is a compact metric space (meaning that every sequence in it has a convergent sub-sequence). A continuous real-valued function defined on a compact metric space always has a maximal value. But S is not compact if the dimension of $\mathcal{H}$ is infinite, and so we lack an important tool for solving the optimization problem. The moral of this paragraph is that a basic problem in the theory of linear operators is a difficult non-linear problem.

Exercise 9.3.3 *Prove that S is not compact if the dimension of $\mathcal{H}$ is infinite.* **Hint:** *Take an orthonormal basis $\mathcal{B}$ of $\mathcal{H}$ and show that there is a sequence in $\mathcal{B}$ that has no convergent sub-sequence.*

For any Hilbert space $\mathcal{H}$ we let

$$\mathcal{L}(\mathcal{H}) := \{T : \mathcal{H} \to \mathcal{H} \mid T \text{ is a bounded operator}\}.$$

Exercise 9.3.4 *Prove that $\mathcal{L}(\mathcal{H})$ is a vector space over the complex numbers, where the sum is defined by $(S + T)\psi := S\psi + T\psi$ and the scalar product by $(\lambda S)\psi := \lambda(S\psi)$ for $S, T \in \mathcal{L}(\mathcal{H})$, $\lambda \in \mathbb{C}$ and $\psi \in \mathcal{H}$. Prove that the operator norm is indeed a norm on the complex vector space $\mathcal{L}(\mathcal{H})$.*

With definition $d(S, T) := \|S - T\|$ for S and T bounded operators, it turns out that d is a metric, and $\mathcal{L}(\mathcal{H})$ is a complete metric space. It can be shown that the norm on $\mathcal{L}(\mathcal{H})$ does not come from an inner product. This means that $\mathcal{L}(\mathcal{H})$ with the operator norm is not a Hilbert space. Rather, $\mathcal{L}(\mathcal{H})$ is an example of a *complete, normed space* which is abbreviated as *Banach space*.

The *identity operator* or *identity map* I, defined by $I\psi := \psi$ for all $\psi \in \mathcal{H}$, is a bounded, linear operator. Therefore $I \in \mathcal{L}(\mathcal{H})$.

Exercise 9.3.5 *Find a condition on $\mathcal{H}$ such that the identity operator I satisfies $||I|| = 1$. Evaluate $||I||$ if $\mathcal{H}$ does not satisfy that condition.*

Exercise 9.3.6 *Let T be a bounded operator acting in the Hilbert space $\mathcal{H}$. Prove that for all $\phi \in \mathcal{H}$ we have*

$$||T\phi|| \le ||T||\,||\phi||.$$

It is important that you understand the meanings of each of the three norms in this inequality.

Definition 9.3.4 *We say that $\sum_{j=0}^{\infty} \phi_j$, is a convergent (infinite) series if the sequence of partial sums $\sum_{j=0}^{n} \phi_j$, where each $\phi_j \in \mathcal{H}$, converges as $n \to \infty$ to a limit $\phi \in \mathcal{H}$ with respect to the standard metric on the Hilbert space $\mathcal{H}$. If this is the case, we denote this as $\phi = \sum_{j=0}^{\infty} \phi_j$.*

Exercise 9.3.7 *Let $T : \mathcal{H} \to \mathcal{H}$ be a linear operator. Show that the following statements are equivalent:*

- *T is bounded.*

- *T is continuous, that is, for any convergent sequence $\psi = \lim_{n\to\infty} \psi_n$ in $\mathcal{H}$, where each $\psi_n \in \mathcal{H}$, we have $T\psi = \lim_{n\to\infty} T\psi_n$. Or, equivalently, if $\phi = \sum_j \phi_j$ is a convergent infinite series in $\mathcal{H}$, where each $\phi_j \in \mathcal{H}$, then we have $T\phi = \sum_j T\phi_j$. (We say that T commutes with limits and with infinite sums.)*

Here is a fundamental result which can be proved with nothing other than the definition of Hilbert space, the definition of its norm and a speck of algebra. We leave the proof to other textbooks or to the industrious reader.

Theorem 9.3.1 (Cauchy-Schwarz inequality) *For $\phi, \psi \in \mathcal{H}$, a Hilbert space, we have that*

$$|\langle \phi, \psi \rangle| \le ||\phi||\,||\psi||. \qquad (9.3.3)$$

Exercise 9.3.8 *Use the Cauchy-Schwarz inequality to show that the inner product, considered as the function $\mathcal{H} \times \mathcal{H} \to \mathbb{C}$ that maps the ordered pair (ϕ, ψ) to its inner product $\langle \phi, \psi \rangle$, is a continuous function. Explicitly, you are asked to prove that if $\psi = \lim_{n\to\infty} \psi_n$ and $\phi = \lim_{n\to\infty} \phi_n$ are convergent sequences in $\mathcal{H}$, then $\langle \psi, \phi \rangle = \lim_{n\to\infty} \langle \psi_n, \phi_n \rangle$. A similar property also holds for the inner product of infinite sums.*

Exercise 9.3.9 *Let $U : \mathcal{H} \to \mathcal{H}$ be a unitary map, where $\mathcal{H}$ is a Hilbert space. Find an extra hypothesis which then implies that $||U|| = 1$. What happens if that hypothesis fails? In any case conclude that $U \in \mathcal{L}(\mathcal{H})$.*

There is a something related to unitary operators. But the reader is advised to note the subtle difference.

Definition 9.3.5 *Let $\mathcal{H}_1, \mathcal{H}_2$ be Hilbert spaces. We say that a linear map $T : \mathcal{H}_1 \to \mathcal{H}_2$ is an* isometry *if T preserves distances between pair of vectors, that is, $||T\psi - T\phi|| = ||\psi - \phi||$ for all $\psi, \phi \in \mathcal{H}_1$. We say that $\mathcal{H}_1$ is* isometric *to $\mathcal{H}_2$ if there exists an isometry $T : \mathcal{H}_1 \to \mathcal{H}_2$ that is also onto (surjective).*

Remark: Note that we do not require an isometry to be onto. Recall that a unitary operator is onto by definition.

Exercise 9.3.10 *Let $T : \mathcal{H}_1 \to \mathcal{H}_2$ be a linear map. Prove:*

- *T is an isometry if and only if T preserves norms, that is, $||T\psi|| = ||\psi||$ for all $\psi \in \mathcal{H}_1$.*

- *If T is unitary, then T is an isometry.*

- *An isometry is one-to-one (injective).*

- *A surjective isometry is unitary. (Use the polarization identity (9.3.7).)*

- *Find an isometry that is not unitary.*

- *Prove that $||T|| = 1$ if T is an isometry and $\mathcal{H}_1 \neq 0$. Find a bounded linear map T with $||T|| = 1$ but T is not an isometry.*

- *The range of an isometry is a closed subspace. Find a bounded linear map T with $||T|| = 1$ but whose range is not closed.*

The norm symbol $|| \cdot ||$ has three different meanings in this exercise. Be sure you understand what they are.

Exercise 9.3.11 *Prove that isometric is a relation between Hilbert spaces that is* reflexive, symmetric, *and* transitive. *(You might need to look up these three words. One says that isometric is an* equivalence relation.*)*

Another important fact is known as the *Riesz representation theorem.* Here is one lengthy way of saying it, though in practice one typically only needs a part of the second paragraph of the following result.

Theorem 9.3.2 (F. Riesz) *Let $f : \mathcal{H} \to \mathbb{C}$ be a bounded linear map, which is called a* (linear) functional. *(Here f bounded means that there exists a real number C such that $|f(\psi)| \leq C||\psi||$ holds for all $\psi \in \mathcal{H}$. Necessarily, $C \geq 0$. Moreover, we define $||f|| := \sup\{|f(\psi)| : ||\psi|| \leq 1\}$.)*

Then there exists a unique element $\phi_f \in \mathcal{H}$ such that $f(\psi) = \langle \phi_f, \psi \rangle$ for every $\psi \in \mathcal{H}$. Also the map $f \mapsto \phi_f$ is an anti-linear, isometric bijection of the dual space $\mathcal{H}'$ onto $\mathcal{H}$, where we define $\mathcal{H}'$ to be the vector space of all bounded functionals $\mathcal{H} \to \mathbb{C}$ together with the norm $\|f\|$ defined above, thereby making it into a complete normed complex vector space (that is, a Banach space).

Finally, this norm on $\mathcal{H}'$ arises from a unique inner product $\langle \cdot, \cdot \rangle_{\mathcal{H}'}$, thereby making $\mathcal{H}'$ into a Hilbert space, and then the map $f \mapsto \phi_f$ is not only anti-linear, but also anti-unitary meaning that for all $f, g \in \mathcal{H}'$ we have

$$\langle \phi_f, \phi_g \rangle_{\mathcal{H}} = \langle g, f \rangle_{\mathcal{H}'}.$$

In just a few words we can summarize this as saying that the dual of a Hilbert space is (almost) isomorphic to itself, and that this is achieved in a 'canonical' way.

For any $T \in \mathcal{L}(\mathcal{H})$ one can prove using the Riesz representation theorem that there exists a unique operator, denoted by $T^* \in \mathcal{L}(\mathcal{H})$ and called the adjoint of T, that satisfies for all $\phi, \psi \in \mathcal{H}$ the condition

$$\langle \phi, T\psi \rangle = \langle T^*\phi, \psi \rangle. \tag{9.3.4}$$

Exercise 9.3.12 Prove for all $T \in \mathcal{L}(\mathcal{H})$ that $T^* \in \mathcal{L}(\mathcal{H})$ and moreover that $\|T^*\| = \|T\|$, where $\| \cdot \|$ denotes the operator norm.

Prove that the mapping $\mathcal{L}(\mathcal{H}) \to \mathcal{L}(\mathcal{H})$ given by $T \mapsto T^*$ is anti-linear and also satisfies $(ST)^* = T^*S^*$ for all $S, T \in \mathcal{L}(\mathcal{H})$. Prove as well $I^* = I$, where I is the identity operator, and $0^* = 0$, where 0 is the zero operator defined by $0(\psi) := 0$, the zero vector, for all $\psi \in \mathcal{H}$.

Definition 9.3.6 Let $T \in \mathcal{L}(\mathcal{H})$. If $T = T^*$, then we say T is self-adjoint.

Exercise 9.3.13 Let $\lambda \in \mathbb{C}$ be an eigenvalue of a self-adjoint operator T, meaning that $T\psi = \lambda\psi$ for some $\psi \neq 0$. Prove that $\lambda \in \mathbb{R}$.

Suppose that $\lambda_1 \neq \lambda_2$ are eigenvalues of a self-adjoint operator T with eigenvectors ϕ_1 and ϕ_2, respectively. Prove that ϕ_1 and ϕ_2 are orthogonal, that is $\langle \phi_1, \phi_2 \rangle = 0$.

Exercise 9.3.14 Suppose that $U \in \mathcal{L}(\mathcal{H})$. Prove that U is unitary if and only if $U^*U = UU^* = I$.

Let $\lambda \in \mathbb{C}$ be an eigenvalue of a unitary operator. Prove that $|\lambda| = 1$.

Definition 9.3.7 Suppose $S, T \in \mathcal{L}(\mathcal{H})$ are self-adjoint operators. Then we define $S \leq T$ to mean that $\langle \psi, S\psi \rangle \leq \langle \psi, T\psi \rangle$ holds for every $\psi \in \mathcal{H}$. Equivalently, we write $T \geq S$.

If $T \geq 0$ we say that T is a positive operator. If $T \leq 0$ we say that T is a negative operator. (In these two statements 0 is the zero operator.)

Exercise 9.3.15 *If $T \in \mathcal{L}(\mathcal{H})$ is self-adjoint, prove that $\langle \psi, T\psi \rangle$ is a real number. (Consequently, the inequality in Definition 9.3.7 is between real numbers, and so makes sense.)*

Let $\lambda \in \mathbb{C}$ be an eigenvalue of a positive operator. Prove that $\lambda \in \mathbb{R}$ satisfies $\lambda \geq 0$. Prove the analogous result for negative operators.

Exercise 9.3.16 *Prove that $\leq$ is a partial order on the set of self-adjoint operators of a Hilbert space $\mathcal{H}$. To say that this is a partial order means that for all self-adjoint operators A, B, C we have these three properties:*

- *Reflexivity: $A \leq A$,*

- *Anti-symmetry: If $A \leq B$ and $B \leq A$, then $A = B$,*

- *Transitivity: If $A \leq B$ and $B \leq C$, then $A \leq C$.*

If $\dim \mathcal{H} \geq 2$, show there exists a pair S, T of self-adjoint operators such that neither $S \leq T$ nor $T \leq S$. (This is why we say that $\leq$ is a partial order.)

Definition 9.3.8 *We say that $P \in \mathcal{L}(\mathcal{H})$ is a projection if $P = P^2 = P^*$.*

The condition $P = P^*$ says that P is self-adjoint. We put the condition $P^2 = P$ into words by saying that P is *idempotent*. Projections are important operators in quantum theory as we shall see.

Exercise 9.3.17 *Prove that the zero operator 0 and the identity operator I are projections.*

A seemingly unrelated concept is defined next. We have already seen this earlier, but it bears repetition.

Definition 9.3.9 *A closed subspace V of $\mathcal{H}$ is an algebraic subspace (that is, $\phi + \psi \in V$ and $\lambda\phi \in V$ whenever $\phi, \psi \in V$, and $\lambda \in \mathbb{C}$) and moreover is topologically closed which means that every limit of a convergent sequence of elements in V is itself in V.*

As noted before, $S^{\perp}$ (see (9.3.1)) is a closed subspace if S is *any* subset of $\mathcal{H}$.

Exercise 9.3.18 *Let V be any (not necessarily closed) subspace of a Hilbert space. Show that $V^{\perp\perp} = \overline{V}$, the closure of V. In particular, if V is a closed subspace, then $V^{\perp\perp} = V$. Here $V^{\perp\perp} := (V^{\perp})^{\perp}$ is the 'perp' of the 'perp'.*

Definition 9.3.10 *Let V_1 and V_2 be closed subspaces of a Hilbert space $\mathcal{H}$ with $V_1 \perp V_2$ and such that every vector $\psi \in \mathcal{H}$ can be written (uniquely) as $\psi = \psi_1 + \psi_2$, where $\psi_1 \in V_1$ and $\psi_2 \in V_2$. Then we say that $\mathcal{H}$ is the direct sum of V_1 and V_2. Notation: $\mathcal{H} = V_1 \oplus V_2$.*

Note that $V_1 \perp V_2$ implies that $V_1 \cap V_2 = 0$, the zero subspace. It follows that the expression $\psi = \psi_1 + \psi_2$ in this definition is necessarily unique. This concept is sometimes called the *internal direct sum* in order to contrast it with another concept known as an *external direct sum*, which we can live without for the time being.

Theorem 9.3.3 (Projection Theorem) *Suppose V is a closed subspace of a Hilbert space $\mathcal{H}$. Then $\mathcal{H} = V \oplus V^\perp$. Moreover, if we take $\psi \in \mathcal{H}$ and write it as $\psi = \psi_1 + \psi_2$, where $\psi_1 \in V$ and $\psi_2 \in V^\perp$, then the function $P_V : \mathcal{H} \to \mathcal{H}$ defined by $P_V \psi := \psi_1$ is an orthogonal projection and, in particular, a bounded linear operator.*

Finally, ψ_1 is the unique nearest vector in the closed subspace V to the vector $\psi \in \mathcal{H}$, that is, $\|\psi - \psi_1\| < \|\psi - \phi\|$ for all $\phi \in V$ with $\phi \neq \psi_1$.

The proof of this theorem depends on the parallelogram identity discussed below. The result of the Projection Theorem does not hold in general for vector spaces with a norm. It is a geometric property of Hilbert spaces.

Exercise 9.3.19 *Given $\psi = \psi_1 + \psi_2$ as in the Projection Theorem, prove the Pythagorean theorem: $\|\psi\|^2 = \|\psi_1\|^2 + \|\psi_2\|^2$.*

Use this result to prove that $\|P_V \psi\| \leq \|\psi\|$ for all $\psi \in \mathcal{H}$ or, in other words, $\|P_V\| \leq 1$.

Next, show that $\|P_V\| = 1$ if and only if $V \neq 0$.

Exercise 9.3.20 *Let $P \in \mathcal{L}(\mathcal{H})$ be a projection.*

- *For all $\phi \in \mathcal{H}$ we have that $\langle \phi, P\phi \rangle$ is a real number and, in particular, $0 \leq \langle \phi, P\phi \rangle \leq 1$. Show that this is equivalent to saying $0 \leq P \leq I$.*

- *Let $\lambda \in \mathbb{C}$ be an eigenvalue of P. Prove that $\lambda = 0$ or $\lambda = 1$.*

- *If $P \neq 0$ and $P \neq I$, then the set of eigenvalues of P is $\{0, 1\}$. What happens in the cases $P = 0$ and $P = I$?*

Given a Hilbert space $\mathcal{H}$ there is a one-to-one and onto correspondence between the set of all of its projections and the set of its closed subspaces. This correspondence maps any closed subspace V to the projection P_V given in the Projection Theorem. And to any projection P we assign its range

$$\operatorname{Ran} P := \{\phi \in \mathcal{H} \mid \exists \psi \in \mathcal{H} \text{ such that } \phi = P\psi\}. \tag{9.3.5}$$

Exercise 9.3.21 *Show that these two mappings are inverses of each other. Prove that $\operatorname{Ran} P$ in (9.3.5) is a closed subspace.*

Now the set of all the closed subspaces has a partial order, namely $V_1 \subset V_2$ means that V_1 is a subset of V_2. And the set of all projections, being self-adjoint operators, also has a partial order. (Recall Exercise 9.3.16.) The next exercise says among other things that this bijection of sets preserves order.

Exercise 9.3.22 Let P_1 and P_2 be projections. Prove that $P_1 \leq P_2$ if and only if $\text{Ran } P_1 \subset \text{Ran } P_2$ if and only if $P_1 P_2 = P_1$ if and only if $P_2 P_1 = P_1$.

Also, the set of closed subspaces becomes an *complete, complemented orthomodular lattice* with respect to its partial order. (Look it up if this is new to you.) For example, the complement of V is $V^\perp$, the orthogonal complement. Therefore, the set of projections has that structure, too. For example, the complement of the projection P is $I - P$, where I is the identity map of $\mathcal{H}$.

Another handy fact is called the *parallelogram identity*, which says that

$$||\phi + \psi||^2 + ||\phi - \psi||^2 = 2 \left(||\phi||^2 + ||\psi||^2 \right) \qquad (9.3.6)$$

holds for all ϕ, ψ in a Hilbert space. This is actually a statement about the smallest subspace containing the vectors ϕ and ψ, that being a subspace of dimension ≤ 2. Thinking about the case when that subspace has dimension exactly 2, we can visualize ϕ and ψ as two sides of a parallelogram, whose two diagonals are then $\phi + \psi$ and $\phi - \psi$. Then the parallelogram identity can be put into words this way: The sum of the squares on the two diagonals of the parallelogram is equal to the sum of the squares on its four sides. Perhaps this wording reminds you of the Pythagorean Theorem. However, you most likely never learned the parallelogram identity in a basic course of Euclidean geometry. I know I didn't.

Exercise 9.3.23 Prove the parallelogram identity (9.3.6).

Here is another useful formula, known as the *polarization identity*. It says

$$\langle \psi, \phi \rangle = \frac{1}{4} \left(||\phi + \psi||^2 - ||\phi - \psi||^2 + i||\phi + i\psi||^2 - i||\phi - i\psi||^2 \right) \qquad (9.3.7)$$

holds for all ϕ, ψ in a complex Hilbert space. Notice that the formula does not even make sense (due to the presence of $i = \sqrt{-1}$) for a real Hilbert space, that is, a Hilbert space which only admits scalar multiplication by real numbers. This identity says that if you know how to evaluate the norm of every vector in the Hilbert space, then you can evaluate the inner product of any pair of vectors. That's quite amazing!

Here is a useful generalization of (9.3.7). Suppose that $T : \mathcal{H} \to \mathcal{H}$ is a bounded, linear operator and $\mathcal{H}$ is a complex Hilbert space. Then for all $\psi, \phi \in \mathcal{H}$ we have

$$\langle \psi, T\phi \rangle = \frac{1}{4} \left(Q_T(\phi + \psi) - Q_T(\phi - \psi) + i Q_T(\phi + i\psi) - i Q_T(\phi - i\psi) \right), \qquad (9.3.8)$$

where $Q_T(\psi) := \langle \psi, T\psi \rangle$ for all $\psi \in \mathcal{H}$ defines the *quadratic form* associated to the linear operator T. This is also called the *polarization identity*. Notice that (9.3.7) is just the special case of (9.3.8) when we take $T = I$, the identity operator.

Exercise 9.3.24 *Prove the polarization identity* (9.3.8).

Exercise 9.3.25 *The parallelogram and polarization identities express a lot of the structure, one might even say geometry, of Hilbert spaces. To see this suppose we have a complex vector space with a norm that satisfies the parallelogram identity* (9.3.6). *Use the polarization identity* (9.3.7) *to define* $\langle \psi, \phi \rangle$ *in terms of the known norms on the right side. Then prove that this is indeed an inner product whose associated norm is the norm we started with.*

9.4   States Revisited (Optional)

For the stationary states $\psi \in L^2(\mathbb{R}^3)$ of a quantum Hamiltonian we now write the normalization condition on ψ as $||\psi|| = 1$. So ψ is a *unit vector*. This says that ψ lies on the *unit sphere* in $L^2(\mathbb{R}^3)$, which is *not* a linear subspace. Moreover, this interpretation says that such a unit vector ψ represents the same physical state as the unit vector $\lambda\psi$ provided that $\lambda \in \mathbb{C}$ satisfies $|\lambda| = 1$, that is, λ lies on the *unit circle* $\mathbb{S}^1 := \{z \in \mathbb{C} \,|\, |z| = 1\}$ in the complex plane $\mathbb{C}$. So, for any Hilbert space $\mathcal{H}$, we let $S(\mathcal{H})$ denote the sphere of its unit vectors. This is a *metric space* where the metric d comes from the norm on $\mathcal{H}$ through the definition $d(\phi_1, \phi_2) := ||\phi_1 - \phi_2||$. Then the quotient topological space $S(\mathcal{H})/\mathbb{S}^1$ is the space of states, or more briefly, the *state space*. Note that this is purely kinematical, that is, the way we describe a quantum system before considering its time evolution. The space $S(\mathcal{H})/\mathbb{S}^1$ is well known in geometry; it is called the *projective space* associated to the Hilbert space $\mathcal{H}$. If we take the complex Hilbert space $\mathbb{C}^{n+1}$ where $n \geq 1$ is an integer, the associated projective space is denoted by $\mathbb{C}P^n$ in some texts and is called the *complex n-dimensional projective space*.

Definition 9.4.1 *A (pure) state for a quantum system described by the Hilbert space $\mathcal{H}$ is an element of $S(\mathcal{H})/\mathbb{S}^1$. In particular, for a single (spin zero) 'particle' in $\mathbb{R}^3$ a state is an element in $S(L^2(\mathbb{R}^3))/\mathbb{S}^1$. (See Chapter 14 for spin zero.)*

   We will not deal with the more general *mixed states* in this book, but only with the pure states, which will simply be called hereafter *states*.

   Another equivalent and standard way of doing this is to say that any non-zero ψ represents the state $\psi/||\psi||$ modulo $\mathbb{S}^1$. Again equivalently, one can construct the projective space as the quotient topological space

$$\frac{L^2(\mathbb{R}^3) \setminus \{0\}}{\mathbb{C} \setminus \{0\}}.$$

This means that the non-zero functions $\phi_1, \phi_2 \in L^2(\mathbb{R}^3))$ are identified if there exists a non-zero $\lambda \in \mathbb{C}$ such that $\phi_1 = \lambda\phi_2$. This alternative definition of the projective space has the virtue that it does not require the existence of a norm and so is applicable in more general algebraic contexts. For example, let $\mathbb{K}$ be any field with finitely many elements. Then for each integer $n \geq 1$ the finite quotient set

$$\frac{\mathbb{K}^{n+1} \setminus \{0\}}{\mathbb{K} \setminus \{0\}}$$

is the projective space over $\mathbb{K}$ of dimension n.

This new definition of a state has an immediate consequence, namely that the set of states is not a vector space. This is rather strange since the set of solutions Ψ_t of the time dependent Schrödinger equation is a vector space under the *point-wise* operations for adding functions and multiplying functions by complex numbers ($\equiv$ scalars):

$$(\Psi_t + \Phi_t)(\mathbf{x}) := \Psi_t(\mathbf{x}) + \Phi_t(\mathbf{x}) \quad \text{and} \quad (\lambda \Psi_t)(\mathbf{x}) := \lambda(\Psi_t(\mathbf{x})),$$

where $\lambda \in \mathbb{C}$ and $\mathbf{x} \in \mathbb{R}^3$. This is called the *superposition principle*, and it is true because the time dependent Schrödinger equation is linear in its unknown. (Parenthetically, let me emphasize that the superposition principle is a property of any *linear* equation, not just the Schrödinger equation.) And a related observation is that the Hilbert space $L^2(\mathbb{R}^3)$ is itself a vector space. But it is better to think of a state $\psi \in L^2(\mathbb{R}^3)$ as being a way of identifying the one dimensional subspace $\mathbb{C}\psi$ of scalar multiples of $\psi \neq 0$.

Now $\psi_1 + \psi_2$ makes sense for any pair of non-zero functions ψ_1 and ψ_2 in $L^2(\mathbb{R}^3)$, and this determines a unique state provided that $\psi_1 \neq -\psi_2$. But this does not define a sum of the subspaces $\mathbb{C}\psi_1$ and $\mathbb{C}\psi_2$. Why not? Well, the non-zero vectors $e^{i\theta_1}\psi_1$ and $e^{i\theta_2}\psi_2$ represent the same pair of states, provided that $\theta_1, \theta_2 \in \mathbb{R}$. (Recall that complex numbers of absolute value 1, such as $e^{i\theta_1}$ and $e^{i\theta_2}$, are called *phases* or *phase factors*. Just to keep you on your toes, sometimes θ_1 and θ_2 are called *phases*.) But their sum is

$$e^{i\theta_1}\psi_1 + e^{i\theta_2}\psi_2 = e^{i\theta_1}(\psi_1 + e^{i(\theta_2 - \theta_1)}\psi_2).$$

The vector $\psi_1 + e^{i(\theta_2 - \theta_2)}\psi_2$ will be non-zero, provided that ψ_1 and ψ_2 represent two distinct states (as we assume from now on), which implies that ψ_1 and ψ_2 determine two distinct one dimensional subspaces and so are linearly independent. And the *overall phase factor* $e^{i\theta_1}$ does not change the state that $\psi_1 + e^{i(\theta_2 - \theta_2)}\psi_2$ represents. However, the *relative phase factor* $e^{i(\theta_2 - \theta_2)}$ does enter in determining the corresponding state. And since θ_1 and θ_2 are arbitrary real numbers, the relative phase factor can be any complex number of absolute value 1.

Exercise 9.4.1 *Suppose that* $\lambda, \mu \in \mathbb{C}$ *satisfy* $|\lambda| = |\mu| = 1$ *and* $\lambda \neq \mu$. *Prove that* $\psi_1 + \lambda\psi_2$ *and* $\psi_1 + \mu\psi_2$ *do not represent the same state. (We are still assuming that* ψ_1 *and* ψ_2 *are linearly independent.)*

Note that we can also write $e^{i\theta_1}\psi_1 + e^{i\theta_2}\psi_2 = e^{i\theta_2}(e^{i(\theta_1 - \theta_2)}\psi_1 + \psi_2)$ whose overall phase factor is $e^{i\theta_2}$. As the reader can check, $e^{i(\theta_1 - \theta_2)}\psi_1 + \psi_2$ and $\psi_1 + e^{i(\theta_2 - \theta_1)}\psi_2$ represent the same state.

Let's now see how this relates to the probability densities $\rho_1(\mathbf{x}) = |\psi_1(\mathbf{x})|^2$ and $\rho_2(\mathbf{x}) = |\psi_2(\mathbf{x})|^2$, where ψ_1 and ψ_2 are two distinct normalized states. Also, for a specific combination of $\psi_\theta := \psi_1 + e^{i\theta}\psi_2$ of ψ_1 and ψ_2 with relative phase factor $\theta \in \mathbb{R}$ we have that the associated probability density function is proportional to

$$|\psi_\theta|^2 = |\psi_1 + e^{i\theta}\psi_2|^2 = (\psi_1 + e^{i\theta}\psi_2)^*(\psi_1 + e^{i\theta}\psi_2) = (\psi_1^* + e^{-i\theta}\psi_2^*)(\psi_1 + e^{i\theta}\psi_2)$$
$$= |\psi_1|^2 + |\psi_2|^2 + 2\operatorname{Re}(e^{i\theta}\psi_1^*\psi_2) = \rho_1 + \rho_2 + 2\operatorname{Re}(e^{i\theta}\psi_1^*\psi_2). \qquad (9.4.1)$$

Recall that the *real part* of $z \in \mathbb{C}$ is defined by $\operatorname{Re} z := (z + z^*)/2$. We still need to divide (9.4.1) by the normalization factor $||\psi_\theta||^2$ to obtain the probability distribution associated to ψ_θ. Nonetheless, (9.4.1) indicates that the relative phase factor θ can be measured modulo 2π, which has been done in experiments. The last term in (9.4.1) can be negative or positive (or zero), depending on the value of the suppressed spatial variable $\mathbf{x}$. In spatial regions where the last term is positive, we say that there is *constructive interference*. And in spatial regions where the last term is negative, we say that there is *destructive interference*. This demonstrates that the probability density of the combination explicitly depends on the relative phase factor $e^{i\theta}$ and is not necessarily the *weighted sum* $(\rho_1 + \rho_2)/2$ of the individual probability densities ρ_1 and ρ_2. The last term in (9.4.1) satisfies this estimate:

$$|2\operatorname{Re}(e^{i\theta}\psi_1^*\psi_2)| \leq 2|e^{i\theta}\psi_1^*\psi_2| = 2|\psi_1^*||\psi_2| \leq |\psi_1^*|^2 + |\psi_2|^2 = \rho_1 + \rho_2.$$

where we used the inequality $2ab \leq a^2 + b^2$ for $a, b \in \mathbb{R}$. Consequently,

$$0 \leq |\psi_1 + e^{i\theta}\psi_2|^2 \leq 2(\rho_1 + \rho_2),$$

an upper and a lower bound which are independent from the relative phase factor, which does appear in the normalization factor $||\psi_\theta||^2$. This says that the destructive interference can reduce the probability density all the way down to zero. On the other hand the constructive interference can increase the probability density up to the weighted sum of the probability densities, $(\rho_1 + \rho_2)/2$. These results arise simply from the *non-linear* relation between a state and its associated probability density function. These results do not come from the so-called 'wave nature' of quantum theory nor 'wave/particle duality', but rather from the facts that the Schrödinger equation is linear and probability densities are real quantities calculated from complex quantities by using a quite explicit non-linear formula. These mathematical considerations help explain the two-slit experiment for those interested in following up on that overly expounded topic.

These considerations defy the way we usually think about the role of equations in physics. Let us be quite clear about this. The Schrödinger equation gives solutions that completely describe a quantum system. But changing that solution by multiplying it by a phase factor provides another solution that also describes completely that quantum system. There is no physical test that can distinguish between these two mathematical solutions;

physically they are the same. But when we take two distinct (that is, linearly independent) solutions, the linearity of the Schrödinger equation implies that their sum is also a solution and so should have a physical meaning. But by changing each summand by a physically insignificant phase factor, these sums give a family of solutions that *do* depend on the relative phase, which is physically significant since it can be and has been measured.

There is a way to specify physically what a state is, namely a *state* of a physical system is the minimal amount of information needed to describe completely that system. More information is not required; less information is not adequate. This does not uniquely *define* the concept of a state, since equivalent information can be used. However, it does *specify* what a state is. In physics, the usual starting point for finding the states of a physical system is a time dependent differential equation for which the Cauchy problem, as we discussed earlier, has a unique solution. The Cauchy problem asks whether a given differential equation with a given *initial condition* has a solution and, if so, whether that solution is unique. This gets into a lot of technical questions about what space of functions the initial condition is to lie in and what space of functions the solution is to lie in. Even the meaning of *solution* can become a technical problem! But however the technical dust settles, the elements of the space of solutions are typically taken to be the states.

For quantum theory the time dependent (possibly partial) differential equation is the Schrödinger equation

$$i\hbar\frac{\partial\psi}{\partial t} = H\psi,$$

where H is a self-adjoint operator acting in a Hilbert space $\mathcal{H}$. When all these words are correctly defined, then the Cauchy problem for an initial condition $\phi \in \mathcal{H}$ and for all $t \in \mathbb{R}$ has the unique solution

$$\psi(t) = e^{-iHt/\hbar}\phi \tag{9.4.2}$$

in the Hilbert space $\mathcal{H}$, where the unitary operators $\{U_t := e^{-iHt/\hbar} \,|\, t \in \mathbb{R}\}$ form a *unitary group*, that is to say $U_s U_t = U_{s+t}$ for all $s, t \in \mathbb{R}$ and $U_0 = I$. (Parenthetically, we note $||\phi|| = 1$ in (9.4.2) implies $||\psi(t)|| = 1$ for all $t \in \mathbb{R}$, since $\psi(t)$ is the image of ϕ by a unitary operator. This is the well known *conservation of probability*. This tells us that the Cauchy solution stays in the sphere $\mathcal{S}(\mathcal{H})$ if it starts in $\mathcal{S}(\mathcal{H})$.)

This is as far as the story goes for the *pure states* of a quantum system. But the physics has led us down a curious rabbit hole! We have learned that the state space for the physical initial value problem is the non-linear complex projective space $\mathbb{CP}(\mathcal{H}) = \mathcal{S}(\mathcal{H})/\mathbb{S}^1$. Moreover, the conservation of probability is automatically built into the definition of $\mathbb{CP}(\mathcal{H})$. There is no physics reason why the solution in $\mathbb{CP}(\mathcal{H})$ should come from a solution of a Cauchy problem associated to a linear differential equation 'upstairs' on the Hilbert space $\mathcal{H}$ nor why this solution should be given in terms of a group of unitary operators. (The second fact is often called *unitarity*.) After all, the physically relevant states for the quantum system are the elements of $\mathbb{CP}(\mathcal{H})$. However, the 'overlying' mathematical structure of a Hilbert space and a linear differential equation is assumed to be necessary for developing quantum theory, to the point that many a physicist gives this mathematical 'over-pinning' a role in the 'intuition' behind (or 'above'?) quantum theory.

Again, my point is that the physically relevant *dynamical system* is the study of a curve (also called a *trajectory*) in the state space $\mathbb{CP}(\mathcal{H})$. And this curve, and nothing else, describes how the state of the quantum system is changing in time. And the state of the system describes completely the quantum system. There is nothing more–or less–to quantum physics than that. So the fact that this trajectory comes from the solution of a linear equation 'upstairs' in the Hilbert space $\mathcal{H}$ seems to be an unnecessary feature of quantum theory. But that's what we have.

9.5   The Spectrum (Optional)

The remaining sections of this chapter are optional. We have already said that the spectrum of a self-adjoint operator consists of all of its eigenvalues plus other numbers that are 'almost' eigenvalues. This half precise and half vague description can serve you well. But for those who want more details, here we will go into a few of those. So without further ado:

Definition 9.5.1 *Let $T \in \mathcal{L}(\mathcal{H})$ be a bounded linear operator, where $\mathcal{H}$ is a Hilbert space. We define the* resolvent set *of T to be*

$$\rho(T) := \{\lambda \in \mathbb{C} \mid A(\lambda I - T) = (\lambda I - T)A \text{ for some } A \in \mathcal{L}(\mathcal{H})\},$$

which can be described as the set of λ's such that $(\lambda I - T)^{-1}$ exists in $\mathcal{L}(\mathcal{H})$. Then the spectrum *of T is defined as $\mathrm{Spec}(T) := \mathbb{C} \setminus \rho(T)$.*

Important facts that are $\mathrm{Spec}(T)$ is a non-empty, closed, and bounded subset of $\mathbb{C}$ for every $T \in \mathcal{L}(\mathcal{H})$, provided that $\mathcal{H} \neq 0$. Moreover, if T is self-adjoint, then $\mathrm{Spec}(T) \subset \mathbb{R}$.

It is not clear at all that Definition 9.5.1 has any relevance to quantum theory or to anything else for that matter! Only as we apply this definition in practice do we begin to understand its importance. The next exercise is the first step in that process.

Exercise 9.5.1 *Show that every eigenvalue of T is in the spectrum of T.*

The next exercise exhibits some new aspects of this *spectral theory*.

Exercise 9.5.2 *In the Hilbert space $\mathcal{H} = L^2([0,2])$, define a linear operator by $Tf(x) := x^3 f(x)$ for all $x \in [0,2]$ and all $f \in L^2([0,2])$. Show that T is a bounded linear operator with operator norm $||T|| = 8$ and $\mathrm{Spec}(T) = [0,8]$. (**Hint:** First find the resolvent set of T.) Prove that T has no eigenvalues. However, prove that any $\lambda \in \mathrm{Spec}(T)$ is an* approximate eigenvalue *in the* *sense that there exists a sequence $f_n \in L^2([0,2])$, where $n \in \mathbb{N}$, such that $\|f_n\| = 1$ for all $n \in \mathbb{N}$ and $\lim_{n \to \infty} \|(T - \lambda)f_n\|_{\mathcal{H}} = 0$. And finally, prove that any $\lambda \in \rho(T)$ is not an approximate eigenvalue.*

Exercise 9.5.3 *Prove that any eigenvalue of a linear operator acting in a Hilbert space is an approximate eigenvalue.*

There is an important theorem due to H. Weyl which says that the spectrum of a self-adjoint operator is exactly its set of approximate eigenvalues. This extends Schrödinger's insight that quantum theory is an eigenvalue problem.

Theorem 9.5.1 (Weyl's Criterion) *Suppose T is a self-adjoint operator acting in the Hilbert space $\mathcal{H}$ and $\lambda \in \text{Spec}\,(T)$. Then there exists a sequence of states $\psi_n \in \mathcal{H}$, that is $\|\psi_n\| = 1$, such that $\lim_{n \to \infty} \|(T - \lambda)\psi_n\| = 0$. Such a sequence is called a* Weyl sequence of approximate eigenvectors *of T.*

This short section is much less than even the tip of an iceberg. The science for identifying the spectrum of a given linear operator is a highly developed and ongoing field in contemporary mathematics, especially for unbounded operators which we will discuss in the next section. See [15] and [30] (and their many references) for a quantum physics perspective on spectral theory. The encyclopedic treatment in the classic [6] is for those who want to see spectral theory in all of its mathematical detail.

9.6 Densely Defined Operators (Optional)

In this section we present some basic facts, again without proof, for this rather technical topic. The idea is to get the reader familiar with some mathematics behind a lot of quantum theory. The most important detail for the beginner to be aware of is the difference between symmetric operators and self-adjoint operators. And why that difference matters in quantum physics! A definite mathematical reference for this topic with complete proofs is [26].

First, a subset $D \subset \mathcal{H}$ is said to be a *dense subspace* if it is closed under the formation of linear combinations of elements of D and if the topological closure of D (i.e., D plus all its limit points) is equal to $\mathcal{H}$. If $\dim \mathcal{H}$ is finite, then the only dense subspace is $\mathcal{H}$ itself. But for $\dim \mathcal{H}$ infinite there is a multitude of distinct, dense subspaces which are *proper* (meaning not equal to $\mathcal{H}$). So the next definition is only non-trivial when $\dim \mathcal{H}$ is infinite.

Definition 9.6.1 *Suppose that $\mathcal{H}$ is a Hilbert space with $D \subset \mathcal{H}$ a dense subspace. A mapping $T : D \to \mathcal{H}$ is said to be a* (linear) densely defined operator *if we have*

$$T(\lambda_1 \phi_1 + \lambda_2 \phi_2) = \lambda_1 T(\phi_1) + \lambda_2 T(\phi_2)$$

for all $\lambda_1, \lambda_2 \in \mathbb{C}$ and all $\phi_1, \phi_2 \in D$. (Note that these conditions imply that $\lambda_1 \phi_1 + \lambda_2 \phi_2 \in D$. So both sides of the previous equation make sense.)

Since the dense subspace D of T is an essential part of this definition, we often write $Dom(T)$ or $D(T)$ instead of D. One says $D = Dom(T) = D(T)$ is the domain *of T.*

In particular we allow the possibility that there exists $C > 0$ such that $||T\phi|| \leq C||\phi||$ holds for all $\phi \in Dom(T)$. If so, we say that T is a bounded operator. *(This agrees with our previous definition of bounded operator in the case when $Dom(T) = \mathcal{H}$.) If no such constant C exists, then we say that T is an* unbounded operator.

A special case is $Dom(T) - \mathcal{H}$. If so, we say that T is globally defined.

In our definition, not all densely defined operators are unbounded. For example, just take any bounded operator defined on $\mathcal{H}$ and restrict it to any dense subspace. The result will be a densely defined, bounded operator. In many texts the only densely defined operators (in their definition) are the unbounded ones. We do not follow that convention.

Here is another property that is related to that of restriction.

Definition 9.6.2 *Suppose $T_j : Dom(T_j) \to \mathcal{H}$ are densely defined operators for $j = 1, 2$ such that $Dom(T_1) \subset Dom(T_2)$ and T_2 restricted to $Dom(T_1)$ is equal to T_1. Then we say that T_2 is an* extension *of T_1.*

If we have $Dom(T_1) \neq Dom(T_2)$ here, then it follows that $T_1 \neq T_2$. This is because a necessary condition for mappings to be equal is that their domains must be equal. Or to put it positively, if $T_1 = T_2$, then we have both $Dom(T_1) = Dom(T_2)$ and $T_1\phi = T_2\phi$ for all $\phi \in Dom(T_1) = Dom(T_2)$.

The basic property of bounded, densely defined operators is given next.

Proposition 9.6.1 *Let $T : Dom(T) \to \mathcal{H}$ be a bounded, densely defined operator which satisfies $||T\phi|| \leq C||\phi||$ for all $\phi \in Dom(T)$. Then T has a unique globally defined extension S which satisfies $||S\phi|| \leq C||\phi||$ for all $\phi \in \mathcal{H}$.*

Because of this result, we often identify T with the unique extension S. Technically, this is incorrect. But the reader can usually see what's happening and so no harm is done.

The fact that we are working in Hilbert space means that we have its inner product at our disposal. This enters into the next definition.

Definition 9.6.3 *Let $T : Dom(T) \to \mathcal{H}$ be a densely defined operator. We define its* adjoint operator, *denoted T^* in the following two steps:*

1. *We first define*

   $$Dom(T^*) := \{\phi \in \mathcal{H} \mid \exists \xi \in \mathcal{H} \text{ such that } \langle \phi, T\psi \rangle = \langle \xi, \psi \rangle \, \forall \psi \in Dom(T)\}$$

   Note that by the density of $Dom(T)$ one can quickly prove that this element ξ is unique for a given $\phi \in Dom(T^*)$. Also, ξ depends on ϕ.

2. *For each $\phi \in Dom(T^*)$, define $T^*\phi := \xi$, where $\xi \in \mathcal{H}$ is defined in the previous part of this definition. (Therefore $\langle \phi, T\psi \rangle = \langle T^*\phi, \psi \rangle$ holds.)*

Exercise 9.6.1 *Given $\phi \in Dom(T^*)$ prove that ξ is unique as claimed.*

Exercise 9.6.2 *Prove that $Dom(T^*)$ is a linear subspace of $\mathcal{H}$ as well as that $T^* : Dom(T^*) \to \mathcal{H}$ is a linear mapping.*

The alert reader will have noticed that it has not been claimed that $Dom(T^*)$ is dense in $\mathcal{H}$. And, yes, there are examples where it is not dense. However, in quantum theory we are interested mainly in certain types of operators. Here is one such type.

Definition 9.6.4 *Let $T : Dom(T) \to \mathcal{H}$ be a densely defined operator that satisfies*

$$\langle \phi, T\psi \rangle = \langle T\phi, \psi \rangle$$

for all $\phi, \psi \in Dom(T)$. Then we say that T is a symmetric operator.

By reading definitions carefully, we see that for a symmetric operator we have that $\phi \in Dom(T)$ implies that $\phi \in Dom(T^*)$ (which is to say, $Dom(T) \subset Dom(T^*)$) and that $T^*\phi = T\phi$ for all $\phi \in Dom(T)$.

In short, T^* is an extension of T if T is symmetric. (Conversely, T is symmetric if T^* is an extension of T.) The inclusion $Dom(T) \subset Dom(T^*)$ together with the assumed density of $Dom(T)$ then implies that the larger subspace $Dom(T^*)$ is also dense. So, the adjoint of a symmetric operator is a densely defined operator.

An even more important type is operator is defined next.

Definition 9.6.5 *Let $T : Dom(T) \to \mathcal{H}$ be a densely defined, symmetric operator that satisfies $Dom(T) = Dom(T^*)$ (and also $T^*\phi = T\phi$ for all $\phi \in Dom(T) = Dom(T^*)$ because it is symmetric). Then we say that T is a* self-adjoint operator.

The role of the domains is the critical factor here. There is a multitude of symmetric operators with $Dom(T) \subset Dom(T^*)$ (by definition), but yet with $Dom(T) \neq Dom(T^*)$. The self-adjoint operators are therefore a very special type of symmetric operator.

One amazing property of self-adjoint operators is that the spectral theory applies to them. This means that they have a Borel functional calculus as well as a spectral representation in terms of a uniquely determined projection valued measure. While this is quite useful, another amazing property is that every self-adjoint, densely defined Hamiltonian operator H has an associated family of unitary operators $\exp(-itH/\hbar)$ for $t \in \mathbb{R}$, which gives the solution ψ_t of the time dependent Schrödinger equation with initial condition ϕ as

$$\psi_t = \exp(-itH/\hbar)\phi.$$

It turns out that the operators $\exp(-itH/\hbar)$ are guaranteed to exist by the Borel functional calculus, but the explicit calculation of them for a specific

H can be quite a chore. These two properties of *self-adjoint* operators are not enjoyed by those *symmetric* operators which are not self-adjoint.

All this theory–and more–is important in quantum theory. For now we lay out the facts such as they are, and let the reader get used to dealing with them. Complete expositions exist in many fine texts, such as [25].

9.7   Dirac Notation (Optional)

One can understand quantum theory without ever using Dirac notation, and I have decided not to use it much in this book. I feel that it is outdated, but that it hangs around for whatever reason. Think of it like Latin, once the universal language of science. But Dirac notation is still found in a lot of the literature and so merits an optional section devoted to it.

We suppose that $\mathcal{H}$ is a Hilbert space. The first idea is that every vector $\psi \in \mathcal{H}$ is denoted as $|\psi\rangle$ and is called a *ket*. Similarly (almost dually) every vector $\phi \in \mathcal{H}$ determines a bounded linear functional (according to the Riesz representation theorem 9.3.2), which is denoted as $\langle\phi|$ and is called a *bra*. The bra $\langle\phi|$ maps the ket $|\psi\rangle$ to the complex number $\langle\phi, \psi\rangle$, which is called the *bracket*. This last sentence can be written as $\langle\phi||\psi\rangle := \langle\phi, \psi\rangle$ or in words as "bra + ket = bracket".

It turns out that the opposite way of combining a bra and a ket is useful (and meaningful!) as well. The notation is $|\psi\rangle\langle\phi|$ and it is defined to be the linear operator mapping $\mathcal{H}$ to itself given by its action on a ket $|\alpha\rangle$ as

$$|\psi\rangle\langle\phi|\,|\alpha\rangle := \langle\phi, \alpha\rangle\psi.$$

The reader can begin to see why this notation is so easy to use. Here are some exercises that re-enforce this remark.

Exercise 9.7.1 *Prove that the composition of two of these operators satisfies*

$$\left(|\psi\rangle\langle\phi|\right) \circ \left(|\psi'\rangle\langle\phi'|\right) = \langle\phi, \psi'\rangle \left(|\psi\rangle\langle\phi'|\right)$$

for all $\psi, \phi, \psi', \phi' \in \mathcal{H}$, where $\circ$ denotes the operation of composition. This formula is usually written compactly as $|\psi\rangle\langle\phi|\,|\psi'\rangle\langle\phi'| = \langle\phi, \psi'\rangle\,|\psi\rangle\langle\phi'|$.

Exercise 9.7.2 *Prove that for all $\psi, \phi \in \mathcal{H}$ the linear operator $|\psi\rangle\langle\phi|$ is bounded and its operator norm is $|| \; |\psi\rangle\langle\phi| \; || = ||\psi||\,||\phi||$. In particular, $|\psi\rangle\langle\phi| = 0$ if and only if either $\psi = 0$ or $\phi = 0$.*

Exercise 9.7.3 *If $|\psi\rangle\langle\phi| \neq 0$, prove $\mathrm{Ran}(|\psi\rangle\langle\phi|) = \mathbb{C}\psi$. In particular, the range of this operator has dimension 1 and is said to be a rank 1 operator.*

Exercise 9.7.4 *The adjoint operator of $|\psi\rangle\langle\phi|$ is $(|\psi\rangle\langle\phi|)^* = |\phi\rangle\langle\psi|$.*

Exercise 9.7.5 *Let $e \in \mathcal{H}$ be a vector with $||e|| = 1$. Prove $P := |e\rangle\langle e|$ is a self-adjoint projection operator whose range is $\mathbb{C}e$, which has dimension 1.*

Conversely, let P be a projection operator whose range has dimension 1. Prove $P := |e\rangle\langle e|$ for any vector e in the range of P which satisfies $||e|| = 1$.

Exercise 9.7.6 *Suppose that $\{e_j \mid j \in J\}$ is an orthonormal basis of a Hilbert space $\mathcal{H}$. Prove that we can write the identity operator I of $\mathcal{H}$ as*

$$I = \sum_{j \in J} |e_j\rangle\langle e_j| \tag{9.7.1}$$

in the sense that $|\psi\rangle = \sum_{j \in J} |e_j\rangle\langle e_j| |\psi\rangle$ holds for every ket $|\psi\rangle$ in $\mathcal{H}$. Here the multiplication of a ket by a complex number from the right means the (usual) scalar multiplication of that ket by that complex number from the left. In other words, old wine in new bottles.

One says that the formula (9.7.1) is a resolution of the identity.

Exercise 9.7.7 *Suppose that $T : \mathcal{H} \to \mathcal{H}$ is a bounded linear operator acting in the Hilbert space $\mathcal{H}$. Suppose that T is* diagonalizable, *meaning that there exists an orthonormal basis $\{e_j \mid j \in J\}$ of the Hilbert space and there exist complex numbers $\{\lambda_j \in \mathbb{C} \mid j \in J\}$ such that $Te_j = \lambda_j e_j$ holds for all $j \in J$. Here J is a countable set of indices. Prove that*

$$T = \sum_{j \in J} \lambda_j |e_j\rangle\langle e_j| \tag{9.7.2}$$

in the sense that the following holds for every ket $|\psi\rangle$ in $\mathcal{H}$:

$$T|\psi\rangle = \sum_{j \in J} \lambda_j |e_j\rangle\langle e_j| |\psi\rangle.$$

We also define $\langle\phi|T|\psi\rangle := \langle\phi, T\psi\rangle = \langle T^*\phi, \psi\rangle$ for $\phi, \psi \in \mathcal{H}$.

In physics texts the formula (9.7.2) is often used instead of the rigorously correct spectral theorem in order to represent an arbitrary, densely defined self-adjoint operator. While it is okay to think of such an operator as being diagonalizable in some *more general* sense, it is not true that it always has a representation as given in (9.7.2). However, *some* densely defined self-adjoint operators (such as the harmonic oscillator Hamiltonian) can be represented as in (9.7.2), in which case $\lambda_j \in \mathbb{R}$, but any number of important physical observables (such as the coordinates of position or momentum) can not be so represented.

9.8 Notes

The quotation of Hilbert translates as: "There would be just one detail. What is a Hilbert space?" It is anecdotal. It may be false, or it may be that he said something else. The story is that one of Hilbert's colleagues, J. von Neumann (a junior colleague since Hilbert had no colleague senior to him), was speaking in a seminar and deliberately used the phrase 'Hilbert space' or rather in German 'Hilbertraum'. Knowing that Hilbert was a stickler for clarity, this was a provocation to get the quoted question. The rest, as they say, is history. Or maybe not.

What is true is that Hilbert and collaborators developed an extensive theory which in modern terms is known as the theory of *bounded operators in Hilbert spaces*. This not only included what we now call spectral theory, but also the word 'spectrum' (in German, of course) was introduced due to a seemingly coincidental relation with the field of spectroscopy in physics. With the advent of quantum theory J. von Neumann realized that a new class of operators acting in the previously known Hilbert space (the *densely defined, self-adjoint, unbounded operators*) was being implicitly used by the physicists. He then developed much of the needed spectral theory, which appeared rather promptly in his famous book on quantum mechanics [32]. And the ultimate irony is that the spectra of this mathematical theory are related to the spectra measured in spectroscopy.

The study of quantum states in the optional Section 9.4 is the beginning of a fascinating theory that includes *entanglement* and the relation of geometry with quantum physics. A nice introduction to this field where physics and mathematics are inseparable is [4].

The mathematics of quantum theory is the mathematics of Hilbert space. But don't be fooled! The physics of quantum theory is very much more than the mathematics of Hilbert space. This is why so many mathematicians (and others with a background only in mathematics) have so much trouble learning quantum theory. The mathematics gets you a good way down the path, but it is not enough to get you all the way to understanding what is going on physically. And since quantum theory is based on a probabilistic interpretation, it is no easy matter to understand it. You do not need a knowledge of classical physics to only feel comfortable with classical thinking. Old habits are hard to break. An analogy: For those who think English is a form of Latin, you have to learn to like splitting infinitives.

Chapter 10

Interpreting ψ: Measurement

> In science, novelty emerges only with difficulty.
> Thomas Kuhn

The physical interpretation in Chapter 8 of the 'wave' function ψ is not complete. We still need to clarify how ψ is related to experiments, that is, to measurements. So we continue this topic. To see the basic idea clearly without getting bogged down in technicalities, we present a simple, though representative, example. Later on in Chapter 16 we will consider this idea in its full generality.

10.1 Some Statements

We start with a linear operator T acting in some Hilbert space $\mathcal{H}$. We suppose that T has the special property that there exists a non-empty orthonormal set $\phi_j \in \mathcal{H}$ with $j \in J$, a finite or countably infinite index set, such that

$$T\phi_j = t_j\,\phi_j,$$

where each $t_j \in \mathbb{R}$ and the set $\{t_j \mid j \in J\}$ is closed in $\mathbb{R}$ (that is, it has no limit points outside of $\{t_j \mid j \in J\}$ itself). To avoid trivial cases we assume that $\dim \mathcal{H} \geq 2$.

Moreover, we assume that every $\phi \in \mathcal{H}$ can be written as

$$\phi = \sum_{j \in J} \alpha_j \phi_j \tag{10.1.1}$$

for unique numbers $\alpha_j \in \mathbb{C}$. Thus $\{\phi_j \mid j \in J\}$ is an *orthonormal basis* of $\mathcal{H}$.

© Springer Nature Switzerland AG 2020
S. B. Sontz, *An Introductory Path to Quantum Theory*,
https://doi.org/10.1007/978-3-030-40767-4_10

Exercise 10.1.1 *Suppose that $\phi \in \mathcal{H}$ is a state. Using the above expansion* (10.1.1) *of ϕ, prove that*

1. $\sum_j |\alpha_j|^2 = 1$.

2. $\alpha_j = \langle \phi_j, \phi \rangle$.

3. $\sum_j |\langle \phi_j, \phi \rangle|^2 = 1$.

4. $0 \leq |\langle \phi_j, \phi \rangle|^2 \leq 1$.

As we have seen, the harmonic oscillator Hamiltonian is such a linear operator. In this more general case we suppose that T represents a physically measurable quantity of some quantum system. Then its possible measured values are given exactly by the closed set of real numbers $\{t_j \in \mathbb{R} \,|\, j \in J\}$, which is the set of eigenvalues of T. As a technical aside for those who know what the following italicized words mean, T determines a unique *densely defined self-adjoint operator* that acts in a dense subspace of $\mathcal{H}$, and the *spectrum* Spec (T) of that operator is Spec $(T) = \{t_j \,|\, j \in J\} \subset \mathbb{R}$.

One final assumption is that $t_j \neq t_k$ for $j \neq k$. This means that each eigenvalue has an associated eigenspace whose dimension is 1, that is,

$$\dim \{\phi \in \mathcal{H} \,|\, T\phi = t_j \,\phi\} = 1 \quad \text{for each } j \in J.$$

Putting $V_j := \{\phi \in \mathcal{H} \,|\, T\phi = t_j \phi\}$ this means that V_j is a one-dimensional space. Clearly, the one element set $\{\phi_j\}$ is a basis of V_j. This assumption is expressed by saying that each eigenvalue t_j has *multiplicity one*. Then the unit eigenvector ϕ_j is a *state* associated to the eigenvalue t_j. We now have:

Measurement Condition: *Suppose $\mathcal{H}$ is a Hilbert space which corresponds to a quantum system $\mathcal{S}$. Suppose $\phi \in \mathcal{H}$ is any state. If an experiment is done on $\mathcal{S}$ when it is in the state ϕ in order to measure the physical quantity associated with T, then the experimental value of that quantity will be t_j with probability $|\alpha_j|^2$, where $\phi = \sum_j \alpha_j \phi_j$ is the unique expansion as given above in* (10.1.1). *It is convenient to write this statement in the notation of quantum probability as follows:*

$$P(T = t_j \,|\, \phi) = |\alpha_j|^2, \tag{10.1.2}$$

which can be read as: "The probability that the observable T when measured yields the value t_j given that the system is in state ϕ is $|\alpha_j|^2$." (This may remind you of a conditional probability, if you have seen such things.) This is a probabilistic affirmation.

Moreover, given that the experimental value is t_j, then $\alpha_j \neq 0$ and the state of the system becomes ϕ_j with no further information about the initial state ϕ of the system. Note that this condition is deterministic. This second affirmation is known as the collapse condition.

A more colloquial reading of equation (10.1.2) is: "The probability that T is t_j, given ϕ, is $|\alpha_j|^2$". Notice that the expression $T = t_j$ when expressed by itself simply means that T is the operator $t_j I$. That is not what that expression means in the above context.

The measurement condition, and especially the collapse condition, is sometimes referred to as the *observer effect*, which is then often confused with the Heisenberg uncertainty inequality. Other common expressions for the collapse condition are *quantum leap*, *quantum jump*, and *reduction of the wave function*.

Notice that the case when dim $\mathcal{H} = 1$ is deterministic, since there is only one eigenvalue (and so only one possible measurable value of the observable). Also, the collapse condition in this case sends the initial (and only) state ϕ_1 to the same state ϕ_1.

Before coming to the controversy surrounding these conditions, we refer back to Part 1 of Exercise 10.1.1, which allows us to construct a unique discrete probability measure μ_ϕ on the subset of eigenvalues $\{t_j \mid j \in J\}$ of T in $\mathbb{R}$ which satisfies $\mu_\phi(\{t_j\}) = |\alpha_j|^2 = |\langle \phi_j, \phi \rangle|^2$ for every $j \in J$. Note that μ_ϕ depends on the observable T as well as the state ϕ. We can extend μ_ϕ to a probability measure on the whole real line $\mathbb{R}$ by putting it equal to zero on the complement of $\{t_j \mid j \in J\}$. Then the *support* of μ_ϕ, that is, the set on which μ_ϕ 'lives', turns out to be $\{t_j \mid \alpha_j \neq 0\} \subset \text{Spec}(T)$. See Appendix A for the definition of the support of a measure.

What is being said here is that the probability measure μ_ϕ is predicted by quantum theory. As with any probability measure, μ_ϕ can have an *average* or *expected value*, $\mathcal{E}(\mu_\phi) := \sum_j |\alpha_j|^2 t_j$ (also predicted by quantum theory if the sum converges absolutely), but this number is not enough information to determine μ_ϕ. This was known as early as 1932 by von Neumann. See [32].

The complex number $\langle \phi_j, \phi \rangle$ is called the *probability amplitude* of the state ϕ_j given the initial state ϕ. Sometimes this is said more colloquially as: The *amplitude* of finding ϕ_j given ϕ is $\langle \phi_j, \phi \rangle$.

Already we can see a difference between the sort of probability that arises in quantum theory and that of probability theory *a la* Kolmogorov. (See Chapter 16 for Kolmogorov's theory.) In the former complex numbers play an essential role while they do not in the latter.

This interpretation generalizes in a non-trivial way to the case of a self-adjoint operator S acting in a Hilbert space. In that case for every state there is a new type of probability measure on the real line $\mathbb{R}$ whose support is a subset of the *spectrum* of S. We will come back to this during the discussion of quantum probability in Chapter 16.

But the measurement condition, and specifically the collapse condition, is controversial among physicists. Before elaborating on the controversy, let's note that these conditions are an essential part of the recent studies of quantum information theory and, more especially, quantum computation theory. Also all experiments on systems with *spin* are consistent with this condition. In short, no experiment has ever falsified this condition while many experiments are consistent with it. We will see this in greater detail after spin is presented in Chapter 14. An unsettling aspect of the collapse condition is that the transition from the initial state before measurement to the final state is not linear. This is clear, since the set of states, as we have seen in Section 9.4, is not a vector space. But this flies in the face of the linearity of the time evolution given by the Schrödinger equation. While this is not a paradox, it is puzzling.

10.2   Some Controversies

So why the controversies? Both aspects of this interpretation are criticized. The first is that it refers directly to measurements, which seems to refer only to experiments (or perhaps just observations) made by humans. While all theories must eventually pass the scrutiny of experimental situations, this principle is not easily applied to the universe at large. How does it apply to nuclear interactions deep inside of rocks on Mars? Can it explain events that occurred billions of years before the creation of the solar system, such as supernova explosions? Does any physicist seriously accept that quantum phenomena can not exist without some interaction with a conscious observer? Unfortunately, the answer to this last seemingly rhetorical question is: Yes.

Some attempts to modify this principle have been proposed. For example, it could refer to all interactions rather than just to measurements. After all measurements are a type of interaction. But usually a measurement is thought of as a deliberate intent to measure a physical quantity associated with a given self-adjoint operator. (Note the hidden conscious observer who has the intent.) But the interactions in nature are not so clearly controlled or intended. However, the Measurement Condition surely does apply to measurements, whatever meaning it might have in other contexts. So it can be checked with experiments. As far as I am aware, it has always been verified in an experiment and never falsified. That is a fine record. Nonetheless, the Measurement, and especially, the Collapse Condition remain a puzzlement.

The Measurement Condition is uncomfortable for those who have become accustomed to deterministic theories of nature. This is understandable, since this condition is not at all deterministic, but rather probabilistic. It says that starting with an identical initial state and performing an identical experiment repeatedly for an observable with two or more possible values in that initially prepared state, one will not get one unique measured result (given a precise enough measurement), but rather a variety of different results. Even spin measurements for spin 1/2 states, which give only two different values, are difficult to accept in a purely probabilistic way. One can be profoundly tempted into believing that the initial conditions were not 'really' identical. This leads one down the trail to the non-laughing place where the nature of 'quantum reality' is discussed *ad nauseum*. When we come back to spin, we shall try to find a place of rational repose, though not a true laughing place for considering this situation.

Another critique is that the collapse of the 'wave' function is not physical. This refers mainly to Hilbert spaces like $L^2(\mathbb{R}^3)$. In such a context a state (or 'wave' function) has an interpretation in terms of the infinitely large space $\mathbb{R}^3$, which engulfs the Moon, the Solar System, and all the galaxies. And the collapse implies an instantaneous change throughout this enormous space to a new 'wave' function. What is the mechanism behind this? Well, maybe that is not a good question. But it does leave one wondering. One argument, which I am unable to support, says that the 'wave' function itself is not physically observable. This argument is then pushed further by positing that being a non-observable, the 'wave' function does not exist physically, that it is merely a convenient mathematical construction. See Section 16.3 for my opinion that states are a special type of *quantum event*, which in turn is a special type of observable. Also, recall that the idea of the physical, rather than the merely mathematical, existence of electric and magnetic fields was slow in being accepted by the scientific community in the 19th century. I leave this for your further consideration. But this is a tough nut to crack.

As is quite usual, in this exposition the collapse condition is considered to be one part of the measurement condition. But the collapse condition is deterministic given the result of the probabilistic measurement condition. Behind all of this is the deterministic time dependent Schrödinger equation. This strange mixture of both deterministic and probabilistic aspects within quantum theory might not be here to stay. But it seems to be here for now.

Since the measurement is a source for human knowledge, one can see how those philosophers who study *epistemology*, the study of knowledge, have gotten interested in quantum theory and so have made contributions to this ongoing discussion. Let me remind the reader that what we call *physics* was once commonly called *natural philosophy*. So we should not be surprised by, nor be unwelcome to, this participation by experts in another academic field.

10.3   Simultaneous Eigenvectors, Commuting Operators

The topic of this section is not typically discussed in the context of the measurement condition. But I find that this is the correct logical place in quantum theory for understanding this phenomenon which has no correlate in classical physics. A basic, if often hidden, assumption in classical physics is that any pair of observables can be simultaneously measured as well as that these two measurements do not interfere with each other in any way. In quantum theory this just isn't so. Please note that time is playing a role here, since the time order of the two measurements is crucial.

To understand this we start by considering two observables represented by the self-adjoint operators A and B acting in some Hilbert space $\mathcal{H}$. To get to the basic idea we will consider a rather special case. First, we assume that the two operators are defined in a common dense subspace V of $\mathcal{H}$ and that V is all finite linear combinations of all the eigenvectors of A as well as being all finite linear combinations of all the eigenvectors of B. Furthermore, we assume that each operator has only a discrete spectrum corresponding to these eigenvectors. Putting this into mathematical notation we let ϕ_j be the eigenstates of A with corresponding eigenvalues a_j, and we let ψ_k be the eigenstates of B with corresponding eigenvalues b_k. Finally, we assume that each eigenvalue for both A and B has multiplicity one, meaning that $A\phi_i = a_i\phi_i$ and $A\phi_j = a_j\phi_j$ with $a_i = a_j$ imply that $\phi_i = \phi_j$. (Recall this quite different and general property: $A\phi_i = a_i\phi_i$ and $A\phi_j = a_j\phi_j$ with $a_i \neq a_j$ imply that $\phi_i \perp \phi_j$.) A similar assumption is made for B. The eigenvectors, or eigenstates, introduced here are time independent states.

Now, as Dirac often noted, the key property (at least for him) of quantum theory was that observables need not commute. And that is exactly what concerns us here. So we define the subspace of V where they do commute:

$$\mathcal{C} := \{\psi \in V \mid AB\psi = BA\psi\}. \qquad (10.3.1)$$

Exercise 10.3.1 *Prove that V is invariant under the actions of A and B because of the above hypotheses. (To say that V is* invariant *under A simply means that $A\psi \in V$ whenever $\psi \in V$. And similarly for B.) Therefore, the expressions $AB\psi$ and $BA\psi$ make sense in the definition* (10.3.1) *of $\mathcal{C}$.*

There are two extreme cases here. In the first case $\mathcal{C} = V$, so that A and B commute on the common domain V. The second extreme case is that $\mathcal{C} = 0$, which means that AB and BA are as different as they possibly could be on V. Clearly, intermediate cases also exist, but again for the sake of simplicity we will not discuss them. We will now consider how measurements work in quantum theory in these two extreme cases.

Say that we are in the first case, so that $AB = BA$ on all of V. Suppose we start with a physical system in some state $\chi \in \mathcal{H}$. Let's make a measurement of A. The result has to be one of the real numbers a_j, since there is no other spectrum, and the system must then collapse into the eigenstate ϕ_j, since the multiplicity of a_j is 1. (Recall that $A\phi_j = a_j\phi_j$.) Next we see that

$$AB\phi_j = BA\phi_j = Ba_j\phi_j = a_jB\phi_j.$$

If $B\phi_j \neq 0$, this says that $B\phi_j$ is an eigenvector of A with eigenvalue a_j. And since a_j has multiplicity 1, that means that $B\phi_j = \lambda_j\phi_j$, for some $\lambda_j \in \mathbb{C} \setminus \{0\}$. On the other hand, $B\phi_j = 0$ implies $B\phi_j = \lambda_j\phi_j$, where $\lambda_j = 0$. Putting these last two sentences together we see that $B\phi_j = \lambda_j\phi_j$ holds, for some $\lambda_j \in \mathbb{C}$. But $\phi_j \neq 0$, since it is a state. Thus, ϕ_j is an eigenvector of B with associated eigenvalue λ_j. However, we know the eigenvalues of B. So, λ_j must be one of them, say $\lambda_j = b_k$, with corresponding eigenstate ψ_k. Consequently, ψ_k must be the same state as ϕ_j. (Of course, this means that $\psi_k = \alpha\phi_j$ for some norm 1 complex number α.)

The story so far for the case $AB = BA$: If we start with the system in an arbitrary state and measure A, then the system collapses into some eigenstate of A *which turns out always to be an eigenstate of B as well.*

Now we continue with a measurement of B which will give the eigenvalue b_k always (that is, with probability 1) and the state collapses from being ψ_k to being ψ_k, which is the state ϕ_j. So, the collapse is trivial in this case and therefore resembles classical physics. If a measurement of A then follows, the value a_j will be found with probability 1 and again the collapse is trivial. So the measurements, excluding the first one, give their results with probability 1 and do not change the state via the collapse. This case is that of commuting operators and their common (or as we also say, simultaneous) eigenstates. Actually, the fact is that the commuting operators A and B, as given in this section, have the same set of eigenstates, but with different eigenvalues in general. And this fact has nothing to do with time, so the terminology 'simultaneous' eigenstate may be misleading.

Exercise 10.3.2 *Understand why A and B have simultaneous (or common) eigenstates without appealing to measurements or time.*

All of this so far should be reminiscent, except for the one non-trivial collapse, of classical physics for those with some background in that subject.

As the reader must have guessed by now, the other extreme case is wildly different. Recall that in this case $C = 0$. As in the first case we start with a physical system in some state $\chi \in \mathcal{H}$ and make a measurement of A. Just as before the result has to be one of the real numbers a_j and the system must then collapse into the eigenstate ϕ_j. So far everything is the same. But next we will make a measurement of B and things are quite different, since ϕ_j is *not* an eigenstate of B. Why not? Well suppose it were, that is, $B\phi_j = b\phi_j$ for some real number b. But then we would have

$$AB\phi_j = Ab\phi_j = bA\phi_j = ba_j\phi_j = a_jb\phi_j = a_jB\phi_j = Ba_j\phi_j = BA\phi_j,$$

which implies that $0 \neq \phi_j \in C$. And that contradicts the hypothesis of this case, namely that $C = 0$. So, as claimed, ϕ_j is not an eigenstate of B.

Now, as indicated above, we proceed to the next step, which is to measure B given that the system is in the state ϕ_j. When we make this measurement the value measured will be some b_k and then the state will also collapse to the corresponding eigenstate ψ_k. But since the starting state ϕ_j is not an eigenstate of B, there are many (i.e., at least two) outcomes with non-zero probability, say $b_k \neq b_l$ with corresponding eigenstates ψ_k and ψ_l and, most importantly, with $k \neq l$. We are face to face with the probabilistic aspect of quantum theory. Given that the system is now in a state that 'has' the value a_k with probability 1 (very much like classical physics), we have that same system in a state that 'has' at least two possible values for the observable B. So we measure B and indeed get one of these values as well as collapsing the system into the corresponding eigenstate. This collapse is a non-trivial operation; it is not the identity map. Now a subsequent measurement of A will give results that depend on which of these various states the system is in after the measurement of B. And by a similar argument none of the eigenstates of B is an eigenstate of A. So again various distinct values for A are possible and the collapse of the eigenstate is non-trivial.

So it goes on and on with alternating measurements of A and B. And at no time do we have the system in a common eigenstate of both A and B. At no time do we have a system that has value a_j for A with probability 1 and at the same time has value b_k for B with probability 1. As noted earlier this can be understood directly as a property of the operators A and B by using an argument that mentions neither measurements nor time. But anyway this now shows that a simultaneous measurement of A and B can not be possible in theory. And yes, I mean simultaneous in time. Because if this were possible the arbitrary initial state would have to collapse into a common eigenstate of A and B. But there is no common eigenstate in this case. But we warned! Experiment always stands present ready to test theory. If an experiment can be performed that actually does measure simultaneously (in time) both A and B, then this aspect of quantum theory is wrong.

Please note that this property of quantum theory is often confused with the Heisenberg uncertainty principle. What is being stated here is that, for observables A and B for which do not commute in the strong sense that $C = 0$, it is completely impossible to have a simultaneous measurement of both A and B. A common 'intuitive' justification for the Heisenberg uncertainty principle is that the lack of commutativity of A and B implies that they can not both be simultaneously measured *with arbitrary precision*. What I am saying is that they can not both be simultaneously measured *in any way, shape or form*; forget about precision! We will discuss this in more detail in Section 16.10 and Section 18.2.

10.4   The Inner Product—Physics Viewpoint

The inner product in a Hilbert space seems to be mathematical–some would even say geometrical–concept. But there is some physics associated to it. First off, physics hints that the inner product $\langle \phi, \psi \rangle$ of two states might have a physical meaning. And it almost does!

But let's back off and think about classical physics, where a (pure) state is a point in a certain space P, called the *phase space*. You don't need to know much about classical physics, just that a point in P completely describes a classical system and that two different points give completely different descriptions of that system. If I consider a system in a particular state $p_1 \in P$ and ask what is the probability that it is in the state $p_2 \in P$, then there are precisely two possible answers. If $p_2 = p_1$, then the probability requested is 1 or, as we sometimes say, 100%. If $p_2 \neq p_1$, then the probability requested is 0 or equivalently 0%. This is so in classical physics because observations of a system never change its state. (Or, more strictly speaking, the change can be made as negligible as one wishes.) So you can now see why we don't ask this question in classical physics. The answer is never a probability strictly between 0 and 1, that is to say, we never get a really interesting probability.

The situation is rather different in quantum physics. If we are given a quantum system in the state ψ, we can again ask what is the probability that it is also in the state ϕ. If $\phi = \psi$ (or even if $\phi = \lambda\psi$ with $|\lambda| = 1$), then we do have that this probability is 1. So far this is like the classical case. But if ψ and ϕ are two unequal states (meaning that they are linearly independent), then the requested probability is $|\langle\phi, \psi\rangle|^2$. All of a sudden the inner product enters into a formula that is experimentally verifiable. And we are facing the counter-intuitive nature of quantum probabilities.

Let's address first the experimental meaning of this probability statement. We suppose that the quantum system is in the state ψ and that we perform a measurement of an observable A which has ϕ as an eigenvector with the eigenvalue a. Also, we suppose the multiplicity of the eigenvalue a is 1. In other words, we have carefully chosen A so that it has these properties with respect to the state ϕ. An example of this is given by taking $A = |\phi\rangle\langle\phi|$, the Dirac notation for the projection onto the subspace $\mathbb{C}\phi$. As we will see in Chapter 16, this choice for A is the *quantum event that ϕ has occurred*. So if we measure a we know that the system collapses to the state ϕ. Also, if we do not measure a, then the system collapses to a state orthogonal to ϕ, since the multiplicity of a is 1. If we measure A and get the value a, we say that the initial state ψ *overlaps* or *transitions* to the state ϕ. And we express this by saying that the state ψ has some probability to be also in the state ϕ. This may seem to be a sloppy way of expressing the results of the measurement process just described, but it is common parlance.

Let's address the issue of calculating this probability. By the Projection Theorem we can decompose ψ as $\psi = c_1\phi + c_2\chi$, the sum of its projection onto the subspace spanned by ϕ plus a vector $c_2\chi$ in the orthogonal complement of ϕ, such that $\chi \perp \phi$, $||\chi||^2 = 1$ and $|c_1|^2 + |c_2|^2 = 1$. It also follows immediately that

$$\langle\phi, \psi\rangle = \langle\phi, c_1\phi\rangle + \langle\phi, c_2\chi\rangle = c_1\langle\phi, \phi\rangle + c_2\langle\phi, \chi\rangle = c_1,$$

since $\langle\phi, \phi\rangle = ||\phi||^2 = 1$ and $\langle\phi, \chi\rangle = 0$. It seems natural to expect that $|c_1|^2$ is the probability of having ψ transition to ϕ. In fact, some physicists take this to be a fundamental rule of quantum physics, although we will see later in Chapter 21 that it can also be understood as a consequence of a more general axiom of quantum theory. In any case we do find that

$$|c_1|^2 = |\langle\phi, \psi\rangle|^2,$$

which is called the *transition probability* for ψ to transition to ϕ. The inner product $\langle\phi, \psi\rangle$ is called the *transition amplitude* for the transition of ψ to ϕ. This is the physics viewpoint of the Hilbert space inner product of states.

Exercise 10.4.1 *Prove that the transition probability is invariant if each state vector is multiplied by a phase factor, but that the transition amplitude does change in general.*

10.5   Notes

All aspects of the topic of this chapter are controversial. Dissenting opinions
are numerous and are held by persons with a lot of knowledge of the matter.
The content of this chapter is my way of expressing what is more or less the
'standard' view. Take it with buckets of salt, if you like. The interested reader
is recommended to search the literature for other viewpoints.

One of my own worries about the measure and collapse condition is that
it is not consistent with Schrödinger's razor, but rather adds something that
does not come from the basic partial differential equation of quantum theory.
More specifically, we end up with a theory that has two distinct ways of
describing time evolution: a deterministic way coming from the Schrödinger
equation and a probabilistic way coming from measurement (or interaction
or whatever). Of course, only the first, deterministic way is consistent with
Schrödinger's razor. So you can see why Schrödinger himself never accepted
the second probabilistic way. As far as I am aware quantum theory is the
only theory in physics that has two ways for describing how things change
with time. And remember that understanding how things evolve in time is
the fundamental goal of physics.

One argument against the collapse condition is that the state (or 'wave'
function) that is collapsing does not really exist, and so it makes no sense
to speak of its collapse. While this may well be a parody of the complete
argument, let me note that the question of what is real (and what is not) is
difficult to decide even in classical physics, where the ether supporting the
vibrations of electromagnetic waves was taken by some in the 19th century
to be a physical reality, though it is now considered to be non-existent. Is
that going to be the fate of the 'wave' function too? On the other hand,
a quantum state is essentially the same as a one-dimensional self-adjoint
projection, which itself 'is' a quantum observable (of a certain type known as
a *quantum event.*) So then one has to address the question whether quantum
observables, or even just quantum events, really exist. It's quite a rabbit hole.

The role of measurement in quantum theory is complicated, certainly
much more so than in classical theory which therefore does not serve at all
as an aid for understanding the quantum theory. So far we have only seen
the tip of an iceberg; a detailed exposition of quantum measurement can be
found in the recent text [8].

I once had the good fortune to hear P. Dirac give a talk. And he did say on
that occasion that the characteristic property of quantum mechanics is that
the observables do not commute. I agree. But I would like to extend this a bit
further by adding that each pair of non-commuting observables should satisfy
a commutation relation that expresses the non-commutativity explicitly in
terms of Planck's constant $\hbar$ and that, moreover, this commutation relation
in the *semi-classical limit* when $\hbar \to 0$ should say that the pair do commute.
This makes $\hbar$ a necessary part of the characteristic property of quantum
mechanics. After all, I think that a theory without $\hbar$ entering in an essential
way should not be considered a theory of quantum phenomena.

Chapter 11

The Hydrogen Atom

> Water, water everywhere.
>
> Samuel Taylor Coleridge

One of the most important considerations of physics, dating to ancient Greek times, was whether the matter of everyday experience is infinitely divisible into ever smaller parts with the same properties or, on the other hand, at some point one would arrive at a piece of matter that is no longer divisible, such an indivisible entity being named an *atom*. The well-established answer is more complicated than this formulation would indicate, but let's note that as matter gets divided into ever smaller pieces one arrives at small structures that correspond to each of the chemical elements in the *Periodic Table*. These are called atoms, even though they are further divisible but not into other atoms. And all the objects of our immediate world are combinations of these atoms. It took over two millennia to settle this question. But now that we know that atoms (at least in the modern sense of that word) do exist, the remaining question is how do they exist! The answer to that question comes from quantum theory. We start with the simplest atom, the first one in the Periodic Table.

11.1 A Long and Winding Road

We return to the study of a particular physical system, the *hydrogen atom*. This is a long and beautiful story that shows the importance of eigenvalue problems in quantum theory. As will become clear shortly, this is the *two body problem* of quantum theory. As such, it corresponds to the *Kepler problem* in classical mechanics of understanding the motion of a planet in orbit around a star. This was one of the first problems solved by quantum theory and so was important in establishing the validity of that new theory.

© Springer Nature Switzerland AG 2020

S. B. Sontz, *An Introductory Path to Quantum Theory*,

https://doi.org/10.1007/978-3-030-40767-4_11

First, here is a bit of *spectroscopy*. Hydrogen gas, that is easily produced in the laboratory using electrolysis of water or is readily available from a gas supply company, is a molecule H_2 consisting of two hydrogen atoms held together by a chemical bond. This molecule can be divided in the laboratory into two unattached hydrogen atoms, which can be studied experimentally. In the 19th century an incredible amount of spectroscopy of visible light (and electromagnetic energy beyond the red end of the visible band) was done. The upshot is that every chemical in the gas phase had its unique *spectrum* for absorption and emission of light. This spectrum is obtained by spreading out the light into its component colors (at first by using prisms, later diffraction gratings) and seeing what one gets. Because of the geometry of the prisms and diffraction gratings, the observed spectrum typically consisted of parallel lines, each of which has its own unique color or frequency. Typically, spectra of emitted light show a discrete set of lines of different colors between which there are gaps with no light at all. These are the *emission lines*.

On the other hand absorption spectra show that *white light* (i.e., light with a full range of colors) when passed through a supposedly transparent gas, has almost the same spectrum as the initial white light, except for certain lines of light that have dimmed considerably, though they still show some light. These are the *absorption lines*. The emission and absorption lines for a given gas are typically at the same frequencies (or actually *wave numbers* to use the jargon of spectroscopy, but never mind). However, the width of these lines (that is, they are not actually perfect lines) as well as their intensities added more data to be explained. For example, the notations S, P, D, F were introduced to qualitatively describe these spectral lines. And the classical physics of the 19th century was not up to the task of providing an explanation for any of this data.

Enough was known that the absorption lines of light from the Sun (and eventually from other stars as well) could be interpreted as indicating the presence of a variety of chemicals, mostly elements, in the photosphere of the Sun. (As indicated earlier sunlight can also be used to measure the temperature of the photosphere). However, there were some absorption lines in sunlight which were unknown in spectroscopy. These were thought to indicate the presence of a chemical in the Sun that was not known on Earth. So this unknown chemical was named *helium*, which is a neologism based on ancient Greek and means the *sun metal*. Subsequently, this chemical was found on Earth, probably in the gas from natural gas wells. This turned out to be the second element in the Periodic Table and is still called helium, even though we now know it is a *noble gas* and not a *metal*. Moreover, it is not so rare since it is the second most abundant element in the universe, only exceeded by hydrogen. All of this shows the importance of the then new science of spectroscopy.

Let's get back to hydrogen atoms. A long and winding experimental road led to the detection in an ordinary matter of the first structures smaller than atoms, namely the *electron* (J.J. Thomson, 1897) and the *nucleus*

(E. Rutherford, 1911). This was the birth of *sub-atomic physics*. The electrons carry negative electric charge while the nuclei (plural of 'nucleus') carry positive electric charge. In a macroscopic sample of matter, these opposite charges cancel out (or very nearly cancel out) and so the electrical nature of ordinary matter was an extremely non-trivial discovery. Next, a planetary model of the hydrogen atom (N. Bohr, 1913) was proposed. It consisted of a single negatively charged electron in an orbit around a positively charged nucleus, which we nowadays know to typically be a single *proton*, even though other nuclei are possible. By the way, Rutherford discovered the proton later in 1917. For our present purposes the charges of the electron and the proton will be taken as -1 and $+1$, respectively. Still, each is a measurable quantity using the standard SI unit of charge, the *coulomb*, denoted C. In terms of the basic SI units of the *ampere* and the *second*, the coulomb is an ampere-second. The charge of an electron in these units is about -1.6×10^{-19} C.

Of course, nowadays we do not think in terms of orbits nor of other sorts of trajectories of 'wave/particles'. And they will not enter the following analysis of the hydrogen atom. So what did Schrödinger know about the hydrogen atom? First off, the masses of these states were known. The electron mass is $m_e \sim .511$ MeV/c^2 and the proton mass is $m_p \sim 938/c^2$ MeV/c^2. Here c is the *speed of light* and MeV is a *mega-electron-volt*, a unit of energy commonly used in high energy physics. (The Einstein equation $E = mc^2$ is used here to convert units of energy into units of mass.) So, $m_p/m_e \sim 1835$, that is, the proton mass is some 3 orders of magnitude larger than the electron mass. Due to the electron mass being very much smaller than that of the first element in the Periodic Table, it was recognized that the electron was a *sub-atomic particle*. The proton is also usually described as being a sub-atomic particle, even though it is exactly the same as most positively ionized hydrogen atoms found in nature. (Perhaps it is worth pointing out parenthetically that the nucleus of some hydrogen atoms is not a proton, but a rather more massive nucleus due to the presence in it of one or two *neutrons*, a sub-atomic particle with zero electric charge. These are the *isotopes* of hydrogen.)

So, the first assumption that we will make is that the proton is a point source of an *electric field* centered on a given point in $\mathbb{R}^3$. For convenience we take that point to be the origin $0 \in \mathbb{R}^3$. The *potential energy* $V(\mathbf{x})$ of the electron in that electric field in our units is obtained from Coulomb's law for the force between electrically charged objects:

$$V(\mathbf{x}) = -\frac{1}{||\mathbf{x}||} = \frac{-1}{\sqrt{x_1^2 + x_2^2 + x_3^2}}.$$

This is the classical potential energy of the attractive force $\mathbf{F}$ between a proton at the origin and an electron at the point $\mathbf{x} = (x_1, x_2, x_3) \in \mathbb{R}^3 \setminus \{0\}$, since

$$\mathbf{F} = -\nabla V(\mathbf{x}) = -\operatorname{grad} V(\mathbf{x}) = -\frac{1}{||\mathbf{x}||^3}\,\mathbf{x}$$

is the classical Coulomb's law, that is, an attractive force that points at the proton (a *central* force) with magnitude (norm) inversely proportional to the distance squared $||\mathbf{x}||^2$, that is, an *inverse square law*.

According to the canonical quantization, the quantum Hamiltonian H for the electron in the hydrogen atom is

$$H = -\frac{\hbar^2}{2m_e}\Delta - \frac{1}{\sqrt{x_1^2 + x_2^2 + x_3^2}}. \qquad (11.1.1)$$

The eigenvalue problem is $H\psi = E\psi$, where the unknowns are $E \in \mathbb{R}$ and a non-zero $\psi \in L^2(\mathbb{R}^3)$. Mathematically, it is not clear whether this problem has any solution at all. If not, then the Schrödinger equation is useless!

But physically, we have the spectroscopic data which had already been interpreted as implying that the hydrogen atom has discrete *energy levels* and that the spectrum of the hydrogen atom was given by $\hbar\omega = E_2 - E_1$. Here $E_1 < E_2$ are two of these energy levels and the transition of the hydrogen atom from the level with energy E_2 down to the level with E_1 is accompanied by the emission of a photon of angular frequency ω and hence, according to Einstein's photon hypothesis, with energy $\hbar\omega$.

Similarly, a hydrogen atom at the energy level E_1 could be excited up to energy level E_2 via the absorption of a photon of angular frequency ω provided that the angular frequency ω is 'tuned' so as to satisfy $\hbar\omega = E_2 - E_1$. So only the light whose frequencies correspond to differences of energy levels are absorbed by the atoms, while light of all other frequencies passes through the gas without change.

So the idea is to see if the eigenvalues of the hydrogen atom Hamiltonian are the same as the experimentally determined energy levels of the hydrogen atom. You might already have guessed that this story has a happy ending, namely that the eigenvalues give the measured energy levels almost exactly. But not all stories have a Hollywood ending. Schrödinger himself had already solved the eigenvalue problem for a more plausible candidate for the hydrogen atom Hamiltonian by using Einstein's special relativity theory as a guide. That path led to what we now call the *Klein-Gordon equation*. And he found that those eigenvalues did not correspond to the measured energy levels. He then tried using a non-relativistic approach, which is what we will now present since it does agree with the experiment.

11.2   The Role of Symmetry

But back to the Hollywood screenplay. The first thing to do with the Hamiltonian H in (11.1.1) is to use its symmetry property. What is that? Well, acting by rotations fixing the origin of $\mathbb{R}^3$ leaves both the Laplacian Δ and the potential energy invariant. This means that if the coordinates $y = (y_1, y_2, y_3)$ are obtained from $x = (x_1, x_2, x_3)$ by a rotation 3×3 matrix

R (that is, $y = Rx$, see Chapter 13), then

$$y_1^2 + y_2^2 + y_3^2 = x_1^2 + x_2^2 + x_3^2,$$
$$\frac{\partial^2}{\partial y_1^2} + \frac{\partial^2}{\partial y_2^2} + \frac{\partial^2}{\partial y_3^2} = \frac{\partial^2}{\partial x_1^2} + \frac{\partial^2}{\partial x_2^2} + \frac{\partial^2}{\partial x_3^2}.$$

So H is also invariant under these rotations, since $\hbar$ and m_e are constants. This indicates that we should change from the standard system of *Cartesian coordinates* (x_1, x_2, x_3) to *spherical coordinates* (r, θ, φ). These coordinate systems are related by this transformation:

$$
\begin{aligned}
x_1 &= r \sin\theta \cos\varphi \\
x_2 &= r \sin\theta \sin\varphi \\
x_3 &= r \cos\theta
\end{aligned}
\tag{11.2.1}
$$

Here $r = +\left(x_1^2 + x_2^2 + x_3^2\right)^{1/2}$. So $r \in [0, \infty)$, i.e., $r \geq 0$. The function $(x_1, x_2, x_3) \mapsto r$ is C^∞ on $\mathbb{R}^3 \setminus \{0\}$, i.e., it has partial derivatives of all orders.

Also, θ is the angle between the ray from the origin passing through the *north pole* $(0, 0, 1)$ on the *unit sphere* $\mathbb{S}^2$ and the ray from the origin passing through the vector (x_1, x_2, x_3), which we assume is not the zero vector. (The *ray* through $0 \neq v \in \mathbb{R}^3$ is the set $\{sv \mid s > 0\}$.) Also, we take $0 \leq \theta \leq \pi$ with $\theta = \pi/2$ corresponding to the *equator* of $\mathbb{S}^2$. Hence, $-1 \leq \cos\theta \leq +1$. The function $(x_1, x_2, x_3) \mapsto \theta := \cos^{-1}(x_3/(x_1^2 + x_2^2 + x_3^2)^{1/2})$ is C^∞ on $\mathbb{R}^3 \setminus Z$, where Z is the z-axis (also called the x_3 axis),

$$Z := \{(0, 0, z) \mid z \in \mathbb{R}\}. \tag{11.2.2}$$

Finally, φ is the angle between a vector (x_1, x_2, x_3) not on Z and the open half-plane which has Z as its boundary and passes through the point $(1, 0, 0)$. We can think of the intersection of that half-plane HP with the unit sphere as defining the *Greenwich meridian* of the spherical coordinates. We take $0 \leq \varphi < 2\pi$. The coordinate φ in defined on $\mathbb{R}^3 \setminus Z$ but is discontinuous on HP, while φ is a C^∞ function of the coordinates (x_1, x_2, x_3) on $\mathbb{R}^3 \setminus (HP \cup Z)$. However, we do not wish to write down a formula now for φ in terms of (x_1, x_2, x_3). We will come back to this detail later.

Notice that the spherical coordinates for the surface of the Earth have different intervals of definition for the two angles θ and φ which are called *latitude* and *longitude*, respectively

Changing the potential energy to spherical coordinates is easy enough: $V = -1/r$, which has a singularity at the origin where $r = 0$. But the spherical coordinates also have a singularity at the origin.

Exercise 11.2.1 *At the origin we have that the variable r is well defined, namely $r = 0$. Understand why there is a singularity at the origin for the spherical coordinates.*

The next thing to do is to write the Laplacian Δ in spherical coordinates. And even though this is nothing other than an application of the chain rule (in dimension 3, of course), that is quite a chore! Here's the answer:

$$\Delta = \frac{1}{r^2}\frac{\partial}{\partial r}\left(r^2\frac{\partial}{\partial r}\right) + \frac{1}{r^2\sin\theta}\frac{\partial}{\partial\theta}\left(\sin\theta\frac{\partial}{\partial\theta}\right) + \frac{1}{r^2\sin^2\theta}\frac{\partial^2}{\partial\varphi^2}. \qquad (11.2.3)$$

Notice that the partial derivative $\partial/\partial r$ means the usual derivative taken with respect to r while the other variables θ and φ are held constant. And *mutatis mutandi* for the other partial derivatives in this formula. For example, in the usual defining formula for the Laplacian, namely

$$\Delta = \frac{\partial^2}{\partial x_1^2} + \frac{\partial^2}{\partial x_2^2} + \frac{\partial^2}{\partial x_3^2}, \qquad (11.2.4)$$

the partial derivative $\partial/\partial x_2$ means the usual derivative taken with respect to x_2 while the other variables x_1 and x_3 are held constant. And so forth. So the meaning of a partial derivative depends on *all* the variables, not just the one that comes in the notation itself. This is a universal abuse of notation that is found throughout the supposedly rigorous mathematical literature. I point this out since beginning students can all too easily misinterpret the meaning of a partial derivative in a given formula. It seems that only physicists, and then only when doing thermodynamics, take the effort to include in the notation of a partial derivative what are the variables being held constant.
Remark: You really should try to prove (11.2.3) in total detail, rather than relying on a formula from a text or a handbook. I have found too many well known books that get this formula wrong.

Exercise 11.2.2 *Prove (11.2.3).*

The formula (11.2.4) looks so much simpler than (11.2.3). For example, (11.2.4) is defined in all of $\mathbb{R}^3$, while (11.2.3) only makes sense in the dense open subset $\mathbb{R}^3 \setminus Z$ of $\mathbb{R}^3$, since Z is the set where either $r = 0$ or $\sin\theta = 0$. Recall that Z is defined in (11.2.2). A rather curious fact is that the variable φ is well defined and discontinuous on $\mathbb{R}^3 \setminus Z$, even though the vector field $\partial/\partial\varphi$ is well defined and continuous on $\mathbb{R}^3 \setminus Z$. (If you don't have *vector field* in your scientific vocabulary, then replace that phrase for now with *partial derivative*.) A technical point is that Z is a subset of measure zero in $\mathbb{R}^3$.

Exercise 11.2.3 *Find the subset of $\mathbb{R}^3$ where $\partial/\partial r$ is well defined. Then do the same for $\partial/\partial\theta$. Understand how $\partial/\partial\varphi$ is well defined on $\mathbb{R}^3 \setminus Z$ and why it is continuous. How do these results relate to the subset where (11.2.3) is well defined?*

Yet for our purposes (11.2.3) is much, much simpler than (11.2.4). How could that be so? Well, we are going to use *separation of variables* using the

variables r, θ, φ in order to study the eigenvalue equation $H\psi = E\psi$. So we
are thinking of ψ as a function of the spherical coordinates, that is $\psi(r, \theta, \varphi)$.
We therefore look for solutions of $H\psi = E\psi$ of the form

$$\psi(r, \theta, \varphi) = F(r)Y(\theta, \varphi).$$

So, using (11.1.1) and (11.2.3), the eigenvalue equation $H\psi = E\psi$ becomes

$$Y(\theta, \varphi)\frac{\partial}{\partial r}\left(r^2\frac{\partial}{\partial r}\right)F(r) + F(r)\left(\frac{1}{\sin\theta}\frac{\partial}{\partial\theta}\left(\sin\theta\frac{\partial}{\partial\theta}\right) + \frac{1}{\sin^2\theta}\frac{\partial^2}{\partial\varphi^2}\right)Y(\theta, \varphi)$$
$$- \frac{2m_e}{\hbar^2}r^2V(r)F(r)Y(\theta, \varphi) = -\frac{2m_e}{\hbar^2}r^2E\,F(r)Y(\theta, \varphi).$$

Next, we divide by $F(r)Y(\theta, \varphi)$ and re-arrange getting

$$\frac{1}{Y(\theta, \varphi)}\left(\frac{1}{\sin\theta}\frac{\partial}{\partial\theta}\left(\sin\theta\frac{\partial}{\partial\theta}\right) + \frac{1}{\sin^2\theta}\frac{\partial^2}{\partial\varphi^2}\right)Y(\theta, \varphi) =$$
$$-\frac{1}{F(r)}\frac{\partial}{\partial r}\left(r^2\frac{\partial}{\partial r}\right)F(r) + \frac{2m_er^2}{\hbar^2}(V(r) - E).$$

The right side does not depend on θ, φ while the left side does not depend
on r. But the two sides are equal, so that each side does not depend on any of
the three variables r, θ, φ. This implies that both sides are the same constant.
So we see that there is a *separation constant*, say $K \in \mathbb{C}$, such that

$$\frac{1}{Y(\theta, \varphi)}\left(\frac{1}{\sin\theta}\frac{\partial}{\partial\theta}\left(\sin\theta\frac{\partial}{\partial\theta}\right) + \frac{1}{\sin^2\theta}\frac{\partial^2}{\partial\varphi^2}\right)Y(\theta, \varphi) = K \qquad (11.2.5)$$

and

$$-\frac{1}{F(r)}\frac{\partial}{\partial r}\left(r^2\frac{\partial}{\partial r}\right)F(r) + \frac{2m_e}{\hbar^2}r^2(V(r) - E) = K. \qquad (11.2.6)$$

We proceed to study these two equations, starting with the first.

11.3   A Geometrical Problem

Notice that (11.2.5) does not involve the energy E nor the potential $V(r)$. So
it is an eigenvalue problem about the unit sphere in terms of its coordinates
θ and φ. This geometrical problem arose from the Schrödinger equation for
the hydrogen atom. But it is an independent problem for the unit sphere;
this problem arises also in the mathematical subject of *harmonic analysis*
without any reference to quantum physics at all. Here is equation (11.2.5)
again, but now looking a bit more like an eigenvalue equation:

$$\Big(\frac{1}{\sin\theta}\frac{\partial}{\partial\theta}\Big(\sin\theta\frac{\partial}{\partial\theta}\Big) + \frac{1}{\sin^2\theta}\frac{\partial^2}{\partial\varphi^2}\Big)Y(\theta,\varphi) = K\,Y(\theta,\varphi) \qquad (11.3.1)$$

By the way, the differential operator of the left side of (11.3.1) can be shown to be a symmetric, negative operator, thereby implying that K is a real number and that $K \leq 0$. We will come back to the exact values for K later.

Also, we did not use the explicit form of the potential energy $V(r)$ to arrive at (11.2.5) and (11.2.6), only the fact that it is a function of r alone. We say that such a function V is a *radial function*. So this separation of variables argument works for any radial potential. This symmetry property alone gives rise to the geometrical eigenvalue problem (11.3.1). We will see in Chapter 12 what this has to do with angular momentum.

Exercise 11.3.1 *Show that the force field* $\mathbf{F} = -\nabla V$ *associated to any radial potential V is a* central force, *that is, the direction of the vector* $\mathbf{F}(x_1, x_2, x_3)$ *points directly at or directly away from the origin* $(0,0,0) \in \mathbb{R}^3$.

And how are we going to solve (11.3.1)? Again, by separation of variables! We write $Y(\theta,\varphi) = Q(\theta)E(\varphi)$ and then substitute into (11.3.1) to get

$$E(\varphi)\Big(\frac{1}{\sin\theta}\frac{\partial}{\partial\theta}\Big(\sin\theta\frac{\partial}{\partial\theta}\Big)\Big)Q(\theta) + Q(\theta)\Big(\frac{1}{\sin^2\theta}\frac{\partial^2}{\partial\varphi^2}\Big)E(\varphi) = K\,Q(\theta)E(\varphi).$$

(Let me make a parenthetical apology for using the letter E for something which is not an energy but, as we shall see, an exponential.) Next we divide by $Q(\theta)E(\varphi)$, multiply by $\sin^2\theta$ and re-arrange to arrive at

$$\frac{1}{E(\varphi)}\frac{\partial^2}{\partial\varphi^2}E(\varphi) = K\sin^2\theta - \frac{1}{Q(\theta)}\Big(\sin\theta\frac{\partial}{\partial\theta}\Big(\sin\theta\frac{\partial}{\partial\theta}\Big)\Big)Q(\theta).$$

The left side does not depend on θ, and the right side does not depend on φ. So both sides are the same constant. We let M be the separation constant such that

$$\frac{\partial^2}{\partial\varphi^2}E(\varphi) = M\,E(\varphi) \qquad (11.3.2)$$

and therefore

$$\Big(\sin\theta\frac{\partial}{\partial\theta}\Big(\sin\theta\frac{\partial}{\partial\theta}\Big)\Big)Q(\theta) + MQ(\theta) = K\,(\sin^2\theta)Q(\theta). \qquad (11.3.3)$$

Technical remark: The eigenfunction $\psi(r,\theta,\varphi) = F(r)Q(\theta)E(\varphi)$ has to be square integrable as a function on $\mathbb{R}^3$, and this imposes square integrability conditions on $F(r)$, $Q(\theta)$ and $E(\varphi)$, each being a function of one variable. From the change of coordinates formula $dx\,dy\,dz = J\,dr\,d\theta\,d\varphi$ for multiple

Riemann (and Lebesgue) integrals, where J is the Jacobian, we get

$$\int_0^\infty dr\, r^2\, |F(r)|^2 < \infty, \quad \int_0^\pi d\theta\, \sin\theta\, |Q(\theta)|^2 < \infty, \quad \int_0^{2\pi} d\varphi\, |E(\varphi)|^2 < \infty,$$

since in this case the Jacobian is $J = r^2 \sin\theta$. (This fact may be difficult for you to see now. However, we will come back to it in Exercise 12.2.2.) These three conditions can also be written as:

$$F \in L^2((0,\infty),\, r^2\, dr), \quad Q \in L^2((0,\pi),\, \sin\theta\, d\theta), \quad E \in L^2((0,2\pi),\, d\varphi).$$

Since we are now working on a purely mathematical problem which is independent from the original physics context, the separation constants K and M can not be guessed by using 'physical intuition'. But mathematics comes to the rescue! Since the original eigenvalue problem is $(H - E)\psi = 0$ and $H - E$ is an *elliptic differential operator* on $\mathbb{R}^3 \setminus \{(0,0,0)\}$, a powerful theorem from partial differential equation (PDE) theory applies. This is known as the *elliptic regularity theorem*. It implies that the solution ψ must be C^∞ on $\mathbb{R}^3 \setminus \{(0,0,0)\}$.

But this forces $E(\varphi)$ to be C^∞ as well. In short, it is necessary that $E(\varphi)$ is defined on $[0, 2\pi]$ and satisfy the *periodic boundary conditions* $E(0) = E(2\pi)$ and $E'(0) = E'(2\pi)$. But, by an integration by parts argument, the operator $\partial^2/\partial\varphi^2$ is a symmetric operator when defined on the dense domain of C^∞ functions in $L^2([0, 2\pi])$ that satisfy these periodic boundary conditions. This implies that M is a real number. Moreover, $\partial^2/\partial\varphi^2 \leq 0$, that is, it is a *negative operator*. Hence, $M \leq 0$.

Exercise 11.3.2 *Prove that $\partial^2/\partial\varphi^2$ is a symmetric, negative operator on the dense domain defined above.*

So, we define $m := +\sqrt{(-M)} \geq 0$. (Warning: This m is not a mass.) By elementary calculus, for $m > 0$ the general solution of (11.3.2) is then $E(\varphi) = ae^{im\varphi} + be^{-im\varphi}$ for complex constants a and b. (We will consider the other case $m = 0$ in a moment.) We next apply the periodic boundary conditions, using the above formula for $E(\varphi)$ and

$$E'(\varphi) = iame^{im\varphi} - ibme^{-im\varphi}.$$

Therefore, the periodic boundary conditions give

$$a + b = ae^{2\pi im} + be^{-2\pi im},$$
$$iam - ibm = iame^{2\pi im} - ibme^{-2\pi im}.$$

We then divide the second equation by $im \neq 0$ giving the pair of equations

$$a + b = ae^{2\pi im} + be^{-2\pi im},$$
$$a - b = ae^{2\pi im} - be^{-2\pi im}.$$

Adding and subtracting these two equation then yields

$$2a = 2ae^{2\pi im} \qquad \text{and} \qquad 2b = 2be^{-2\pi im}. \qquad (11.3.4)$$

Now the choice of integration constants $a = b = 0$ gives us the zero solution $E(\varphi) \equiv 0$, implying $\psi(r, \theta, \varphi) \equiv 0$ which has been excluded for lacking a physical interpretation. But if either $a \neq 0$ or $b \neq 0$, then at least one of these last two equations (11.3.4) tells us that $e^{2\pi im} = 1$. Then by using a standard identity, this means the real number $m > 0$ satisfies

$$\cos(2\pi m) + i \sin(2\pi m) = 1.$$

This is satisfied if and only if $m > 0$ is an integer, in which case we have two linearly independent solutions of the eigenvalue problem, namely the eigenfunctions $e^{im\varphi}$ and $e^{-im\varphi}$, each having the eigenvalue $-m^2 < 0$. Also, each of them *is* C^∞ on $\mathbb{R}^3 \setminus Z$. In short, $\{e^{im\varphi} \mid m \in \mathbb{Z} \setminus \{0\}\}$ is an infinite family of linearly independent eigenfunctions.

We now return to the case $m = 0$. In this case the general solution of the differential equation (11.3.2) is

$$E(\varphi) = a + b\varphi$$

for $a, b \in \mathbb{C}$. The boundary condition $E(0) = E(2\pi)$ says that $a = a + 2\pi b$, which implies that $b = 0$. Moreover $E(\varphi) = a$ solves the differential equation, since $M = -m^2 = 0$. And in this case $e^{im\varphi} = e^0 = 1$, which is not only a solution but also a C^∞ function. So the *complete solution* of the eigenvalue problem (11.3.2) *with C^∞ eigenfunctions* is that the eigenfunctions are given (up to a non-zero multiplicative factor) by the infinite family

$$\{E_m(\varphi) := e^{im\varphi} \mid m \in \mathbb{Z}\} \qquad (11.3.5)$$

and the corresponding eigenvalue M in (11.3.2) is $-m^2$. So we have

$$\frac{\partial^2}{\partial \varphi^2} E_m(\varphi) = -m^2 E_m(\varphi). \qquad (11.3.6)$$

While a physics text will arrive at the same 'answer' for this eigenvalue problem, often the derivation is not as detailed as that given here. But this extra mathematical care to detail is important, since we do not want to find just *some* of the solutions of (11.3.2) but rather *all* of them. Why? Because each valid solution tells us something about the *physics* of the situation. Therefore this is, or should be, a textbook example of the interdependence of mathematics and physics.

The reader may have already noticed that the family of functions (11.3.5) plays a central role in *Fourier analysis*. The next exercise considers this.

Exercise 11.3.3 *Prove that* (11.3.5) *is a family of orthogonal functions in* $L^2[0, 2\pi]$. *Find the normalization factor for each of those functions so that we obtain an orthonormal set.*
Remark: *A fundamental result in Fourier analysis is that this orthonormal set is actually an orthonormal basis of* $L^2[0, 2\pi]$.

Now we go back to (11.3.3) and put $M = -m^2$ with $m \in \mathbb{Z}$ to get

$$\left(\sin\theta \frac{\partial}{\partial\theta} \left(\sin\theta \frac{\partial}{\partial\theta} \right) \right) Q(\theta) - m^2 Q(\theta) = K \left(\sin^2\theta \right) Q(\theta). \qquad (11.3.7)$$

So for each $m \in \mathbb{Z}$ we have an eigenvalue problem whose unknowns are the separation constant K and the function $Q(\theta)$, where $0 \le \theta \le \pi$. Note that (11.3.7) is a second order, linear ordinary differential equation with variable coefficients. Since the coefficient of the highest order derivative (i.e., the second derivative) is $\sin^2\theta$, this differential equation is singular at the values of θ where $\sin^2\theta = 0$, that is, for $\theta = 0$ and $\theta = \pi$. So we restrict ourselves to solving (11.3.7) on the domain $\mathbb{R}^3 \setminus Z$, where θ is a well defined variable and $0 < \theta < \pi$. Again by the elliptic regularity theorem, the solution $Q(\theta)$ will be C^∞ in $\theta \in (0, \pi)$. But we only want those solutions that give a solution $Q \in L^2((0, \pi), \sin\theta\, d\theta)$.

Solving (11.3.7) is a dicey business, and we will not do that in total detail. However, we will re-write it by changing variables to get a form that is more frequently used in the literature. So, we let $t = \cos\theta$ for $\theta \in (0, \pi)$. Hence $t \in (-1, +1)$. Then

$$\frac{\partial}{\partial\theta} = \frac{\partial t}{\partial\theta} \frac{\partial}{\partial t} = (-\sin\theta) \frac{\partial}{\partial t}.$$

which implies that

$$\sin\theta \frac{\partial}{\partial\theta} = (-\sin^2\theta) \frac{\partial}{\partial t} = (t^2 - 1) \frac{\partial}{\partial t}.$$

We now apply this change of variable to (11.3.7) and write $P(t) = Q(\cos^{-1} t)$ to get

$$(t^2 - 1) \frac{\partial}{\partial t} \left((t^2 - 1) \frac{\partial}{\partial t} P(t) \right) - m^2 P(t) = K(1 - t^2) P(t). \qquad (11.3.8)$$

We are using the standard branch $\cos^{-1} : (-1, +1) \to (0, \pi)$, which is C^∞. So Q being C^∞ on $(0, \pi)$ implies that P is C^∞ on $(-1, +1)$ by the chain rule.

Moreover, the condition $Q \in L^2((0, \pi), \sin \theta \, d\theta)$ under the change of variables $t = \cos \theta$ becomes $P \in L^2((-1, 1), dt)$.

Finally, let's divide (11.3.8) by $1 - t^2$ to obtain an eigenvalue equation

$$\frac{\partial}{\partial t} \left((1 - t^2) \frac{\partial}{\partial t} P(t) \right) - \frac{m^2}{1 - t^2} P(t) = K \, P(t).$$

Recall that $1 - t^2 \neq 0$ for $t \in (-1, +1)$. This is a second order linear ordinary differential operator with variable coefficients. Using the standard notation for functions of one variable, we can also write this eigenvalue equation as

$$(1 - t^2) P''(t) - 2t P'(t) - \frac{m^2}{1 - t^2} P(t) = K \, P(t). \tag{11.3.9}$$

There are still more techniques which one can now apply here. It seems that A.-M. Legendre did just that in the very productive 19th century. Or one can turn to the handbooks to see this is a well known differential equation. It seems that Schrödinger depended a lot on H. Weyl for help with differential equations. The material in the classic [9] by R. Courant and D. Hilbert was apparently also rather useful. We will follow that strategy here, and let the interested reader consult the mathematical literature on *Sturm-Liouville equations* in general, and this equation in particular, for the justification of what we are about to assert about (11.3.9) in the next paragraph.

The left side of (11.3.9) defines a self-adjoint operator (on an appropriate domain) in $L^2(-1, +1)$ whose spectrum is given by

$$\{-l(l + 1) \mid l \in \mathbb{N}, \, l \geq |m|\}.$$

This tells us the allowed values of K in (11.3.9). Each of these eigenvalues has multiplicity 1 and a corresponding eigenfunction is given for $t \in (-1, +1)$ and $|m| \leq l$ by the *associated Legendre function*

$$P_l^m(t) := (-1)^m (1 - t^2)^{m/2} \frac{d^m}{dt^m} \left(P_l(t) \right),$$

where $P_l(t)$ is the *Legendre polynomial* of degree l that itself is defined for $t \in (-1, +1)$ and $l \in \mathbb{N}$ by

$$P_l(t) := \frac{1}{l! \, 2^l} \frac{d^l}{dt^l} (t^2 - 1)^l.$$

One finds other normalization conventions in the literature for the Legendre polynomials and the associated Legendre functions. The reader should be aware that Schrödinger's razor has been applied here, since we are discarding all solutions of (11.3.9) that are not in the Hilbert space $L^2(-1, 1)$.

Therefore, the eigenvalue problem (11.3.1) for the function $Y(\theta, \varphi)$ of the angular variables is solved by the eigenfunctions

$$Y_{l,m}(\theta,\varphi) = e^{im\varphi}\, P_l^m(\cos\theta)$$

for all pairs of integers $-l \leq m \leq l$ with the associated eigenvalue K being $-l(l+1)$ for every integer $l \geq 0$. Putting this into (11.3.7) we get

$$\frac{1}{\sin^2\theta}\left[\left(\sin\theta\frac{\partial}{\partial\theta}\left(\sin\theta\frac{\partial}{\partial\theta}\right)\right) - m^2\right]Y_{l,m}(\theta,\varphi) = -l(l+1)\,Y_{l,m}(\theta,\varphi) \quad (11.3.10)$$

Of course, l is a *quantum number* associated to this geometric, as well as physical, problem. One inelegant name for this is the *azimuthal quantum number*. We will eventually give it a better name: the *angular momentum quantum number*.

The functions $Y_{l,m} : S^2 \to \mathbb{C}$ on the unit sphere S^2 are usually multiplied by some normalization constant (depending on l and m) and are then called the *spherical harmonics*, fundamental objects in *harmonic analysis*.

The integer m is called the *magnetic quantum number*. In the next section when we come to the radial problem, we will see immediately that m does not enter into that problem, although l does. Thus, the energy eigenvalues E do not depend on m. So how do we know about m experimentally? Well, as its name suggests, when a hydrogen atom is placed in a magnetic field the degenerate levels associated with an energy E are split into distinct energy levels (the *Zeeman effect*), and we use m to label these levels. This also happens for other quantum systems that interact with magnetic fields. From the theoretical viewpoint this amounts to changing the Hamiltonian of the system and using *perturbation theory* to learn how the energy levels change. This is the topic of the amazingly well written text [17] by T. Kato.

11.4   The Radial Problem

Next we attack the eigenvalue equation (11.2.6) for the radial function $F(r)$. To do this we will use the explicit formula $V(r) = -1/r$ for the Coulomb potential energy. Recall that we have identified the separation constant as $K = -l(l+1)$ for $l \in \mathbb{N}$ in the angular part of the problem. So we obtain

$$-\frac{\partial}{\partial r}\left(r^2\frac{\partial}{\partial r}\right)F(r) + \frac{2m_e}{\hbar^2}r^2\left(-\frac{1}{r} - E\right)F(r) = -l(l+1)F(r). \quad (11.4.1)$$

The equation (11.4.1) contains all the physics of this problem: the mass of the electron m_e, Planck's constant $\hbar$, the energy E of the hydrogen atom, and the potential energy. Note that the electric charges of the nucleus and the electron are included in the constant -1 in the potential energy.

The solution of (11.4.1) must combine with the spherical harmonic Y_{lm} to give a solution of $H\psi = E\psi$ in $L^2(\mathbb{R}^3)$, where $\psi(r,\theta,\varphi) = F(r)Y_{lm}(\theta,\varphi)$. As noted earlier $F(r)$ must satisfy $F \in L^2(0, \infty, r^2 \, dr)$. This condition, which is a consequence of Schrödinger's razor, excludes many solutions of (11.4.1) and will impose some condition on the energy E, as we shall see. Recall that (11.4.1) is a linear ordinary differential equation of second order and so has two linearly independent solutions for every energy E.

Expanding formula (11.4.1) in units with $m_e = 1$ and $\hbar = 1$ then yields

$$r^2 F''(r) + 2r F'(r) + 2(r + Er^2) F(r) - l(l+1) F(r) = 0. \qquad (11.4.2)$$

Recall that the last two equations come from the original eigenvalue equation $H\psi = E\psi$, where E is the eigenvalue. In particular, that means that we have to solve for all admissible values of E. Of course, (11.4.2) is also an eigenvalue equation with eigenvalue E, if we re-arrange the terms and divide by $2r^2 > 0$. So we have two unknowns: E and the eigenfunction $F(r)$.

By just looking at equation (11.4.2), we expect E to depend non-trivially on the integer $l \geq 0$, though the magnetic quantum number m has totally disappeared from this part of the theory. The particular formula for the potential energy $V(r) = -1/r$ gives us a problem that we can solve explicitly. For general radial potentials $V(r)$ that is not the case. Another unexpected consequence is that the energy eigenvalue E does *not* depend on l. This rather curious property of the Coulomb potential energy will emerge later.

Now we are going to play with (11.4.2) in a non-rigorous, intuitive way to see if we can understand it better. We are going to speculate that $F(r) = r^\alpha$ is a solution. Then, as the reader can check, the left side of (11.4.2) gives us

$$-\alpha(\alpha+1)r^\alpha - 2(r^{\alpha+1} + Er^{\alpha+2})$$

while the right side gives $-l(l+1)r^\alpha$. This clearly is not a solution! However, it does show that near $r = 0$ it is approximately correct if we take $\alpha = l$, since the terms proportional to r^{l+1} and to r^{l+2} go to zero faster than the terms proportional to r^l as $r \to 0$. So we expect, though we do not know, that the solution of (11.4.1) looks like r^l as $r \to 0$. This sort of statement is said to be about the *asymptotic behavior* of the solution as $r \to 0$.

What can we say about the asymptotic behavior of the solution of (11.4.2) as $r \to \infty$? Well, again we will deal in intuition. We assume (hope!) that $F(r)$ and its derivatives go to zero very rapidly as $r \to \infty$, so that the only terms in (11.4.1) that matter for very large r are those proportional to r^2, since this is the highest power of r that we see in that equation. Dropping the 'lower order terms' and dividing by r^2, we assert that as $r \to \infty$ the dominant terms in (11.4.2) give us $F''(r) + 2EF(r) = 0$. We want the solution of this to be square integrable near $r = +\infty$, so this forces $E < 0$ with two linearly independent solutions

$$F(r) = e^{+\sqrt{-2E}\, r} \qquad \text{and} \qquad F(r) = e^{-\sqrt{-2E}\, r}.$$

But only the second of these functions is square integrable near $r = +\infty$. So we hope that the asymptotic behavior of the solution of (11.4.1) as $r \to \infty$ is $e^{-\lambda r}$, where $E < 0$ and $\lambda := \sqrt{-2E} > 0$.

We really do not know if any of this helps. And so it goes when doing research; one looks for clues which sometimes help. But not always. Anyway, we use these ideas to re-write the solution as

$$F(r) = r^l e^{-\lambda r} G(r). \tag{11.4.3}$$

Rigorously speaking, this is merely a definition of a new function $G(r)$, of course. This changes the problem from that of finding $F(r)$ to that of finding $G(r)$. The condition on $G(r)$ is that $F(r)$ as given in (11.4.3) must be in $L^2((0, \infty), r^2\, dr)$. When we have solved this problem we will see that $G(r)$ is typically an unbounded function. But that will not matter provided that the decaying exponential factor in (11.4.3) is enough to guarantee that $F(r)$ is in $L^2((0, \infty), r^2\, dr)$.

This new problem is not exactly equivalent to the original one, since we have imposed the condition $E < 0$ on the eigenvalue of the original problem $H\psi = E\psi$. This restriction on the energy can be understood both physically and mathematically. Physically, it means we have limited ourselves to look only for the *bound states* of the hydrogen atom, though there are also *scattering states* with positive energy. Mathematically, it means we are not considering the *continuous spectrum* $[0, \infty)$ of the hydrogen atom but only its *point spectrum* in $(-\infty, 0)$.

Now we have to find the differential equation that $G(r)$ satisfies. To do this we first compute derivatives:

$$F'(r) = \left(lr^{l-1}G(r) - \lambda r^l G(r) + r^l G'(r) \right) e^{-\lambda r}$$

$$F''(r) = -\lambda \left(lr^{l-1}G(r) - \lambda r^l G(r) + r^l G'(r) \right) e^{-\lambda r}$$
$$+ \left(l(l-1)r^{l-2}G(r) + lr^{l-1}G'(r) - \lambda lr^{l-1}G(r) \right.$$
$$\left. - \lambda r^l G'(r) + lr^{l-1}G'(r) + r^l G''(r) \right) e^{-\lambda r}.$$

Combining terms we find that

$$e^{\lambda r}F''(r) = \left(l(l-1)r^{l-2} - 2\lambda lr^{l-1} + \lambda^2 r^l \right) G(r)$$
$$+ \left(-2\lambda r^l + 2lr^{l-1} \right) G'(r) + r^l G''(r).$$

Going back to (11.4.2) and substituting in these formulas we obtain

$$r^{l+2}G''(r) + \left(-2\lambda r^{l+2} + 2lr^{l+1}\right)G'(r)$$
$$+ \left(l(l-1)r^l - 2\lambda lr^{l+1} + \lambda^2 r^{l+2}\right)G(r)$$
$$+ \left(2lr^l G(r) - 2\lambda r^{l+1}G(r) + 2r^{l+1}G'(r)\right)$$
$$+ \left(2(Er^2 + r)r^l - l(l+1)r^l\right)G(r) = 0.$$

The three terms that are scalar multiples of $r^l\,G(r)$ cancel. So we can divide out by r^{l+1} without introducing singularities. There remain three terms with $G'(r)$ and four terms with $G(r)$. Here is what we now have:

$$rG''(r) + (-2\lambda r + 2l + 2)G'(r) + (-2\lambda l + \lambda^2 r - 2\lambda + 2Er + 2)G(r) = 0.$$

Since $\lambda^2 = -2E$, the two terms proportional to r in the coefficient of $G(r)$ cancel, giving

$$rG''(r) + (-2\lambda r + 2l + 2)G'(r) + (-2\lambda l - 2\lambda + 2)G(r) = 0. \qquad (11.4.4)$$

This is a type of second order linear ordinary differential equation which was studied in the 19th century by E. Laguerre. We now rely on several facts about the solutions of this equation. First, we define the *Laguerre polynomials* by the *Rodrigues formula*

$$L_j(x) := e^x \frac{d^j}{dx^j}(x^j e^{-x})$$

for all integers $j \geq 0$ and $x \in \mathbb{R}$. Clearly, $L_j(x)$ is a polynomial of degree j. Next, we define the *associated Laguerre polynomials* by

$$L_j^k(x) := \frac{d^k}{dx^k}\left(L_j(x)\right)$$

for all integers $j, k \geq 0$ and $x \in \mathbb{R}$. Clearly, for $k \leq j$ this is a polynomial of degree $j - k$. Also, it is non-zero if and only if $k \leq j$. Then the *associated Laguerre differential equation*

$$xy'' + (k + 1 - x)y' + (j - k)y = 0$$

has $y = L_j^k(x)$ as a non-zero solution for integers $0 \leq k \leq j$.
It follows immediately that $G(x) = L_j^k(ax)$ satisfies

$$xG''(x) + (k + 1 - ax)G'(x) + a(j - k)G(x) = 0.$$

Consequently, $G(x)$ will satisfy (11.4.4) provided that $a = 2\lambda$, $k = 2l + 1$ and

$$a(j - k) = -2\lambda l - 2\lambda + 2.$$

These three equations imply that $j = l + 1/\lambda$ as the reader can check.

Now we appeal to the mathematical literature (but without proof here) for this critical fact about the associated Laguerre differential equation: The only solutions of (11.4.4) that give a non-zero eigenfunction in $L^2(\mathbb{R}^3)$ for the original problem $H\psi = E\psi$ are the associated Laguerre polynomials. This is proved by expanding $G(x)$ in a power series and doing a lot of analysis. One consequence is that the degree d of the polynomial $G(x) = L_j^k(2\lambda x)$ is $d = j - k \geq 0$.

Another amazing consequence is that $j - l = 1/\lambda$ must be a both positive (since $\lambda > 0$) and an integer (since j and l are integers). So we introduce the notation $n := j - l = 1/\lambda \geq 1$. Recalling that $\lambda = \sqrt{-2E} > 0$, we immediately get this formula for the energy E, which is the eigenvalue in the original problem of the hydrogen atom:

$$E = -\frac{1}{2n^2} =: E_n,$$

where $n \geq 1$ is an integer. We also have that $\lambda = 1/n$ and $j = l + n$.

The moral of this long story is that the eigenvalues of the hydrogen atom form a discrete sequence of strictly negative numbers which have a unique accumulation point at $0 \in \mathbb{R}$. We say that n is the *principal quantum number* for the eigenfunctions ψ satisfying $H\psi = E_n\psi$. The relation of n with the angular quantum number $l \geq 0$ is $2l + 1 = k \leq j = l + n$, which is equivalent to $l \leq n - 1$.

Putting this all together we see that the eigenfunctions for the hydrogen atom are given by

$$\psi_{nlm}(r, \theta, \varphi) := r^l e^{-r/n} L_{l+n}^{2l+1}(2r/n) P_l^m(\cos\theta) e^{im\varphi}$$

for integers n, l, m satisfying $n \geq 1$, $0 \leq l \leq n - 1$ and $-l \leq m \leq +l$. They satisfy $H\psi_{nlm} = E_n\psi_{nlm}$. (Notice that L_{l+n}^{2l+1} is a polynomial of degree $n-l-1$. However, be aware that other notations are used in the literature for this polynomial.) These are non-zero functions, but they are not *normalized*, that is, they do not have norm 1 in the Hilbert space $L^2(\mathbb{R}^3)$. In applications one would multiply these functions by a positive constant in order to get norm 1 elements in the Hilbert space, namely, states of the hydrogen atom. But there are scattering states in the Hilbert space that are orthogonal to all of these eigenfunctions, a point that we leave to the reader's further consideration. However, we do have the next result.

Exercise 11.4.1 *Prove that the family ψ_{nlm} with integers l, m, n as given above is orthogonal.*

Given the constraints on the the three quantum numbers n, l, m we can count how many eigenfunctions there are for each fixed value of energy E_n. For each fixed value of $l \geq 0$ the quantum number m has $2l + 1$ allowed

integer values in the interval $[-l, +l]$. So the total number of eigenfunctions
having energy E_n is

$$\sum_{l=0}^{n-1}(2l + 1) = n^2,$$

a standard formula that can be proved by induction on $n \geq 1$. Since these
eigenfunctions are orthogonal, they span a subspace of dimension n^2. One
says that the *degeneracy* or *multiplicity* of the energy level E_n is n^2.

The case $n = 1$ has special interest, since that gives the lowest (negative)
energy of all the values of E_n. Also, the degeneracy is $n^2 = 1$, that is, the
state is unique up to a scalar multiple. Moreover, in this case $l = m = 0$ and
the corresponding (normalized) eigenfunction Φ is called the *ground state*
and is given by $\Phi(r, \theta, \varphi) = c\psi_{100}(r, \theta, \varphi) = ce^{-r}$ for some normalization
constant $c > 0$. It follows that the ground state is a strictly positive function.
As expected from elliptic regularity, it is a C^∞ function on $\mathbb{R}^3 \setminus \{0\}$.

It is important to note how Scrödinger's razor has worked in this problem.
At no point did we impose an *ad hoc* condition in order to get the quantum
numbers l, m, n that define the eigenfunction. Nor did we use anything except
the condition that the solution must be square integrable in order to conclude
that the bound states have energies that lie in a discrete subset of the real
line. By finding the solutions of the Schrödinger equation that are in $L^2(\mathbb{R}^3)$
we have automatically found all the so-called 'quantum numbers'.

The units we have used were convenient up to now, but to compare with
experiment we have to introduce the standard values in physical units of
Planck's constant $\hbar$, the mass of the electron m_e and the electric charge e
of the electron, three constants of nature known from experiments quite
independent from the atom of hydrogen. We then find for $n \geq 1$ that

$$E_n = -\frac{Ry}{n^2},$$

where $Ry > 0$ is the *Rydberg energy* and is given by $Ry = m_e e^4/2\hbar^2$. (We
are using *Gaussian units* here for the electric charge.) By using the measured
values of $\hbar$, m_e and e, one can evaluate the predicted value of $Ry \sim 13.6\,\text{eV}$,
where eV is an electron-volt, a unit of energy that is not the basic energy
unit, the *joule*, of the SI. The electron-volt is the amount of energy gained by
an electron when it is accelerated through an electrical potential difference
of one volt and with no other forces acting on the electron.

The dimension-less variable r in our previous analysis becomes r/a_0 in
physical units, where a_0 is the *Bohr radius* defined by

$$a_0 := \frac{\hbar^2}{m_e e^2}.$$

Again, this formula is in Gaussian units. The approximate value, using
independently measured values of $\hbar$, m_e and e, is $a_0 = 5.29 \times 10^{-11}$ meters or

$a_0 = 0.529$ Å, where Å is the called the Ångström, a common unit of length in atomic physics, (1 Å $= 10^{-10}$ meters). Thus the exponential factor in the ground state of the hydrogen atom becomes e^{-r/a_0}. This says roughly that the 'size' of the hydrogen atom is a_0. Other atoms have comparable sizes. While atoms seem incredibly small compared to sizes of objects in our everyday lives, a_0 is amazingly larger than the Planck distance of $\sim 1.6 \times 10^{-35}$ meters. So the unsolved problem is to explain why atoms are so very large.

Notice that despite the original expectation, the predicted energy levels do *not* depend on l. This is a very special property of the Coulomb potential energy, and not a general property of radial potentials. The discrete spectrum $\{-Ry/n^2 \,|\, n \geq 1\}$ given by this quantum theory agrees quite well with the experimental values of the energy levels of the hydrogen atom. To date all discrepancies between this quantum theory and experiment can be explained using a version of quantum theory that incorporates the theory of *special relativity*. The Schrödinger equation is consistent with the symmetries of classical Newtonian mechanics (embodied in the *Galilean group*), while the group of symmetries of special relativity is the *Poincaré group*.

11.5 Spherical Coordinates (Optional)

This section concerns subtle details of the spherical coordinate system. This is not directed to those who have a good geometrical notion of how these coordinates work on and near the surface of (the approximately spherical) Earth. Rather it is directed at those with some background in differential geometry and how coordinate systems work in that context. The point is that spherical coordinates, as used in physics, are not quite what one is studying in differential geometry. But, almost. So, heads up, mathematicians! The physicists have got this right, even if they do not spell it out in the detail you might like. We shall now present this material in that detail.

Using our conventions, the three vector fields $\partial/\partial r$, $\partial/\partial\theta$, and $\partial/\partial\varphi$ of the spherical coordinate system correspond to three well known physical directions near the Earth's surface. And a complete set of directions always gives *directional derivatives*, namely, hold all variables constant for all the directions except one and then vary the remaining variable (in the remaining direction) to get a partial derivative in that remaining direction. So what are these directions for the Earth? First, $\partial/\partial r$ is 'up'. Second, $\partial/\partial\theta$ is 'south'. And finally, $\partial/\partial\varphi$ is 'east'. The coordinate r is used near the Earth's surface, for example for the positions of airplanes. However, r is that case is usually measured with respect to mean sea level. But that does not change the meaning of $\partial/\partial r$. In fact, thinking mathematically, $\partial/\partial r$ is well defined on $\mathbb{R}^3 \setminus \{0\}$. And $r = +(x_1^2 + x_2^2 + x_3^2)^{1/2}$ is a C^∞ function of the Euclidean coordinates (x_1, x_2, x_3) on the domain $\mathbb{R}^3 \setminus \{0\}$.

Exercise 11.5.1 *Prove that the three directions up, south, east in that order form a right handed basis at each point for which all three directions are defined.*

Hint: *Let* **u**, **s** *and* **e** *represent the unit vectors in these three directions, respectively. Then the assertion of this exercise is equivalent to* $\mathbf{u} \times \mathbf{s} = \mathbf{e}$.

The coordinate θ is also well defined on $\mathbb{R}^3 \setminus \{0\}$. Here is a way to think of θ geometrically. The set where θ is a constant in the interval $[0, \pi]$ is a cone of rays (without vertex) emitting from the origin such that all of these rays have the same angle with respect of the ray $(0, 0, z)$ with $z > 0$, i.e., the positive z-axis. When $\theta = \pi/2$, this cone is actually the x_1, x_2-plane (minus the origin). The x_1, x_2-plane itself is usually called the *equatorial plane*. This particular cone for $\theta = \pi/2$ is not a singularity of the spherical coordinate system. However, the cones for $\theta = 0$ and $\theta = \pi$ reduce to a single ray (the positive and negative z-axis, respectively), and these are singularities. Mathematically, what is going on is that the change of 'coordinates'

$$\theta = \cos^{-1}\left(\frac{x_3}{+(x_1^2 + x_2^2 + x_3^2)^{1/2}} \right), \tag{11.5.1}$$

using the principal branch of $\cos^{-1}$, is a well defined continuous function $\mathbb{R}^3 \setminus \{0\} \to [0, \pi]$, but it is not differentiable on the z-axis, that is to say, for $x_1 = x_2 = 0$. So, even though the 'coordinate' θ is well defined on this domain $\mathbb{R}^3 \setminus \{0\}$, it is not a C^∞ coordinate as required in differential geometry. However, $\theta : \mathbb{R}^3 \setminus Z \to (0, \pi)$ as defined in (11.5.1) is a C^∞ function. Notice that the direction $\partial/\partial\theta$ (south) on the Earth is not well defined at the poles. At the North Pole all directions are south, while at South Pole there is no direction south. Nonetheless, the function $\mathbb{R}^3 \setminus \{0\} \to [-1, +1]$ given by

$$(x_1, x_2, x_3) \mapsto \frac{x_3}{+(x_1^2 + x_2^2 + x_3^2)^{1/2}} = \cos\theta$$

is C^∞, being the quotient of C^∞ functions with a non-zero denominator. In this last formula the non-smooth 'coordinate' θ lies in the closed interval $[0, \pi]$. We say that the function $\cos$ *removes the singularity* in the variable θ.

Exercise 11.5.2 *Prove that the function* $\sin\theta$ *is also* C^∞ *on* $\mathbb{R}^3 \setminus \{0\}$.

The coordinate φ also requires very careful consideration. The perfectly correct formula $\tan\varphi = x_2/x_1$ has multiple built-in difficulties. First, it does not make sense on the plane $x_1 = 0$. Second, it only defines φ implicitly and to get an explicit formula for φ in terms of the Euclidean coordinates is not so easy, since no branch of $\tan^{-1}$ will do the job. The expression $\tan^{-1}(x_2/x_1)$ only equals φ in an open half-space, no matter which branch of $\tan^{-1}$ one chooses. To see that this is so, just consider the two distinct points (x_1, x_2, x_3) and $(-x_1, -x_2, x_3)$ for $x_1 \neq 0$. Nonetheless, the geometrical definition of φ gives a well-defined function $\varphi : \mathbb{R}^3 \setminus Z \to [0, 2\pi)$, which is discontinuous along the half-plane where $\varphi = 0$, the Greenwich meridian plane G. However, $\varphi : \mathbb{R}^3 \setminus \overline{G} \to (0, 2\pi)$ is a C^∞ function. Here, $\overline{G}$, the closure of G, includes the z-axis. Proving that φ is C^∞ is a bit tricky, since it is not clear how to write *one* formula for φ as a function of (x_1, x_2, x_3) on this domain. Hint: We do not need one formula. Various formulas on open subsets that agree on overlaps is sufficient.

To see what is happening geometrically, let's fix some $r > 0$ and some $\theta \in (0, \pi)$. This gives us the intersection of a cone with vertex at the origin and the sphere of radius r. So this is a circle. Such a circle on the Earth is known as a *circle of constant latitude*. We have excluded $\theta = 0$ and $\theta = \pi$ for which values this circle degenerates to a point, namely, one of the poles. The coordinate φ locates the points on this circle but is not a globally defined continuous coordinate (let alone C^∞), while the directional derivative $\partial / \partial \varphi$ (east) is perfectly well defined at every point on that circle.

It is well known that a circle has no global coordinate system (in the sense of differential geometry), although it does have discontinuous global coordinates. In our case, φ is such a discontinuous coordinate. But the 2π periodic function $e^{i\varphi}$ is C^∞ on $\mathbb{R}^3 \setminus Z$ as are its real and imaginary parts, $\cos \varphi$ and $\sin \varphi$. One says that these functions remove the singularity of the variable φ. Also the vector field $\partial / \partial \varphi$ extends uniquely to a continuous, and even C^∞, vector field on the unit circle in the (x_1, x_2) plane as well as on $\mathbb{R}^3 \setminus Z$. This is proved using at least two coordinate systems for the circle or for the spherical coordinates on $\mathbb{R}^3 \setminus Z$.

Using the definition of coordinate system from differential geometry, the spherical coordinates form a *chart* κ defined on the open subset $\mathbb{R}^3 \setminus \overline{G}$ of $\mathbb{R}^3$, given by $\kappa : \mathbb{R}^3 \setminus G \to (0, \infty) \times (0, \pi) \times (0, 2\pi)$, which sends the point $(x_1, x_2, x_3) \in \mathbb{R}^3 \setminus \overline{G}$ to $\kappa(x_1, x_2, x_3) = (r, \theta, \varphi)$. We already have formulas for r and θ that exhibit them as C^∞ functions of $(x_1, x_2, x_3) \in \mathbb{R}^3 \setminus \overline{G}$. As we commented earlier, this is the 'tricky bit':

Exercise 11.5.3 *Prove that φ is a C^∞ function of $(x_1, x_2, x_3) \in \mathbb{R}^3 \setminus \overline{G}$.*

But what is the domain of the definition of the spherical coordinates as used in physics? That depends on what you want to do! The main point of this section is that this can be a tricky business, but if one shuts one's eyes and does what comes 'naturally', then things could be okay. But a better policy is to keep one's eyes open.

When studying the hydrogen atom, we have a potential that is singular only at the origin, while the spherical coordinates introduce singularities all along the z-axis Z. But at the end of the day, elliptic regularity tells us that the solutions of the time independent Schrödinger equation for the hydrogen atom must be C^∞ on $\mathbb{R}^3 \setminus \{0\}$. So, even though we initially work on $\mathbb{R}^3 \setminus Z$ or on $\mathbb{R}^3 \setminus \overline{G}$, eventually we must extend our solutions to C^∞ functions on $\mathbb{R}^3 \setminus \{0\}$. This is the principle, an idea originally due to Schrödinger, that an equation tells us what its solutions are. As you already know, I have dubbed this idea *Schrödinger's razor*. This idea is very appealing for mathematicians, but seems to have been overlooked by some physicists.

11.6   Two-body Problems (Optional)

The two-body problem of quantum theory is usually called the hydrogen atom, while the two-body problem of Newtonian gravitation is called the Kepler problem. In both cases one can reformulate the problem as a one-body problem for the less massive object moving in a potential energy field generated by the more massive object. But why don't we study this as a one-body problem for the more massive object moving in the potential energy field generated by the less massive object? What is good for the goose is good for the gander, and all that. But, no! Either way we are making an approximation, but one way the approximation is very, very good while the other way it is wildly bad.

In both cases we have to start with $\mathbb{R}^6$ as the *configuration space*: three spatial variables for each of the two bodies. (There are some comments on configuration spaces in the Notes for this Chapter.) Let's denote these variables by $\mathbf{x} = (x_1, x_2, x_3)$ for the body with the bigger mass M and by $\mathbf{y} = (y_1, y_2, y_3)$ for the body with the smaller mass m. For the quantum case the appropriate Schrödinger operator acting in $L^2(\mathbb{R}^6)$ is

$$H_{atom} = -\frac{\hbar^2}{2M}\Delta_{\mathbf{x}} - \frac{\hbar^2}{2m}\Delta_{\mathbf{y}} - \frac{1}{\|\mathbf{x} - \mathbf{y}\|}.$$

This is the total kinetic energy operator plus the potential energy operator. While the variables $\mathbf{x}$ and $\mathbf{y}$ enter symmetrically in the potential energy, they do not have the same 'weight' in the kinetic energy terms. One way to think about this is to let $M \to +\infty$ and see that the first term goes to zero (in some sense which is not clearly specified). But this is not such a good idea, because we are also interested in other two-body problems. One such is the positron-electron 'atom'. This is described by the above atomic Hamiltonian, but now with $M = m =$ the common mass of the positron and the electron. (The *positron* is the anti-particle of the electron. It turns out that every particle has an associated anti-particle with the same mass.) The analogous gravitational situation prevails in a binary star system (that is, two stars rotating around each other) with stars of equal mass.

In both the classical and the quantum cases we make a change of variables to the *center of mass* coordinates. These are defined by

$$X_{cm} := \frac{1}{M+m}(M\mathbf{x} + m\mathbf{y}), \qquad X_{rel} := \mathbf{x} - \mathbf{y}.$$

This change of variables allows us to write the original problem in terms of two uncoupled Schrödinger operators. For the variables X_{cm} the Schrödinger operator acting in $L^2(\mathbb{R}^3)$ is

$$-\left(\frac{1}{M+m}\right)\Delta_{X_{cm}}, \qquad (11.6.1)$$

which is the Hamiltonian for a free massive 'particle' with mass $M + m$. Here *free* means that the 'particle' is subject to zero potential energy. This is called a *one-body problem*, even though there is no 'body' here with mass $M + m$, because of the mathematical form of the Schrödinger operator (11.6.1), that is, it acts on functions whose domain is $\mathbb{R}^3$.

Next, for the variables X_{rel} the Schrödinger operator acting in $L^2(\mathbb{R}^3)$ is

$$-\frac{\hbar^2}{2m_{red}} \Delta_{X_{rel}} - \frac{1}{||X_{rel}||},$$

where the *reduced mass* m_{red} is defined implicitly by the formula

$$\frac{1}{m_{red}} = \frac{1}{M} + \frac{1}{m}. \tag{11.6.2}$$

So the Schrödinger operator in the relative coordinates X_{rel} is the one–body Schrödinger operator for a state with reduced mass m_{red} subject to the potential energy field of an attractive central Coulomb force. Hence, in all our calculations for the hydrogen atom, we should replace the electron mass with m_{red}. Moreover when $M \gg m$, as is the case of the proton mass in comparison with the electron mass, we can approximate (11.6.2) by

$$\frac{1}{m_{red}} \sim \frac{1}{m}.$$

This says that $m_{red} \sim m$, that is, the reduced mass is equal to the smaller mass in very good approximation. Clearly, the reduced mass is not well approximated by the larger mass. For the two-body positron-electron system we have that $m_{red} = (1/2)m_e$, where m_e is the electron mass, which is equal to the positron mass. The energies of the bound states of the positron-electron system, known as *positronium*, have been measured experimentally and agree quite well with this theory. We will discuss known discrepancies between experiment and this theory in the Notes. Of course, the experiment wins in all cases.

11.7 A Moral or Two

There is a moral or two to the long story told in this chapter. And one has to do with the discipline and perseverance of the scientific researcher. Schrödinger knew that publishing an abstract partial differential equation by itself would not be sufficient to justify a new quantum theory. He needed to show that it could explain significant experimental results. This gave him more than enough motivation to solve the highly non-trivial eigenvalue problem for the hydrogen atom. So when one sees that the only obstacle to getting an important result is a tedious calculation, then one must roll up one's sleeves and do that calculation. But doing long calculations just for the sake of demonstrating one's virtuosity is not worth much esteem. At least, that is my opinion of what one moral of this story is.

But even the long and winding road of the calculations of this chapter still is an incomplete rendition of the strict mathematical development of the theory of the hydrogen atom. Believe it or not, there have been gaps in our presentation. Just to cite one such gap, we never proved that there are no positive eigenvalues of the hydrogen Hamiltonian. The curious reader can find a fully rigorous mathematical treatment of the hydrogen atom in Teschl's book [30] for example, but only after more than 200 pages of highly technical mathematical details. My point of view is that those details would not be worth the effort to learn without the fantastic payoff of a thorough understanding of something as fundamental in physics as the hydrogen atom. So another moral to this story is that to get a truly complete and rigorous mathematical description of even a rather simple quantum system is no easy matter at all, but that the physics behind it motivates the effort to achieve that. But on the other hand the presentation in this chapter gives a rather good idea of what is going on.

There is also a moral about the technique of separation of variables, which replaced one partial differential equation in three spatial variables with three ordinary differential equations (ODE). The point is that the study of ODE is more than a mathematical exercise; it helps us understand the physics of what happens in three-dimensional space. And as noticed previously, this technique works well for any radial potential whatsoever. So this motivates the study of a wide class of ODE on the half-line $[0, \infty)$.

11.8   Notes

Since $\mathbb{R}^3$ models well (at least locally) the space we inhabit, it seems more or less reasonable that the Hilbert space $L^2(\mathbb{R}^3)$ should enter in quantum theory. Since quantum theory was developed by physicists all of whom knew classical mechanics, it was natural for them to also think of $\mathbb{R}^3$ in classical mechanical terms, namely as the configuration space for a single point mass. When interest arose in studying quantum systems with $n \geq 2$ particles for which the classical mechanical configuration space is $\mathbb{R}^{3n}$, it was not clear whether the Hilbert space for the quantum theory should still be $L^2(\mathbb{R}^3)$ or rather $L^2(\mathbb{R}^{3n})$. Opinion was divided! Of course, this is part of the mystery of first quantization of choosing 'correctly' the Hilbert space as well as the self-adjoint operators acting in it. We now know that $L^2(\mathbb{R}^{3n})$ is the 'correct' Hilbert space because it gives us results that agree with experiment. But I have tried to avoid justifying such choices by appealing to classical mechanics, since when all is said and done quantum theory should stand on its own legs.

The hydrogen atom and other two particle states have been a testing ground for quantum theory. Heisenberg found the spectrum for the negative energy states using his matrix mechanics in 1925. Schrödinger did the same using his version of quantum theory in 1926. So both theories passed a key test successfully. The degenerate levels were split with external fields (electric or magnetic) and the corresponding *perturbation theory* was able to explain them also in terms of quantum theory. Effects were detected and explained with relativistic corrections to the non-relativistic theory we have presented. A grandson of Charles Darwin was involved with that! The quantum field theory known as QED (quantum electrodynamics) has explained *hyperfine splittings* in the hydrogen spectrum. Other two particle systems such as proton and negative muon, positron and electron, anti-proton and positron have been studied experimentally. And to date quantum theory in general and QED in particular hold true.

Chapter 12

Angular Momentum

> The miracle of the appropriateness of
> the language of mathematics for the
> formulation of the laws of physics is
> a wonderful gift which we neither
> understand nor deserve.
> Eugene Wigner

The symmetry property of the hydrogen atom or of any quantum system with a radial potential is related to the mathematics of angular momentum, as we shall see.

12.1 Basics

There are various quantities from classical mechanics that we have yet to *quantize*, that is, give them their corresponding quantum theory. So far we have only considered *linear momentum*, which we simply called *momentum*. Now we wish to consider *angular momentum*. In classical mechanics the definition of the angular momentum of a point particle with mass $m > 0$ is

$$\mathbf{L} := \mathbf{x} \times \mathbf{p},$$

where $\mathbf{x} = (x_1, x_2, x_3) \in \mathbb{R}^3$ is the particle's position and $\mathbf{p} = (p_1, p_2, p_3) \in \mathbb{R}^3$ is its *(linear) momentum*. Here $\mathbf{p} = m\mathbf{v}$, where $\mathbf{v}$ is the *velocity vector* of the particle. Also $\times$ denotes the usual *cross product* (or *vector product*) of vectors in $\mathbb{R}^3$. Then using the definition of the cross product, $\mathbf{L} = (L_1, L_2, L_3)$ is given by

$$L_1 = x_2 p_3 - x_3 p_2, \qquad L_2 = x_3 p_1 - x_1 p_3, \qquad L_3 = x_1 p_2 - x_2 p_1. \quad (12.1.1)$$

© Springer Nature Switzerland AG 2020
S. B. Sontz, *An Introductory Path to Quantum Theory*,
https://doi.org/10.1007/978-3-030-40767-4_12

Exercise 12.1.1 *Verify the formulas in* (12.1.1). *Look up the definition of cross product, if needed.*

Each L_j in (12.1.1) is a function of $(x_1, x_2, x_3, p_1, p_2, p_3) \in \mathbb{R}^6$, the *phase space* of classical mechanics. Now the product of any two such functions does not depend on their order, that is, the multiplication of functions is commutative. (By the way, real-valued functions defined on the phase space are known as *observables* in classical mechanics.) In particular, in classical mechanics, all the x's commute with all the p's, since multiplication in $\mathbb{R}$ is commutative. But that is not so in quantum theory where the x's and p's are operators and satisfy the *canonical commutation relations* (5.1.2), (5.1.3), and (5.1.4).

Remark: In these relations and the subsequent material the domain of definition of the operators will not be specified. One can use either a vector space of functions for which all expressions are defined or a vector subspace of the appropriate Hilbert space of (equivalence classes of) functions for which, again, all expressions are defined.

Exercise 12.1.2 *Compute the dimensions and units of angular momentum in the SI. Compare this with the dimensions of Planck's constant $\hbar$.*

By (5.1.4) $p_j x_k = x_k p_j$ for $j \neq k$ in quantum theory, and so the canonical quantization is unambiguous and gives these angular momentum operators:

$$L_1 = \frac{\hbar}{i}\Big(x_2\frac{\partial}{\partial x_3} - x_3\frac{\partial}{\partial x_2}\Big),$$

$$L_2 = \frac{\hbar}{i}\Big(x_3\frac{\partial}{\partial x_1} - x_1\frac{\partial}{\partial x_3}\Big),$$

$$L_3 = \frac{\hbar}{i}\Big(x_1\frac{\partial}{\partial x_2} - x_2\frac{\partial}{\partial x_1}\Big). \tag{12.1.2}$$

These are first order partial differential operators. We do not use notation to distinguish between the classical and quantum expressions for angular momentum, since from now on we will only consider the quantum version.

Exercise 12.1.3 *Verify that L_1, L_2, and L_3 are symmetric operators by a formal argument (or, if these words make sense to you, on some appropriate dense vector subspace of $L^2(\mathbb{R}^3)$, such as $C_0^\infty(\mathbb{R}^3)$, the space of all complex-valued C^∞ functions of compact support).*

Exercise 12.1.4 *Verify the commutation relations for the operators of the angular momentum:*

$$[L_1, L_2] = i\hbar L_3, \qquad [L_2, L_3] = i\hbar L_1, \qquad [L_3, L_1] = i\hbar L_2.$$

This implies that no two of these operators correspond to simultaneously measurable observables.

Exercise 12.1.5 *Let* A, B, C *be linear operators acting on a Hilbert space. Prove that*

$$[A, B\,C] = [A, B]\,C + B\,[A, C].$$

Compare this with the Liebniz rule in calculus.

Exercise 12.1.6 *Define*

$$L^2 := L_1^2 + L_2^2 + L_3^2.$$

Beware: L^2 *is defined as one thing, namely, a certain second order partial differential operator. It is not defined as the square of something else. The notation* L^2 *is universally used, and so I would be doing you, my kind reader, a disservice if I were to change it.*

   Prove these commutation relations:

$$[L^2, L_1] = 0, \qquad [L^2, L_2] = 0, \qquad [L^2, L_3] = 0.$$

Physically, this means that L^2 *and any one (but not two) of the* L_j's *are simultaneously measurable.*

We call L^2 the *total angular momentum*. We say that L^2 is a *Casimir operator* because it has the property of commuting with all algebraic combinations (sums and products in any order) of L_1, L_2, L_3.

Exercise 12.1.7 *Prove the last statement.* **Hint:** *Use Exercise 12.1.5.*

12.2 Spherical Symmetry

While all this is fine and well, it might be worthwhile to write these three operators in spherical coordinates, which were introduced in (11.2.1). Why? Because, if we have a function $\psi(r, \theta, \varphi)$ and we apply to it the first order differential operator L_1, the chain rule gives

$$(L_1 \psi)(r, \theta, \varphi) = \frac{\partial \psi}{\partial r} L_1 r + \frac{\partial \psi}{\partial \theta} L_1 \theta + \frac{\partial \psi}{\partial \varphi} L_1 \varphi.$$

Let's focus on the coefficient of $\partial \psi / \partial r$. For this we have

$$L_1 r = \frac{\hbar}{i} \left(x_2 \frac{\partial r}{\partial x_3} - x_3 \frac{\partial r}{\partial x_2} \right) = \frac{\hbar}{i} \left(x_2 \frac{x_3}{r} - x_3 \frac{x_2}{r} \right) = 0.$$

Here we have used the following exercise.

Exercise 12.2.1 *Prove for* $j = 1, 2, 3$, *and* $0 < r = (x_1^2 + x_2^2 + x_3^2)^{1/2}$ *that*

$$\frac{\partial r}{\partial x_j} = \frac{x_j}{r}.$$

Thus we have seen that $L_1\psi$ only depends on the two angular partial derivatives of ψ. By similar arguments we see that $L_2\psi$ and $L_3\psi$ also depend on the two angular partial derivatives. This gives some hint that spherical coordinates are a good thing for studying this topic. And this is a hint that angular momentum is important for the hydrogen atom.

Let's start with the change of coordinates at the infinitesimal level. Using (11.2.1), the chain rule gives us

$$\frac{\partial}{\partial r} = \frac{\partial x_1}{\partial r}\frac{\partial}{\partial x_1} + \frac{\partial x_2}{\partial r}\frac{\partial}{\partial x_2} + \frac{\partial x_3}{\partial r}\frac{\partial}{\partial x_3}$$

$$= \sin\theta\cos\varphi\frac{\partial}{\partial x_1} + \sin\theta\sin\varphi\frac{\partial}{\partial x_2} + \cos\theta\frac{\partial}{\partial x_3}.$$

Continuing with the partial with respect to θ we have

$$\frac{\partial}{\partial\theta} = \frac{\partial x_1}{\partial\theta}\frac{\partial}{\partial x_1} + \frac{\partial x_2}{\partial\theta}\frac{\partial}{\partial x_2} + \frac{\partial x_3}{\partial\theta}\frac{\partial}{\partial x_3}$$

$$= r\cos\theta\cos\varphi\frac{\partial}{\partial x_1} + r\cos\theta\sin\varphi\frac{\partial}{\partial x_2} - r\sin\theta\frac{\partial}{\partial x_3}.$$

Finally, for φ we see that

$$\frac{\partial}{\partial\varphi} = \frac{\partial x_1}{\partial\varphi}\frac{\partial}{\partial x_1} + \frac{\partial x_2}{\partial\varphi}\frac{\partial}{\partial x_2} + \frac{\partial x_3}{\partial\varphi}\frac{\partial}{\partial x_3}$$

$$= -r\sin\theta\sin\varphi\frac{\partial}{\partial x_1} + r\sin\theta\cos\varphi\frac{\partial}{\partial x_2}.$$

These three formulas for the partial derivatives with respect to the three spherical coordinates are true, but employ a 'mixed' notation. What I mean by this is that the expressions on the right sides of these equations have coefficients written in spherical coordinates while the partial derivatives are with respect to the Cartesian coordinates. So in this sense we are mixing the notations from two different coordinate systems in one expression. This is perfectly acceptable. However, eventually we want expressions entirely written in just one coordinate system.

We next write these three equations as one matrix equation

$$\begin{pmatrix} \partial/\partial r \\ \partial/\partial\theta \\ \partial/\partial\varphi \end{pmatrix} = \begin{pmatrix} \sin\theta\cos\varphi & \sin\theta\sin\varphi & \cos\theta \\ r\cos\theta\cos\varphi & r\cos\theta\sin\varphi & -r\sin\theta \\ -r\sin\theta\sin\varphi & r\sin\theta\cos\varphi & 0 \end{pmatrix} \begin{pmatrix} \partial/\partial x_1 \\ \partial/\partial x_2 \\ \partial/\partial x_3 \end{pmatrix}.$$

Let M denote the 3×3 matrix in this formula.

Exercise 12.2.2 *Prove that* $\det M = r^2\sin\theta$, *where* det *is the determinant of a matrix.*

So $\det M \neq 0$ on the complement of the z-axis Z, that is, on $\mathbb{R}^3 \setminus Z$, where $Z := \{(0, 0, z)\,|\,z \in \mathbb{R}\}$. On that set, we have that the inverse of the matrix

M is given by

$$M^{-1} = \frac{1}{r^2 \sin \theta} \begin{pmatrix} r^2 \sin^2 \theta \cos \varphi & r \cos \theta \sin \theta \cos \varphi & -r \sin \varphi \\ r^2 \sin^2 \theta \sin \varphi & r \cos \theta \sin \theta \sin \varphi & r \cos \varphi \\ r^2 \cos \theta \sin \theta & -r \sin^2 \theta & 0 \end{pmatrix}.$$

Exercise 12.2.3 *Verify this formula for* M^{-1}. **Hint:** *Cramer's rule if you know it. Otherwise, verify by matrix multiplication that* $M^{-1}M = I$.

So, we obtain

$$\begin{pmatrix} \partial/\partial x_1 \\ \partial/\partial x_2 \\ \partial/\partial x_3 \end{pmatrix} = \begin{pmatrix} \sin \theta \cos \varphi & \cos \theta \cos \varphi \, / \, r & -\sin \varphi \, / \, r \sin \theta \\ \sin \theta \sin \varphi & \cos \theta \sin \varphi \, / \, r & \cos \varphi \, / \, r \sin \theta \\ \cos \theta & -\sin \theta \, / \, r & 0 \end{pmatrix} \begin{pmatrix} \partial/\partial r \\ \partial/\partial \theta \\ \partial/\partial \varphi \end{pmatrix}.$$

Note that the expressions on the right side here are now written exclusively in spherical coordinates. Of course, the entries in this matrix are the nine partials of the spherical coordinates r, θ, φ with respect to the Cartesian coordinates x_1, x_2, x_3. One could calculate this matrix by calculating those partials. The tricky bit is writing the spherical coordinate φ as a well-defined function of the Cartesian coordinates on appropriate subsets of $\mathbb{R}^3 \setminus Z$. This can be done, but is not as easy as you might think.

In the next exercise we come back to a point we emphasized earlier.

Exercise 12.2.4 *Find explicit formulas for the spherical coordinate* φ *in terms of the Cartesian coordinates* (x_1, x_2, x_3).

Warning: *This is not so easy. In many texts one finds* $\varphi = \tan^{-1}(x_2/x_1)$, *which is correct on a proper subset of* $\mathbb{R}^3 \setminus Z$, *but not on all of* $\mathbb{R}^3 \setminus Z$. *And remember that* φ *had better be a* C^∞ *function of the Cartesian coordinates on all of* $\mathbb{R}^3 \setminus Z$.

Now we can write L_3 in spherical coordinates. We obtain

$$i\hbar^{-1} L_3 = \left(x_1 \frac{\partial}{\partial x_2} - x_2 \frac{\partial}{\partial x_1} \right)$$

$$= (r \sin \theta \cos \varphi) \left(\sin \theta \sin \varphi \frac{\partial}{\partial r} + \frac{\cos \theta \sin \varphi}{r} \frac{\partial}{\partial \theta} + \frac{\cos \varphi}{r \sin \theta} \frac{\partial}{\partial \varphi} \right)$$

$$- (r \sin \theta \sin \varphi) \left(\sin \theta \cos \varphi \frac{\partial}{\partial r} + \frac{\cos \theta \cos \varphi}{r} \frac{\partial}{\partial \theta} - \frac{\sin \varphi}{r \sin \theta} \frac{\partial}{\partial \varphi} \right)$$

$$= \frac{\partial}{\partial \varphi}$$

where the last equality is straightforward. Therefore, on $\mathbb{R}^3 \setminus Z$ we have

$$L_3 = \frac{\hbar}{i} \frac{\partial}{\partial \varphi}. \tag{12.2.1}$$

We already knew that this first order linear operator was going to be a linear combination of $\partial/\partial\theta$ and $\partial/\partial\varphi$ with possibly variable coefficients. So this result is consistent with that expectation, but gives us only one of the partial derivatives. And its coefficient is constant. So our expectations are exceeded. However, the partial derivative ($\equiv$ vector field) $\partial/\partial\varphi$ does not exist on Z. Intuitively, the direction 'east' does not exist for the points on Z. But the first order differential operator L_3 is singular for those points x_1, x_2, x_3 satisfying $x_1 = x_2 = 0$, that is along the axis Z, since by (12.1.2) on the axis Z we have $L_3 - 0$, which is not a direction. Similar comments hold for L_1 and L_2.

The operator L_3 was defined for functions whose domain is $\mathbb{R}^3$, but the spherical coordinates introduce singularities. It is a curious and important fact that the partial derivative (also known as a *vector field*) $\partial/\partial\varphi$ is well defined on $\mathbb{R}^3 \setminus Z$, even though the coordinate φ is only defined as a C^∞ function on the complement of the closed half-plane $x_1 \geq 0$ in the coordinate x_1, x_3 plane. Also, the functions $\cos\varphi$ and $\sin\varphi$ are C^∞ functions on $\mathbb{R}^3 \setminus Z$.

The calculations for L_1 and L_2 do not work out anywhere as nicely. Let's start with L_1. We find that

$$i\hbar^{-1}L_1 = \left(x_2\frac{\partial}{\partial x_3} - x_3\frac{\partial}{\partial x_2}\right)$$

$$= (r\sin\theta\sin\varphi)\left(\cos\theta\frac{\partial}{\partial r} - \frac{\sin\theta}{r}\frac{\partial}{\partial\theta}\right)$$

$$- (r\cos\theta)\left(\sin\theta\sin\varphi\frac{\partial}{\partial r} + \frac{\cos\theta\sin\varphi}{r}\frac{\partial}{\partial\theta} + \frac{\cos\varphi}{r\sin\theta}\frac{\partial}{\partial\varphi}\right)$$

$$= -\sin\varphi\frac{\partial}{\partial\theta} - \cot\theta\cos\varphi\frac{\partial}{\partial\varphi}.$$

Finally, here's L_2:

$$i\hbar^{-1}L_2 = \left(x_3\frac{\partial}{\partial x_1} - x_1\frac{\partial}{\partial x_3}\right)$$

$$= (r\cos\theta)\left(\sin\theta\cos\varphi\frac{\partial}{\partial r} + \frac{\cos\theta\cos\varphi}{r}\frac{\partial}{\partial\theta} - \frac{\sin\varphi}{r\sin\theta}\frac{\partial}{\partial\varphi}\right)$$

$$- (r\sin\theta\cos\varphi)\left(\cos\theta\frac{\partial}{\partial r} - \frac{\sin\theta}{r}\frac{\partial}{\partial\theta}\right)$$

$$= \cos\varphi\frac{\partial}{\partial\theta} - \cot\theta\sin\varphi\frac{\partial}{\partial\varphi}.$$

We are now in a good position to solve the eigenvalue problem for L_3,

$$L_3\psi = M\psi.$$

Since L_3 is a linear, first order partial differential operator, this equation has solutions for every complex number M. But in quantum theory we want solutions in $L^2(\mathbb{R}^3)$. And since L_3 is a symmetric operator acting in that Hilbert space, M must be a real number. Using spherical coordinates we have

$$\frac{\hbar}{i}\frac{\partial\psi}{\partial\varphi} = M\psi,$$

whose general solution is

$$\psi(\varphi) = a\, e^{iM\varphi/\hbar},$$

where $a \in \mathbb{C}$. The argument given in Chapter 11, which shows that ψ and ψ' must have periodic boundary conditions on $(0, 2\pi)$ applies here because L_3 reduces to the ordinary derivative $d/d\varphi$ of one variable (modulo a non-zero constant), which is an elliptic operator. Therefore, as we saw in Chapter 11, the set of eigenvalues of L_3 is

$$\hbar\mathbb{Z} = \{m\hbar \mid m \in \mathbb{Z}\}.$$

An eigenfunction corresponding to $m\hbar$ is $\psi_m(\varphi) := e^{im\hbar\varphi/\hbar} = e^{im\varphi}$ for $\varphi \in [0, 2\pi)$. Since $\hbar\mathbb{Z}$ is an unbounded subset of $\mathbb{R}$, it follows by functional analysis that any self-adjoint extension of the symmetric operator L_3 is an unbounded operator.

Strange as it may seem the operators L_1 and L_2 are like L_3 in disguise. What could that possibly mean? Well, in coordinates analogous to the usual spherical coordinates, except with the z-axis replaced by the x-axis or y-axis, respectively, the angular coordinate (say φ_x and φ_y, respectively) measuring rotations about that axis will correspond to L_1 and L_2 via

$$L_1 = \frac{\hbar}{i}\frac{\partial}{\partial\varphi_x} \quad \text{and} \quad L_2 = \frac{\hbar}{i}\frac{\partial}{\partial\varphi_y},$$

respectively. So now I hope that it is clear that the eigenvalues of L_1 and L_2 are the same as those for L_3, while the eigenfunctions will be $e^{im\varphi_x}$ and $e^{im\varphi_y}$, respectively, where $m \in \mathbb{Z}$.

12.3 Ladder Operators

This section concerns some very helpful, but rather technical, operators that will remind us of the raising and lowering operators of the harmonic oscillator. So, we introduce the *ladder operators*

$$L_+ := L_1 + iL_2 \quad \text{and} \quad L_- := L_1 - iL_2.$$

Since L_1 and L_2 are symmetric operators, the adjoint of L_+ (resp., L_-) is L_- (resp., L_+). So neither L_+ nor L_- is a symmetric operator, and therefore neither represents a physical observable. Nonetheless, they are quite useful.

Exercise 12.3.1 *Prove these identities:*

$$L_+L_- = L^2 - L_3^2 + \hbar L_3 \qquad L_-L_+ = L^2 - L_3^2 - \hbar L_3 \qquad [L_+, L_-] = 2\hbar L_3$$
$$[L_3, L_+] = \hbar L_+ \qquad\qquad [L_3, L_-] = -\hbar L_-$$

Exercise 12.3.2 *Suppose that $L_3\psi = m\hbar\psi$ for some real number m and some $\psi \neq 0$. (This simply says that ψ is an eigenvector of L_3 with eigenvalue $m\hbar$. We already know that $m \in \mathbb{Z}$, but that fact does not enter into this exercise.)*

- *Prove that $L_+\psi = 0$ or else $L_+\psi$ is an eigenvector of L_3 with eigenvalue $(m+1)\hbar$.*

- *Prove that $L_-\psi = 0$ or else $L_-\psi$ is an eigenvector of L_3 with eigenvalue $(m-1)\hbar$.*

These results are often stated under the hypothesis that we have chosen units such that $\hbar = 1$ and $\hbar$ is dimensionless. This exercise shows why we call L_+ the raising ladder operator *and L_- the* lowering ladder operator.

As mentioned above, these ladder operators should remind you a little bit of the raising and lowering operators associated with the harmonic oscillator. However, they are *not* the same.

Exercise 12.3.3 *Derive these formulas for the ladder operators written in spherical coordinates:*

$$L_+ = \hbar e^{i\varphi}\left(\frac{\partial}{\partial\theta} + i\cot\theta\frac{\partial}{\partial\varphi}\right) \quad \text{and} \quad L_- = \hbar e^{-i\varphi}\left(-\frac{\partial}{\partial\theta} + i\cot\theta\frac{\partial}{\partial\varphi}\right).$$

We next use this machinery to compute L^2 in spherical coordinates, which is the whole point of this section. But be prepared; the calculation is messy. We start off with this:

$$e^{-i\varphi}\hbar^{-2}L_+L_- = \left(\frac{\partial}{\partial\theta} + i\cot\theta\frac{\partial}{\partial\varphi}\right)\left[e^{-i\varphi}\left(-\frac{\partial}{\partial\theta} + i\cot\theta\frac{\partial}{\partial\varphi}\right)\right]$$

$$= e^{-i\varphi}\left(-\frac{\partial^2}{\partial\theta^2} - i\frac{1}{\sin^2\theta}\frac{\partial}{\partial\varphi} + i\cot\theta\frac{\partial^2}{\partial\theta\partial\varphi}\right) \quad \text{(action of } \partial/\partial\theta)$$

$$+ e^{-i\varphi}\cot\theta\left(-\frac{\partial}{\partial\theta} + i\cot\theta\frac{\partial}{\partial\varphi}\right) \quad \text{(action of } i\cot\theta\,\partial/\partial\varphi \text{ on } e^{-i\varphi})$$

$$+ e^{-i\varphi}i\cot\theta\left(-\frac{\partial^2}{\partial\varphi\partial\theta} + i\cot\theta\frac{\partial^2}{\partial\varphi^2}\right) \quad \text{(rest of action of } i\cot\theta\,\partial/\partial\varphi)$$

$$= e^{-i\varphi}\left(-\frac{\partial^2}{\partial\theta^2} - i\frac{1}{\sin^2\theta}\frac{\partial}{\partial\varphi} - \cot\theta\frac{\partial}{\partial\theta} + i\cot^2\theta\frac{\partial}{\partial\varphi} - \cot^2\theta\frac{\partial^2}{\partial\varphi^2}\right)$$

So, this gives

$$L_+L_- = \hbar^2\left(-\frac{\partial^2}{\partial\theta^2} - i\frac{1}{\sin^2\theta}\frac{\partial}{\partial\varphi} - \cot\theta\frac{\partial}{\partial\theta} + i\cot^2\theta\frac{\partial}{\partial\varphi} - \cot^2\theta\frac{\partial^2}{\partial\varphi^2}\right).$$

Then from this we calculate

$$L^2 = L_+L_- + L_3^2 - \hbar L_3$$

$$= \hbar^2\left(-\frac{\partial^2}{\partial\theta^2} - i\frac{1}{\sin^2\theta}\frac{\partial}{\partial\varphi} - \cot\theta\frac{\partial}{\partial\theta} + i\cot^2\theta\frac{\partial}{\partial\varphi}\right.$$

$$\left. - \cot^2\theta\frac{\partial^2}{\partial\varphi^2} - \frac{\partial^2}{\partial\varphi^2} + i\frac{\partial}{\partial\varphi}\right)$$

$$= \hbar^2 \left(-\frac{\partial^2}{\partial\theta^2} - \cot\theta \frac{\partial}{\partial\theta} - \cot^2\theta \frac{\partial^2}{\partial\varphi^2} - \frac{\partial^2}{\partial\varphi^2} \right)$$

$$= \hbar^2 \left(-\frac{\partial^2}{\partial\theta^2} - \cot\theta \frac{\partial}{\partial\theta} - \frac{1}{\sin^2\theta} \frac{\partial^2}{\partial\varphi^2} \right)$$

$$= -\hbar^2 \left(\frac{1}{\sin\theta} \frac{\partial}{\partial\theta} \left(\sin\theta \frac{\partial}{\partial\theta} \right) + \frac{1}{\sin^2\theta} \frac{\partial^2}{\partial\varphi^2} \right). \tag{12.3.1}$$

In the third equality the three terms with $\partial/\partial\varphi$ canceled out, since

$$\frac{-1}{\sin^2\theta} + \cot^2\theta + 1 = \frac{-1}{\sin^2\theta} + \frac{\cos^2\theta}{\sin^2\theta} + 1 = \frac{-1 + \cos^2\theta + \sin^2\theta}{\sin^2\theta} = 0.$$

The fourth equality also follows from a trigonometric identity.

12.4 Relation to Laplacian on $\mathbb{R}^3$

We have seen the expression (12.3.1) for L^2 before, though in a context where $\hbar$ was not present. See the left side of (11.3.1). Also see the formula (11.2.3) for the Laplacian on $\mathbb{R}^3$ in spherical coordinates which now becomes

$$\Delta = \frac{1}{r^2} \frac{\partial}{\partial r} \left(r^2 \frac{\partial}{\partial r} \right) + \frac{1}{r^2 \sin\theta} \frac{\partial}{\partial\theta} \left(\sin\theta \frac{\partial}{\partial\theta} \right) + \frac{1}{r^2 \sin^2\theta} \frac{\partial^2}{\partial\varphi^2}$$

$$= \frac{1}{r^2} \frac{\partial}{\partial r} \left(r^2 \frac{\partial}{\partial r} \right) - \frac{1}{\hbar^2 r^2} L^2.$$

Exercise 12.4.1 *Prove from its definition that $L^2 \geq 0$.*

However, Laplacians are generally ≤ 0. But $-L^2 \leq 0$ and so we get a Laplacian type operator associated with the unit sphere. This is actually the Euclidean Laplacian Δ on $\mathbb{R}^3$ restricted to the unit sphere $\mathbb{S}^2$, which is a *Riemannian manifold* with respect to the restriction of the Euclidean inner product on $\mathbb{R}^3$ to each of its *tangent spaces*, that is the planes tangent to each of the points of $\mathbb{S}^2$. Also, $-L^2$ is a special case of a *Laplace-Beltrami operator* associated to a Riemannian manifold.

All of this geometry was relevant to the study of the hydrogen atom due to the invariance of its Hamiltonian H under rotations. This leads to these commutation relations:

$$[H, L_j] = 0 \text{ for } j = 1, 2, 3 \quad \text{and} \quad [H, L^2] = 0. \tag{12.4.1}$$

The last commutation relation in (12.4.1) tells us that the total angular momentum is *conserved* in the quantum system of the hydrogen atom. This means we can find states which are eigenfunctions of both H and L^2. This is what we did in Chapter 11 where we identified those states to be the ψ_{nlm}'s.

Exercise 12.4.2 *Prove that the commutation relations (12.4.1) holds for any Schrödinger operator $H = -\Delta + V(r)$, where the potential energy V depends* *only on the radial variable r. So, angular momentum is conserved for such an operator.*

This exercise is an instance of *Noether's theorem* that says roughly that symmetries imply conservation laws. However, in many common examples L^2 does not commute with the Hamiltonian, and one has to include *spin*, a new type of angular momentum, in order to maintain this conservation law.

Our solution for the negative energy states ψ_{nlm} of the hydrogen atom gave simultaneous eigenfunctions for H, L^2 as well as L_3. But note, that in general these are not eigenfunctions of L_1 nor of L_2. This is a common feature of Hamiltonians satisfying (12.4.1).

Exercise 12.4.3 *Evaluate $L_1\psi_{nlm}$. Find necessary and sufficient condition on the integers n, l, m such that ψ_{nlm} is an eigenfunction of L_1. Do the same for $L_2\psi_{nlm}$.*

Suppose that ψ is a simultaneous eigenfunction of L^2 and L_3, which are commuting operators. By the way, in the solution of the angular part of the hydrogen atom we have shown that such a ψ does exist, namely the spherical harmonic $\psi = Y_{lm}$ and, according to (11.3.10), (12.3.1), (11.3.6), and (12.2.1) we have

$$L^2 \psi = \hbar^2 l(l+1)\,\psi \qquad \text{and} \qquad L_3\,\psi = \hbar m\,\psi, \qquad (12.4.2)$$

where $l \geq 0$ is an integer and m is an integer with $|m| \leq l$. (Note that these formulas have the dimensionally correct powers of $\hbar$. In the angular part of the hydrogen atom problem we considered only dimension-less quantities.) Also, these are the only solutions which are square integrable with respect to the angular variables θ and φ. Therefore, contrary to a possibly naïve expectation, the eigenvalues of L^2 are not proportional to l^2, but rather to $l^2 + l$. Nonetheless, this is still close enough to motivate calling l the *(orbital) angular momentum quantum number*, although as already noted in Chapter 11 it is also called the *azimuthal quantum number*.

The point of this discussion is that the relation $|m| \leq l$ satisfied by the quantum numbers l and m comes from conservation of angular momentum of the hydrogen Hamiltonian and so works equally well for any quantum two-body problem with a central force.

12.5   Notes

The quantum theory of angular momentum gives a good learning experience, since we have little intuition, only operators, equations, and commutation relations. And so this is a great topic for strengthening the skill set needed in quantum theory. This becomes second nature, which is a good news, though it should not be confused with intuition. This fits in well with the philosophy of "shut up and compute." However, we also become accustomed to working with the very useful ladder operators, which are not self-adjoint and so do not correspond to observables. Still we think of them as pushing states up and down, as if those were physical processes. Ladder operators show us how amazingly useful mathematics can be, but we should not always consider such mathematics to be physics, but rather as aids for the comprehension of not-so-intuitive quantum physics.

Chapter 13

The Rotation Group $SO(3)$

<blockquote>
What goes around comes around.

Proverb
</blockquote>

We are now going to see how the geometry of physical space is related to its rotations. This allows us to understand angular momentum as a topic in geometry. This is part of a larger program of understanding various physics theories in terms of geometry, including spin in Chapter 14. Often this is achieved by finding a Lie group of symmetries associated with the theory. In this chapter we study the geometry and linear algebra of the Lie group $SO(3)$ and how that is related to angular momentum. Familiarity with matrices (especially matrix multiplication, determinants, and traces) is assumed.

13.1 Basic Definitions

We consider the Euclidean space $\mathbb{R}^3$ with its standard inner product and norm. This space is important in physics since it models (at least locally) the three-dimensional spatial world in which we live. A linear operator $A : \mathbb{R}^3 \to \mathbb{R}^3$ that preserves the inner product on $\mathbb{R}^3$, and hence distances and angles as well, is a geometrical symmetry of this space. So we define

$$O(3) := \{A \in Mat(3 \times 3; \mathbb{R}) \,|\, \langle Av, Aw \rangle = \langle v, w \rangle \quad \text{for all } v, w \in \mathbb{R}^3\}.$$

(Recall (9.1.2) for $n = 3$.) Here $Mat(3 \times 3; \mathbb{R})$ is the vector space over $\mathbb{R}$ of all 3×3 matrices with real entries. Since there are 9 independent entries in a 3×3 matrix,

$$Mat(3 \times 3; \mathbb{R}) \cong \mathbb{R}^9,$$

© Springer Nature Switzerland AG 2020
S. B. Sontz, *An Introductory Path to Quantum Theory*,
https://doi.org/10.1007/978-3-030-40767-4_13

a Euclidean space. Clearly, $O(3)$ is a subset of $Mat(3 \times 3; \mathbb{R})$ and so inherits its topology, since the usual metric on $\mathbb{R}^9$ allows us to define what the limit of a sequence (of matrices!) is. But $O(3)$ is not a vector subspace of $Mat(3 \times 3; \mathbb{R})$. We say that $O(3)$ is the *orthogonal group* of $\mathbb{R}^3$ and that a matrix in $O(3)$ is an *orthogonal matrix*. We will see presently that $O(3)$ is indeed a group under the binary operation of matrix multiplication.

The topological space $O(3)$ can also be given local C^∞ coordinates in order to make it into a *differential manifold* and into a *Lie group*.

We note that the following statements are equivalent by some elementary results in linear algebra:

- $\langle Av, Aw \rangle = \langle v, w \rangle$ for all $v, w \in \mathbb{R}^3$.

- $\langle A^t Av, w \rangle = \langle v, w \rangle$ for all $v, w \in \mathbb{R}^3$, where A^t is the *transpose* of A.

- $A^t Av = v$ for all $v \in \mathbb{R}^3$.

- $A^t A = I$, the *identity matrix*. (1's on the diagonal, 0's off the diagonal)

Therefore, an alternative description of $O(3)$ is

$$O(3) = \{A \in Mat(3 \times 3; \mathbb{R}) \mid A^t A = I\}.$$

This is often used as the definition of $O(3)$, but then its geometric significance is not so obvious.

Exercise 13.1.1 *Prove that $O(3)$ is a group, where its binary operation is matrix multiplication. In particular, identify which matrix serves as the identity element. (Recall Definition 5.2.1.)*

Exercise 13.1.2 *Prove that $-I \in O(3)$.*

The condition $A^t A = I$ tells us something about the *determinant*, denoted as det, of an orthogonal matrix A, namely that

$$1 = \det I = \det A^t A = \det A^t \det A = \det A \det A = (\det A)^2.$$

Here we used standard properties of the determinant. Hence, $\det A = \pm 1$.

Exercise 13.1.3 *Find orthogonal matrices A_1 and A_2 such that $\det A_1 = 1$ and $\det A_2 = -1$.*

The last exercise proves that the function $\det : O(3) \to \{-1, +1\}$ is onto.

Exercise 13.1.4 *Prove that $\det : O(3) \to \{-1, +1\}$ is continuous.*

Exercise 13.1.5 *Prove that* $O(3)$ *is not a connected topological space.*

Remark: *If you do not know what* connected *means in this context, prove that there exist orthogonal matrices A and B such that there is no continuous path lying entirely in* $O(3)$ *that starts at A and ends at B. This means that* $O(3)$ *is not* pathwise connected.

Since the value of the determinant seems to play a role in this theory, we are lead to make the next definition.

Definition 13.1.1 *We define the* special orthogonal group *of* $\mathbb{R}^3$ *to be*

$$SO(3) := \{A \in O(3) \mid \det A = 1\}.$$

This is also called the rotation group *of* $\mathbb{R}^3$.

Exercise 13.1.6 *Prove that* $SO(3)$ *is a subgroup of* $O(3)$, *which means that* $A, B \in SO(3)$ *implies* $AB \in SO(3)$ *and* $A^{-1} \in SO(3)$. *In particular* $SO(3)$ *with matrix multiplication is a group in and of itself.*

13.2 Euler's Theorem (Optional)

Euler's Theorem justifies our calling $SO(3)$ the *rotation group* of $\mathbb{R}^3$. Though it is a mathematical interlude, the reader should understand what it says, but the lengthy proof can be skipped.

First, we recall that a *ray* in $\mathbb{R}^3$ is a half-line $\mathbb{R}^+ v$ for some $0 \neq v \in \mathbb{R}^3$, where $\mathbb{R}^+ = (0, +\infty)$. Since the ray determined by $v \neq 0$ and the ray determined by λv for $\lambda > 0$ are the same ray, we can identify rays with points on the unit sphere $S^2 := \{\mathbf{x} \in \mathbb{R}^3 \mid \|\mathbf{x}\| = 1\} \subset \mathbb{R}^3$. For each ray the corresponding point on S^2 is called the *direction* of the ray. For any point $v \in S^2$ we say that v and $-v$ (the latter of which also lies on S^2) are *antipodal* points on S^2 and that the corresponding rays are *antipodal* rays.

Here is an intuitive, but not totally rigorous, definition. A *rotation* of $\mathbb{R}^3$ is an invertible linear map $R : \mathbb{R}^3 \to \mathbb{R}^3$ such that either $R = I$ or there is exactly one subspace of dimension one that is pointwise fixed by R and the set complement of that subspace undergoes a rigid motion that revolves around that subspace. It is clear that a rotation R, being a rigid motion, preserves distances and angles. Therefore R preserves the inner product and so $R \in O(3)$. Also, R can be connected by a continuous curve to the identity $I \in SO(3)$ and so $R \in SO(3)$. (If R rotates around a subspace with the angle θ, the curve consists of rotations around the same subspace by the angle $t\theta$ for $t \in [0, 1]$.) It is not obvious that the composition of two rotations is necessarily a rotation, although this is an immediate, non-trivial consequence of Euler's Theorem.

Theorem 13.2.1 (Euler's Theorem) *Let $A \in SO(3)$. Then A can be viewed as the rotation in the counter-clockwise direction by a unique angle $\theta \in [0, \pi]$ around some ray in $\mathbb{R}^3$.*

Here are the various cases of this statement:

- *$A = I$: In this case any ray will do and $\theta = 0$.*

- *$A \neq I$ and $A^2 \neq I$: In this case the ray is unique and $\theta \in (0, \pi)$.*

- *$A \neq I$ and $A^2 = I$: In this case there are exactly two possible rays, which are antipodal, and $\theta = \pi$.*

Remark: We will describe what *counter-clockwise* means rather than give a fully rigorous mathematical definition. So, suppose that $0 \neq v \in \mathbb{R}^3$ so that v determines the ray $\mathbb{R}^+ v := \{rv \mid r > 0\}$. Then the orthogonal complement

$$W := \{v\}^\perp = \{w \in \mathbb{R}^3 \mid \langle v, w \rangle = 0\}$$

is a two-dimensional vector subspace of $\mathbb{R}^3$. Notice that $v \notin W$. Now suppose we are at the point $v \in \mathbb{R}^3$, and we look towards the origin $(0, 0, 0)$ in the Euclidean plane W. Then a rotation in $\mathbb{R}^3$ around the ray $\mathbb{R}^+ v$ produces a rotation in the plane W.

Then by using our common experience about how the hands of a clock rotate, we understand what is a rotation in the plane W in the opposite, that is counter-clockwise, sense. Notice that by viewing the plane W from the point $-v \notin W$ the rotation which is counter-clockwise when viewed from v now becomes a clockwise rotation. This finishes this description, which is hoped to be intuitive.

Proof: Before discussing the three cases, we make some general observations. First, let's suppose that $Av = \lambda v$ for some $0 \neq v \in \mathbb{C}^3$ and some $\lambda \in \mathbb{C}$. This simply means that λ is one of the eigenvalues of A. Then we have

$$\langle Av, Av \rangle = \langle \lambda v, \lambda v \rangle = \lambda^* \lambda \langle v, v \rangle = |\lambda|^2 \langle v, v \rangle,$$

where we are using the standard inner product on $\mathbb{C}^3$. But the identity $A^t A = I$ still holds when these matrices act on $\mathbb{C}^3$ instead of on $\mathbb{R}^3$. Since A has real entries, its adjoint matrix satisfies $A^* = (\overline{A})^t = A^t$, where $\overline{A}$ denotes the complex conjugate of the matrix A. So, we also have

$$\langle Av, Av \rangle = \langle A^* Av, v \rangle = \langle A^t Av, v \rangle = \langle v, v \rangle.$$

We conclude that $|\lambda|^2 \langle v, v \rangle = \langle v, v \rangle$. But $v \neq 0$, and therefore $|\lambda|^2 = 1$. So, $\lambda \in S^1$, the unit circle in the complex plane.

We next define $p(x) := \det(xI - A)$, which is known as the *characteristic polynomial of A*. This is a polynomial of degree 3 with *real* coefficients, since the entries of A are real numbers. By the theory of polynomials we know that this polynomial has exactly three *roots* (i.e., solutions of $p(x) = 0$) in the complex numbers $\mathbb{C}$, provided that we count each root with its multiplicity. We denote these roots by $\lambda_1, \lambda_2, \lambda_3$, realizing there can be repetitions in this list. These are the eigenvalues of A considered as an operator mapping $\mathbb{C}^3$ to itself. A fundamental theorem of linear algebra says that any orthogonal matrix A is *diagonalizable*, meaning there exists a unitary map $U : \mathbb{C}^3 \to \mathbb{C}^3$ such that $U^*AU = \mathrm{diag}\{\lambda_1, \lambda_2, \lambda_3\}$, where $\mathrm{diag}\{a, b, c\}$ is the diagonal matrix with the complex numbers a, b, c along the diagonal and zeros elsewhere.

The eigenvectors corresponding to these eigenvalues are in $\mathbb{C}^3$. We claim that at least one of these eigenvalues is $+1$ and that there is a corresponding eigenvector lying in $\mathbb{R}^3$. This claim has to depend on the fact that A is orthogonal with determinant 1, since it is not true for arbitrary 3×3 matrices with real entries.

Well, we know by the general theory of polynomials that the eigenvalues of A (which are exactly the distinct roots of $p(x)$) satisfy exactly one of these two cases:

1. All the eigenvalues $\lambda_1, \lambda_2, \lambda_3$ are real numbers.

2. Exactly one eigenvalue is a real number, say λ_1 while the other two eigenvalues lie in $\mathbb{C} \setminus \mathbb{R}$ and are a pair of conjugate complex numbers, that is, $\lambda_2 = \lambda_3^*$.

In case 1, we have that each eigenvalue lies both on the unit circle and on the real line. So $\lambda_j = \pm 1$ for each $j = 1, 2, 3$. But $1 = \det A = \lambda_1 \lambda_2 \lambda_3$. There are exactly two possibilities: All the eigenvalues are $+1$ or exactly two of them are -1 and the remaining eigenvalue is $+1$. Using the diagonalization of A, one sees that the first possibility corresponds to $A = I$, while for the second possibility we have $A \neq I$ but $A^2 = I$. However, for both possibilities we have that at least one eigenvalue is equal to $+1$.

In case 2, one of the two non-real eigenvalues lies in the upper half-plane, say λ_2. Then there exists $\theta \in (0, \pi)$ such that $\lambda_2 = e^{i\theta}$ and so $\lambda_3 = \lambda_2^* = e^{-i\theta}$. From this it follows that

$$1 = \det A = \lambda_1 \lambda_2 \lambda_3 = \lambda_1 e^{i\theta} e^{-i\theta} = \lambda_1.$$

So $\lambda_1 = +1$ is an eigenvalue as are both λ_2 and λ_3. Also, the matrix A^2 has eigenvalues $+1, e^{2i\theta}, e^{-2i\theta}$. But $e^{\pm 2i\theta} \neq +1$ for $\theta \in (0, \pi)$ and so $A^2 \neq I$.

In short, as claimed, there is always at least one eigenvalue equal to $+1$. Also, if $A \neq I$, we have seen that the other two eigenvalues are distinct from $+1$. It remains to show that there is an eigenvector in $\mathbb{R}^3$ corresponding to the eigenvalue $+1$. But we do know that there exists $0 \neq w \in \mathbb{C}^3$ which is an eigenvector for the eigenvalue $+1$, that is $Aw = w$. Now since the entries of A are real, we have by taking the real part that

$$A \, Re(w) = Re(Aw) = Re(w).$$

Similarly, by taking the imaginary part, we get

$$A\,Im(w) = Im(Aw) = Im(w).$$

Next, we remark that $Re(w), Im(w) \in \mathbb{R}^3$. Also, $w \neq 0$ implies that either $Re(w) \neq 0$ or $Im(w) \neq 0$. Picking v to be one of these two vectors and to be non-zero, we see that $0 \neq v \in \mathbb{R}^3$ and that $Av = v$. This was the second part of our claim above.

Now we are ready to analyze the three cases as stated in the theorem itself. First, if $A = I$, then clearly the rotation around any ray by the angle $\theta = 0$ is the identity I. Also, rotation by $\theta \in (0, \pi]$ around any ray is not the identity I. So, $\theta = 0$ is the unique angle in this case.

If $A \neq I$ and $A^2 \neq I$, then we are in the case of three eigenvalues $1, \lambda_2, \lambda_3$ with $\lambda_2^* = \lambda_3$ and $|\lambda_2| = |\lambda_3| = 1$. Let $0 \neq v \in \mathbb{R}^3$ be an eigenvector of A corresponding to the eigenvalue $+1$.

We claim that the orthogonal complement $W = \{v\}^{\perp} \subset \mathbb{R}^3$ is invariant under the action of A. To show this let $w \in W$ be arbitrary. So we have $\langle v, w \rangle = 0$. We have to consider $\langle v, Aw \rangle = \langle A^t v, w \rangle$. We next want to find a formula for $A^t v$. But we know that $A v = v$ and so by acting on each side with A^t we get $v = A^t A v = A^t v$ since $A^t A = I$. Substituting this back into our equation we obtain $\langle v, Aw \rangle = \langle A^t v, w \rangle = \langle v, w \rangle = 0$. In other words, $Aw \in W$ as claimed.

Let $\{\eta_1, \eta_2, \eta_3\}$ be any orthonormal basis of $\mathbb{R}^3$ such that $\eta_1 = v/\|v\|$. It follows that $W = \text{span}\{\eta_2, \eta_3\}$. Then the matrix of A with respect to this orthonormal basis is represented by the 3×3 matrix

$$A' = \begin{pmatrix} 1 & 0 & 0 \\ 0 & b_{11} & b_{12} \\ 0 & b_{21} & b_{22} \end{pmatrix} = \begin{pmatrix} 1 & 0 \\ 0 & B \end{pmatrix}$$

in block matrix form, where $B = (b_{ij})$ is the 2×2 matrix which represents A restricted to W. Here $b_{ij} \in \mathbb{R}$ for $1 \leq i, j \leq 2$. From this we have that $1 = \det A = \det A' = 1 \cdot \det B = \det B$. Clearly, B is also orthogonal, that is, $B^t B = I$. So we have reduced this situation to studying the Lie group

$$SO(2) := \{T \in Mat(2 \times 2; \mathbb{R}) \mid T^t T = I, \det T = 1\},$$

which is called the *special orthogonal group in dimension* 2. Next, we write an arbitrary element $T \in SO(2)$ as

$$T = \begin{pmatrix} a & b \\ c & d \end{pmatrix} \quad \text{for } a, b, c, d \in \mathbb{R}. \tag{13.2.1}$$

For the next exercise recall from linear algebra that $T^t T = I$ implies $T T^t = I$.

Exercise 13.2.1 *Using the notation (13.2.1), prove $T^t T = I$ is equivalent to these three equations:*

$$a^2 + c^2 = 1, \quad b^2 + d^2 = 1, \quad ab + cd = 0.$$

Similarly show that $TT^t = I$ is equivalent to these three equations:

$$a^2 + b^2 = 1, \quad c^2 + d^2 = 1, \quad ac + bd = 0.$$

We learn from this exercise that each column and each row of T is a unit vector. Moreover, we learn that the two columns are orthogonal vectors and as well that the two rows are orthogonal vectors. We have learned a lot!

We start out with the first column of T. Since it is a unit vector, there is a unique $\theta \in [-\pi, \pi)$ such that $(a, c) = (\cos\theta, \sin\theta)$, because (a, c) lies on the unit circle in the Euclidean plane. This forces the second column vector of T, which is a unit vector orthogonal to the first column, to be one of these two possibilities: $(b, d) = \pm(-\sin\theta, \cos\theta)$. Which sign will it be? To decide this we use $\det T = 1$. Clearly, by choosing the sign $+$ we obtain this matrix with determinant 1:

$$T = \begin{pmatrix} \cos\theta & -\sin\theta \\ \sin\theta & \cos\theta \end{pmatrix} \quad \text{for } \theta \subset [-\pi, \pi). \tag{13.2.2}$$

And choosing the sign $-$ gives determinant -1. So every matrix $T \in SO(2)$ has the form (13.2.2). A matrix of the form (13.2.2) when acting on $\mathbb{R}^2$ is called a *rotation* of $\mathbb{R}^2$ by the angle θ. This is a counter-clockwise rotation if and only if $\theta > 0$, and it is a clockwise rotation if and only if $\theta < 0$.

Exercise 13.2.2 *Prove that every matrix of the form (13.2.2) is in $SO(2)$. Prove that $SO(2)$ is isomorphic as a group to the group (whose product is complex multiplication)*

$$\mathbb{S}^1 := \{ z \in \mathbb{C} \mid |z| = 1 \},$$

the unit circle in the complex plane.

This ends our detour for studying $SO(2)$. We now apply this to the 2×2 matrix B. So, B is a rotation of the plane W by some unique angle $\theta \in [-\pi, \pi)$. But we know in this particular case that $\theta \neq 0$ and $\theta \neq -\pi$. Thus, we have $\theta \in (0, \pi)$ or $\theta \in (-\pi, 0)$. The first case is a counter-clockwise rotation by $\theta \in (0, \pi)$ with respect to one of the two rays lying on the line $\mathbb{R}\eta_1$. The second case is also a counter-clockwise rotation, but now by $-\theta \in (0, \pi)$ with respect to the antipodal ray. (Recall that η_1 is an eigenvector of A with eigenvalue $+1$, that is, $\mathbb{R}\eta_1$ is fixed pointwise by A.)

We now know that A fixes the one-dimensional subspace $\mathbb{R}\eta_1$ pointwise and rotates its orthogonal subspace passing through the origin by an angle

$\theta \in (0, \pi)$. The reader should understand that by the linearity of A this implies that A rotates $\mathbb{R}^3$ counter-clockwise around one of the rays lying on $\mathbb{R}\eta_1$ by the same angle $\theta \in (0, \pi)$. This finishes the argument for this case.

The last case remaining is $A \neq I$ but $A^2 = I$. We have seen that in this case the three eigenvalues are $+1, -1, -1$, including multiplicity. The eigenvector $0 \neq v \in \mathbb{R}^3$ corresponding to the eigenvalue $+1$ then has as before the orthogonal complement $W = \{v\}^\perp \subset \mathbb{R}^3$. In this case A restricted to the plane W is $-I_W$, where I_W is the identity map of W. But $-I_W$ is the rotation by the unique angle $\theta = \pi$ in the counter-clockwise direction when viewed from v and also when viewed from $-v$. These are the only possible rays for A. Again, the reader is advised to draw a picture of this situation. While that picture is not a rigorous mathematical argument, it is worth a thousand words of such arguments. Please draw it!

This exhausts all the cases and therefore concludes the proof. ∎

Here is a non-trivial result whose proof is now amazingly easy.

Corollary 13.2.1 *Let R_1 and R_2 be rotations of $\mathbb{R}^3$. Then their product $R_1 R_2$ is also a rotation of $\mathbb{R}^3$.*

Proof: By the comments at the beginning of this section $R_1, R_2 \in SO(3)$. But that implies $R_1 R_2 \in SO(3)$, since $SO(3)$ is a group. Consequently $R_1 R_2$ is a rotation by Euler's theorem. ∎

We are now in a position for describing $SO(3)$ in terms of coordinates. We define the solid, closed ball $B(\pi)$ in $\mathbb{R}^3$ centered at the origin and of radius π by
$$B(\pi) := \{v \in \mathbb{R}^3 \mid ||v|| \leq \pi\}.$$
The boundary $\partial B(\pi)$ of this ball is the two-dimensional sphere $S(\pi)$ centered at the origin with radius π:

$$S(\pi) = \partial B(\pi) = \{v \in \mathbb{R}^3 \mid ||v|| = \pi\}.$$

Now we use Euler's Theorem to identify elements in $SO(3)$ with points in $B(\pi)$. So consider $A \in SO(3)$.

- If $A = I$, we identify A with the origin $(0, 0, 0) \in B(\pi)$.

- If $A \neq I$ and $A^2 \neq I$, then we have a unique angle $\theta \in (0, \pi)$ and a unique unit vector $v \in \mathbb{R}^3$ such that A is the counter-clockwise rotation by the angle θ around the ray determined by v. Then we identify A with the vector $w = \theta v \in B(\pi)$. In this case $||w|| = \theta$ and so $w \neq 0$ and $w \notin S(\pi)$.

- If $A \neq I$ but $A^2 = I$, then we identify A with the *two* points

$$\pi v, -\pi v \in S(\pi) \subset B(\pi),$$

where A is the rotation in the counter-clockwise direction by the angle π with respect to the unit vectors v and $-v$.

This paragraph uses some topology but is worth skimming. Each point in the interior of $B(\pi)$ corresponds to a unique element in $SO(3)$, while each pair of antipodal points on $S(\pi)$ corresponds to a unique element in $SO(3)$. In all cases the norm of the point(s) in $B(\pi)$ is the angle of rotation of the corresponding element in $SO(3)$. So, we can identify $SO(3)$ as the quotient space of $B(\pi)$ when we divide out by the equivalence relation $\pi v \cong -\pi v$ for points on the boundary $S(\pi)$ of $B(\pi)$, where $v, -v$ are antipodal unit vectors. It is to be devotedly wished that this quotient topological space of $B(\pi)$ is homeomorphic to the topological space $SO(3)$. (*Homeomorphic* means isomorphic as topological spaces.) Given that this is so (and it is), we see that $SO(3)$ is a compact, connected topological space being the quotient of a compact, connected space. Also, $\dim SO(3) = 3$ follows.

13.3 One-parameter Subgroups

While $SO(3)$ is not a vector space, but rather a compact group, we would still like to understand its elements in terms of three basic elements, where each element corresponds in some sense to one of three 'independent' directions in $SO(3)$. Actually, we are going to construct three *families* of elements of $SO(3)$. For every $\theta \in \mathbb{R}$ we define these matrices:

$$R_x(\theta) := \begin{pmatrix} 1 & 0 & 0 \\ 0 & \cos\theta & -\sin\theta \\ 0 & \sin\theta & \cos\theta \end{pmatrix} \tag{13.3.1}$$

$$R_y(\theta) := \begin{pmatrix} \cos\theta & 0 & \sin\theta \\ 0 & 1 & 0 \\ -\sin\theta & 0 & \cos\theta \end{pmatrix} \tag{13.3.2}$$

$$R_z(\theta) := \begin{pmatrix} \cos\theta & -\sin\theta & 0 \\ \sin\theta & \cos\theta & 0 \\ 0 & 0 & 1 \end{pmatrix} \tag{13.3.3}$$

Exercise 13.3.1 *Prove that $R_x(\theta), R_y(\theta), R_z(\theta) \in SO(3)$ for every $\theta \in \mathbb{R}$.*

For $\theta \geq 0$ each of $R_x(\theta), R_y(\theta)$, and $R_z(\theta)$ is a rotation in the counter-clockwise direction by the angle θ with respect to the corresponding positive semi-axis (which is a ray), namely the positive x-axis, the positive y-axis or the positive z-axis. We assume throughout that these positive axes form a right-handed orientation on $\mathbb{R}^3$. (The southpaws among my readers should not feel imposed upon. It is perfectly fine to consistently use the left-handed orientation on $\mathbb{R}^3$ throughout.)

In particular $R_x(0) = R_y(0) = R_z(0) = I$, the 3×3 identity matrix.

Exercise 13.3.2 *Draw pictures to convince yourself that each of these three families consists of counter-clockwise rotations by the angle θ around the appropriate positive semi-axis (x, y or z) provided that $\theta \geq 0$.* **Hint:** *Doing this for $\theta = \pi/2$ should be enough to convince you.*
Having done that, it is an easy next step to see that these are clockwise rotations by the angle $-\theta$ around the same positive semi-axis for $\theta \leq 0$.

Exercise 13.3.3 *The previous exercise should have convinced you also that $R_x(\theta_1 + \theta_2) = R_x(\theta_1)R_x(\theta_2)$ holds. Prove that is so for any $\theta_1, \theta_2 \in \mathbb{R}$. Similar identities hold for the families $R_y(\theta)$ and $R_z(\theta)$. Prove those identities too.*

Exercise 13.3.4 *Prove $R_x(\theta_1)R_x(\theta_2) = R_x(\theta_2)R_x(\theta_1)$ for all $\theta_1, \theta_2 \in \mathbb{R}$.*

Note that $R_x : \mathbb{R} \to SO(3)$ is a C^∞ function in the sense that each matrix entry in $R_x(\theta)$ is a C^∞ real-valued function of $\theta \in \mathbb{R}$. Also, Exercise 13.3.3 says that this is a group morphism from the (abelian) group $\mathbb{R}$ with $+$ as its group operation to the (non-abelian) group $SO(3)$. One says that R_x is a *one-parameter subgroup of $SO(3)$*. This is an unfortunate terminology, since it is the range of R_x that is a subgroup of $SO(3)$, while R_x itself is a morphism of groups. Similar statements hold for the functions R_y and R_z. For the record here is a more general definition:

Definition 13.3.1 *We say $R : \mathbb{R} \to SO(3)$ is a* one-parameter subgroup of *$SO(3)$ if R is a differentiable function and a group morphism.*

We can also think of each of R_x, R_y, and R_z as a curve in $SO(3)$. And at each point of a differentiable curve, we have an associated direction given by its *derivative*, which in geometry is also called its *tangent vector* and in physics is called its *velocity vector*. Let's evaluate these tangent vectors at $\theta \in \mathbb{R}$. For example, applying elementary calculus to each of the 9 entries of R_x we see that

$$R_x'(\theta) = \begin{pmatrix} 0 & 0 & 0 \\ 0 & -\sin\theta & -\cos\theta \\ 0 & \cos\theta & -\sin\theta \end{pmatrix}. \tag{13.3.4}$$

This is called the *infinitesimal rotation* around the positive x-axis at the 'point' $R_x(\theta)$ in $SO(3)$. Now believe it or not, it is not heresy to speak of infinitesimals. However, this matrix is definitely not a rotation matrix, despite its name. In particular it has 0 as an eigenvalue or, in other words, its *kernel* (also called its *null space*) is non-zero. Also its determinant is zero. All of this is quite different from the properties of a rotation matrix. We now compute the other two infinitesimal rotations of interest:

$$R_y'(\theta) = \begin{pmatrix} -\sin\theta & 0 & \cos\theta \\ 0 & 0 & 0 \\ -\cos\theta & 0 & -\sin\theta \end{pmatrix}, \tag{13.3.5}$$

$$R'_z(\theta) = \begin{pmatrix} -\sin\theta & -\cos\theta & 0 \\ \cos\theta & -\sin\theta & 0 \\ 0 & 0 & 0 \end{pmatrix}. \tag{13.3.6}$$

These are also matrices with non-zero kernel and determinant zero.

Here is an important result about one-parameter subgroups. Suppose that $R : \mathbb{R} \to SO(3)$ is any one-parameter subgroup. So, R could be one of the three one-parameter subgroups introduced above, or it could be the family of rotations, parameterized by the angle θ of rotation, around any given one-dimensional subspace of $\mathbb{R}^3$. In particular, we have that $R(0) = I$. The properties we will now use are that $R(\theta)$ is differentiable at $\theta = 0$ and is a group morphism. We compute its infinitesimal rotation by first considering the usual difference quotient for $0 \neq \alpha \in \mathbb{R}$ from a calculus course:

$$\frac{R(\theta + \alpha) - R(\theta)}{\alpha} = \frac{R(\alpha)R(\theta) - R(\theta)}{\alpha} = \frac{R(\alpha) - R(0)}{\alpha} R(\theta).$$

The quotient on the right side has limit $R'(0)$ by hypothesis. Therefore, by taking the limit as $\alpha \to 0$ we obtain

$$R'(\theta) = \lim_{\alpha \to 0} \frac{R(\theta + \alpha) - R(\theta)}{\alpha} = R'(0)R(\theta). \tag{13.3.7}$$

This is an amazing result! It says that the infinitesimal generator at an arbitrary angle θ depends only on the infinitesimal generator at $\theta = 0$ and the rotation itself at the arbitrary angle θ. But even more remarkably (13.3.7) is a first order, ordinary differential equation (ODE) for $R(\theta)$ with the constant coefficient $G := R'(0)$, which is called the *infinitesimal generator* of the one-parameter subgroup. Also, the initial condition for this ODE is $R(0) = I$.

Of course, this is a differential equation for *matrix valued* functions of θ and not for *scalar valued* functions, which one studies in elementary calculus. If all is for the best in the best of all possible worlds, then the unique solution should be $R(\theta) = \exp(\theta G)$, where the exponential of a matrix M is defined by the usual infinite series:

$$\exp(M) := I + M + \frac{1}{2!}M^2 + \frac{1}{3!}M^3 + \cdots + \frac{1}{k!}M^k + \cdots \tag{13.3.8}$$

Another common notation for this is e^M. As the reader may have already guessed, all of this does hold rigorously. One technical point is that the infinite series (known as *power series*) for e^M does converge. Also, the infinite series e^{tM} can be differentiated term by term with respect to a variable t when M is a matrix that does not depend on t. Here t can be a real or complex variable.

Exercise 13.3.5 *Let M be a matrix. Prove that e^M commutes with M and hence with any power M^n with integer $n \geq 0$. (Note that $M^0 := I$.) Also, if M is invertible, prove that e^M commutes with $(M^{-1})^n$ for any integer $n \geq 1$. Finally, differentiate the infinite series for e^{tM} term by term to see that*

$$\frac{d}{dt} e^{tM} = M e^{tM} = e^{tM} M.$$

The moral of this little story is that a one-parameter subgroup of $SO(3)$ is determined by its infinitesimal generator. Of course, the infinitesimal generator is determined by the one-parameter subgroup; one just evaluates the derivative of the one-parameter subgroup at $\theta = 0$. This relation between one-parameter *semigroups* and their infinitesimal generators turns out to be a major industry in functional analysis. (See [28] for example.) But you have already seen the basic ideas in a concrete example.

Let's evaluate the infinitesimal rotations at $\theta = 0$ (also known as the infinitesimal generators) for R_x, R_y, and R_z. So we evaluate the expressions (13.3.4), (13.3.5), and (13.3.6) at $\theta = 0$. We get the following:

$$R_1 := R_x'(0) = \begin{pmatrix} 0 & 0 & 0 \\ 0 & 0 & -1 \\ 0 & 1 & 0 \end{pmatrix} \tag{13.3.9}$$

$$R_2 := R_y'(0) = \begin{pmatrix} 0 & 0 & 1 \\ 0 & 0 & 0 \\ -1 & 0 & 0 \end{pmatrix} \tag{13.3.10}$$

$$R_3 := R_z'(0) = \begin{pmatrix} 0 & -1 & 0 \\ 1 & 0 & 0 \\ 0 & 0 & 0 \end{pmatrix} \tag{13.3.11}$$

Exercise 13.3.6 *To check that our understanding of this is really right, prove by direct calculation what we already know must be true, namely that $R_x(\theta) = \exp(\theta R_1)$ for all $\theta \in \mathbb{R}$. (You will need to use the power series representations of $\cos \theta$ and $\sin \theta$.)*

We pause for some definitions and an exercise.

Definition 13.3.2 *For any $n \times n$ matrix $A = (a_{ij})$ with $a_{ij} \in \mathbb{C}$ we define its adjoint matrix to be $A^* := (a_{ji}^*)$, that is, the transpose matrix complex conjugated. Here $1 \leq i, j \leq n$. We say that A is symmetric (or self-adjoint or hermitian) if $A^* = A$ and that it is anti-symmetric if $A^* = -A$.*

Exercise 13.3.7 *Let $A = (a_{ij})$ be an $n \times n$ matrix with entries $a_{ij} \in \mathbb{R}$. Let $Tr(A) := \sum_{i=1}^n a_{ii}$ denote the trace of A and $\det(A)$ its determinant. Prove that if A is anti-symmetric, then $Tr(A) = 0$. Moreover, if n is an odd integer, show that $\det(A) = 0$.*

Now rotations about distinct axes will not commute in general. This is sometimes referred to as a textbook exercise because of the following exercise with a textbook.

Exercise 13.3.8 *Take a textbook and consider the rotations about its three symmetry axes. (The plural of 'axis' is 'axes'.) To your own satisfaction prove that in general AB is not equal to BA if A is a rotation about one of these axes and B is a rotation about another of these axes. It is easier to see this with rotations by $\pi/2$ radians (colloquially known as $90°$) but you can try other angles too. This is basically an experimental exercise, though you might try to describe what happens in words. (Suggestion: For 21st century readers who do not have printed textbooks, use your laptop instead.)*

13.4 Commutation Relations, $so(3)$ and All That

Well, if rotations about distinct axes do not commute in general, what about infinitesimal rotations about distinct axes?

Exercise 13.4.1 *If R_1 commutes with R_2, prove that every rotation $R_x(\theta)$ commutes with every rotation $R_y(\tau)$. Conclude that R_1 does not commute with R_2. Recall that R_1 and R_2 are defined in (13.3.9) and (13.3.10).*

Of course, one can simply compute $R_1 R_2$ and compare with the computed value of $R_2 R_1$, something which we are about to do anyway. The point of the previous exercise is to give an intuition behind this lack of commutativity. So we are motivated to evaluate the commutators $[R_1, R_2] = R_1 R_2 - R_2 R_1$, and so forth.

Exercise 13.4.2 *Prove the following commutation relations:*

$$[R_1, R_2] = R_3, \qquad [R_2, R_3] = R_1, \qquad [R_3, R_1] = R_2.$$

In words we can say that the infinitesimal rotations R_1 and R_2 around the x-axis and y-axis, respectively, fail to commute by the infinitesimal rotation R_3 around the z-axis. Similar remarks hold for the other two commutation relations.

Try to verify this using your textbook or laptop. Rotate it by a small angle θ first around the y-axis and then the x-axis in that order and then compare that with rotations by the small angle $-\theta$ and again first around the y-axis and then the x-axis in that order. Be careful! All of these must be counter-clockwise rotations around the appropriate positive semi-axis. The result of these four rotations should be equal approximately to a very, very small counter-clockwise rotation around the z-axis.

These three commutation relations in the previous exercise look rather familiar. They are almost the same as the commutation relations for the angular momentum operators L_1, L_2, and L_3. But not exactly! Recall that for $\hbar = 1$, we have that $[L_1, L_2] = iL_3$ and its cyclic permutations. So the difference is a factor of $i = \sqrt{-1}$. This difference has to do with the difference between symmetric operators and anti-symmetric operators. Here, L_1, L_2, and L_3 are symmetric operators while R_1, R_2, and R_3 are anti-symmetric matrices (and therefore anti-symmetric operators).

For example, if we define $M_j := (1/i)L_j$ for $j = 1, 2, 3$, then we have the commutation relations $[M_1, M_2] = M_3$ and cyclic permutations for the anti-symmetric operators M_1, M_2, and M_3. In general the difference between symmetric and anti-symmetric operators which act on *complex* vector spaces is simply a factor of $i = \sqrt{-1}$.

You might think that you can not multiply the matrices R_1, R_2, R_3 with real entries by $i = \sqrt{-1}$. But of course you can! And what will happen is that the resulting 3×3 matrices will not have real entries and therefore will not represent operators acting on $\mathbb{R}^3$. But they will represent operators acting on $\mathbb{C}^3$. In fact, the original matrices R_1, R_2, R_3 also represent operators acting on $\mathbb{C}^3$ by using the same formula for their action on $\mathbb{R}^3$.

Exercise 13.4.3 *Suppose that A and B are symmetric operators acting in a complex vector space with inner product. Prove that $[A, B]$ is anti-symmetric, and therefore $i[A, B]$ and $(1/i)[A, B]$ are symmetric.*
Suppose that C and D are anti-symmetric matrices. Prove that $[C, D]$ is anti-symmetric. (For this part of the problem the vector space may be real or complex. Notice that real vector spaces are making an appearance here in the quantum world. More on this in a moment.)

The moral of this exercise is that the symmetric operators are closed under the binary operation $(1/i)[\cdot, \cdot]$, while the anti-symmetric matrices are closed under the binary operation $[\cdot, \cdot]$. The first operation only makes sense for vector spaces over the complex numbers $\mathbb{C}$, while the second operation makes sense for vector spaces over the real numbers $\mathbb{R}$ as well as over the complex numbers $\mathbb{C}$. In quantum physics the basic field is $\mathbb{C}$ and so there is no obvious advantage in using anti-symmetric operators together with the usual commutator $[\cdot, \cdot]$. And in fact, in quantum physics one typically only considers symmetric operators. But in mathematics one often wishes to study the purely real case including only anti-symmetric operators. Of course, such a mathematical study sometimes leads to useful results for physics. So it is best to keep both points of view in mind.

Right now, let's consider the matrices R_1, R_2, and R_3 as defined above. We define a vector space over the *real numbers* by

$$so(3) := \text{span}_{\mathbb{R}}\{R_1, R_2, R_3\} = \{aR_1 + bR_2 + cR_3 \mid a, b, c \in \mathbb{R}\}.$$

We remark that this an important *real* vector space in quantum theory. By its very definition we have that $\dim_{\mathbb{R}} so(3) \leq 3$.

Exercise 13.4.4 *Prove that the three matrices R_1, R_2, and R_3 are linearly independent. Consequently, $\dim_{\mathbb{R}} so(3) = 3$.*

Exercise 13.4.5 *Prove that any anti-symmetric 3×3 matrix can be written uniquely in the form $aR_1 + bR_2 + cR_3$, where $a, b, c \in \mathbb{R}$. Conclude that $so(3)$ is the real vector space of all anti-symmetric 3×3 matrices.*
Suppose that $R : \mathbb{R} \to SO(3)$ is a one-parameter subgroup. Prove that $R'(0)$ is an anti-symmetric matrix and therefore $R'(0) \in so(3)$.
Suppose that $M \in so(3)$. Prove that M is the infinitesimal generator of the one-parameter subgroup $\mathbb{R} \ni t \mapsto e^{tM} \in SO(3)$.

The upshot is that the set of one-parameter subgroups of $SO(3)$ is in bijective correspondence with the elements of the vector space $so(3)$.

Since the matrices R_1, R_2, and R_3 are tangent to the space $SO(3)$ at the point $I \in SO(3)$, we call $so(3)$ the *tangent space* to $SO(3)$ at the point $I \in SO(3)$. While $SO(3)$ is some sort of three-dimensional compact space, $so(3)$ is a three-dimensional real vector space that infinitesimally, and it turns out even locally, approximates the space $SO(3)$ at the point $I \in SO(3)$. But $so(3)$ has even more structure. For any two elements $R, S \in so(3)$ we define $[R, S] := RS - SR$, the *commutator* of matrices.

Exercise 13.4.6 *Prove that the binary operation $[\cdot, \cdot]$ so defined on $so(3)$ satisfies the following properties for $R, S, T \subset so(3)$:*

- *(Closure) $[R, S] \in so(3)$.*

- *(Bilinear) The binary operation $R, S \mapsto [R, S]$ is bilinear.*

- *(Anti-symmetry) $[R, S] = -[S, R]$.*

- *(Jacobi's identity) $[R, [S, T]] + [S, [T, R]] + [T, [R, S]] = 0$.*

This leads to an extremely important definition.

Definition 13.4.1 *Let $\mathcal{G}$ be any vector space over the real numbers $\mathbb{R}$ (resp., the complex numbers $\mathbb{C}$) with a binary operation $\mathcal{G} \times \mathcal{G} \to \mathcal{G}$, denoted by $X, Y \mapsto [X, Y]$ for $X, Y \in \mathcal{G}$, which is bilinear, anti-symmetric, and satisfies Jacobi's identity. Then we say that $\mathcal{G}$ together with this Lie bracket $[\cdot, \cdot]$ is a real (resp., complex) Lie algebra. We often say simply Lie algebra when the field of scalars is implicitly known.*

This structure is named in honor of the mathematician Sophus Lie. We will not dwell on the much more technical definition of a *Lie group*. However, we do remark that $SO(3)$ is a Lie group. It turns out that every Lie group has a *tangent space* at the identity, and it is given by differentiating one-parameter subgroups. And that tangent space is always a finite-dimensional Lie algebra, because I am supposing that the Lie group is itself a finite-dimensional space. An arbitrary finite-dimensional Lie algebra over the field $\mathbb{R}$ always comes in this manner from some Lie group, but that Lie group need not be unique.

So we have seen that $so(3)$ is the Lie algebra of the Lie group $SO(3)$.

Exercise 13.4.7 *a) Prove that $Mat(n \times n; \mathbb{R})$, the vector space of all $n \times n$ matrices with real entries, equipped with the commutator as its Lie bracket, forms a Lie algebra. b) Prove that $\mathbb{R}^3$ with the* cross product *(also known as the* vector product*) is a Lie algebra isomorphic to $so(3)$.*

13.5   Notes

Euler did indeed prove something like Euler's Theorem. However, matrices came much after his time. He formulated this result in terms of the geometry of Euclidean three-dimensional space. Thought of in that way it is quite remarkable that rotations form a group. That is to say, given first a rotation by some angle about some axis passing through some point p and following that by another rotation by some other angle about some other axis passing through the same point p, then it is amazing that the combination of these two rotations is itself a rotation about yet another axis, but still passing through the same point p. After all, there must be a way (an algorithm!) for finding the axis and angle of the combined rotation from that data for the two individual rotations. This may be intuitively obvious to you, but it leaves me wondering. Yet this is what Euler proved. Explicitly, he showed that any rigid motion of a solid sphere that leaves the center fixed will always leave (at least) one diameter fixed pointwise and so is a rotation about that diameter. Since the combination of two such rigid motions is again such a rigid motion, Euler clearly knew what was going on.

The modern form of Euler's Theorem identifies the rotations as the group of matrices $SO(3)$. This thereby allows us to study geometry by using linear algebra.

Chapter 14

Spin and $SU(2)$

> The wheel is come full circle.
> Edmund in *King Lear*, William Shakespeare

In quantum theory the commutation relations of pairs of observables play a central role. In the last two chapters we saw the commutation relations of the three components of the angular momentum operators and the three infinitesimal rotations R_1, R_2, and R_3. These commutation relations are more important than the particular operators that satisfy them, believe it or not! This is one of the great insights in the development of quantum theory. This is all part and parcel of the role played in quantum theory by *Lie groups* and *Lie algebras*. This aspect of quantum theory was originally developed by H. Weyl, W. Pauli, and E.P. Wigner. Some work by E. Noether around 1918 in classical physics foreshadowed developments in quantum theory.

14.1 Basics of $SU(2)$

The theory of spin can begin with the Lie group

$$SU(2) := \{U \in Mat(2 \times 2; \mathbb{C}) \mid U^*U = I \text{ and } \det U = 1\}.$$

This is called the *special unitary group* for $\mathbb{C}^2$. For the sake of completeness we note that this is a group which itself is a subgroup of

$$U(2) := \{U \in Mat(2 \times 2; \mathbb{C}) \mid U^*U = I\},$$

which is called the *unitary group* for $\mathbb{C}^2$. Of course, $Mat(2 \times 2; \mathbb{C})$ denotes the complex vector space of all 2×2 matrices with complex entries.

Exercise 14.1.1 *Show that $U \in U(2)$ if and only if $\langle Uw, Uz \rangle = \langle w, z \rangle$ for all $w = (w_1, w_2)$ and $z = (z_1, z_2)$ in $\mathbb{C}^2$, where the inner product $\langle \cdot, \cdot \rangle$ is defined by $\langle w, z \rangle := w_1^* z_1 + w_2^* z_2$. We say such a matrix U is* unitary.

© Springer Nature Switzerland AG 2020
S. B. Sontz, *An Introductory Path to Quantum Theory*,
https://doi.org/10.1007/978-3-030-40767-4_14

Alternatively, the theory of spin can begin with the Lie algebra $su(2)$ associated to the Lie group $SU(2)$. This involves finding the infinitesimal generator of every one-parameter subgroup of $SU(2)$. Now, a one-parameter subgroup is a differentiable morphism of groups $h : \mathbb{R} \to SU(2)$. So h satisfies these two conditions for every $t \in \mathbb{R}$:

$$h(t)^* h(t) = I \quad \text{and} \quad \det h(t) = 1.$$

Taking derivatives at t of the first condition gives

$$h'(t)^* h(t) + h(t)^* h'(t) = 0. \tag{14.1.1}$$

Next evaluating at $t = 0$ yields

$$h'(0)^* + h'(0) = 0,$$

since $h(0) = I$. We set $G := (1/i)h'(0)$. So this says that G is a self-adjoint matrix. Also, $h(t) = e^{itG}$ for all $t \in \mathbb{R}$ is the unique solution of (14.1.1) satisfying $h(0) = I$ as can be seen by using Exercise 13.3.5.

The condition on the determinant gives us

$$1 = \det h(t) = \det e^{itG} = e^{it\,Tr(G)}$$

for all $t \in \mathbb{R}$, where $Tr(G)$ is the trace of G. The last equality is an identity in linear algebra. The only way this can hold for all $t \in \mathbb{R}$ is if $Tr(G) = 0$.

Exercise 14.1.2 *Conversely, prove that any self-adjoint, trace zero, 2×2 matrix A with complex entries exponentiates to the one-parameter subgroup $e^{itA} \in SU(2)$ for all $t \in \mathbb{R}$. Thus, the set of one-parameter subgroups of $SU(2)$ is in bijection with the real vector space defined by*

$$su(2) := \{A \in Mat(2 \times 2; \mathbb{C}) \mid A^* = A \text{ and } Tr(A) = 0\}.$$

N.B. A matrix $A \in su(2)$ is allowed to have entries which are complex numbers. Nevertheless, the vector space $su(2)$ of all such matrices is only a vector space over the real numbers, not over the complex numbers.

Next, we address the problem of identifying all self-adjoint 2×2 matrices.

$$\text{If} \quad A = \begin{pmatrix} \alpha & \beta \\ \gamma & \delta \end{pmatrix}, \quad \text{then } A^* = \begin{pmatrix} \alpha^* & \gamma^* \\ \beta^* & \delta^* \end{pmatrix}.$$

So, $A = A^*$ if and only if $\alpha, \delta \in \mathbb{R}$ and $\beta = \gamma^*$. Therefore,

$$A = \begin{pmatrix} a & b - ic \\ b + ic & d \end{pmatrix}$$

where $a, b, c, d \in \mathbb{R}$. We conclude that the *real* vector space of 2×2 self-adjoint matrices has dimension 4. A vector space basis of this space of matrices is given by the identity matrix I together with the three *Pauli matrices*:

$$\sigma_1 = \begin{pmatrix} 0 & 1 \\ 1 & 0 \end{pmatrix} \quad \sigma_2 = \begin{pmatrix} 0 & -i \\ i & 0 \end{pmatrix} \quad \sigma_3 = \begin{pmatrix} 1 & 0 \\ 0 & -1 \end{pmatrix} \tag{14.1.2}$$

This particular choice of basis is taken, since we then have that all of the Pauli matrices are in $su(2)$, while $I \notin su(2)$. Therefore, $\dim_{\mathbb{R}} su(2) = 3$. One immediate consequence from manifold theory is that the Lie group $SU(2)$ is a real differential manifold with dimension 3.

Exercise 14.1.3 *Directly from the definition of $SU(2)$, prove that $SU(2)$ is homeomorphic (that is, isomorphic as topological space) to the 3-sphere S^3. Recall that $S^3 := \{v \in \mathbb{R}^4 \mid \|v\| = 1\}$.*

The Pauli matrices have many elementary, but important properties.

Exercise 14.1.4 *Prove that*

$$\sigma_1^2 = \sigma_2^2 = \sigma_3^2 = I$$

and that

$$\sigma_1\sigma_2 = i\sigma_3 = -\sigma_2\sigma_1 \qquad \sigma_2\sigma_3 = i\sigma_1 = -\sigma_3\sigma_2 \qquad \sigma_3\sigma_1 = i\sigma_2 = \sigma_1\sigma_3.$$

Prove that $\mathrm{Spec}(\sigma_1) = \mathrm{Spec}(\sigma_2) = \mathrm{Spec}(\sigma_3) = \{-1, +1\}$, *where* $\mathrm{Spec}(M)$ *means the spectrum of M, that is the set of all eigenvalues of the matrix M.*

It follows immediately from this exercise that

$$[\sigma_1, \sigma_2] = \sigma_1\sigma_2 - \sigma_2\sigma_1 = 2i\sigma_3.$$

Also, we obtain two more identities by cyclically permuting the three sub-indices in this identity. But these identities are not quite those of the angular momentum operators. To get those commutation relations we define the *spin matrices* by $S_j := (1/2)\sigma_j$ for $j = 1, 2, 3$.

Exercise 14.1.5 *Prove that $[S_1, S_2] = iS_3$ plus cyclic permutations of this identity. Also show that* $\mathrm{Spec}(S_1) = \mathrm{Spec}(S_2) = \mathrm{Spec}(S_3) = \{-1/2, +1/2\}$.

Clearly, the set $\{S_1, S_2, S_3\}$ is a basis of $su(2)$. Also, $su(2)$ is a real Lie algebra under the bilinear operation $M, N \mapsto \{M, N\} := (1/i)[M, N]$ for $M, N \in su(2)$.

The most important consequence is that the commutation relations of the S_j in $su(2)$ with respect to its Lie bracket $\{\cdot, \cdot\}$ are exactly the same as those for the R_j in $so(3)$ as given in Exercise 13.4.2. The conclusion is that the real Lie algebras $so(3)$ and $su(2)$ are isomorphic as Lie algebras.

14.2   A Crash Course on Spin

In this section we present a telegraphic discussion of spin. Please consult a text on *representation theory* if you need to flesh out this overview.

The isomorphism of $so(3)$ and $su(2)$ might lead one to suspect that the Lie groups $SO(3)$ and $SU(2)$ are isomorphic as Lie groups. Now that is not right, but very nearly. What happens is that there is a group morphism

$$p : SU(2) \to SO(3)$$

that is a C^∞ function, too. The map p is onto, but it is not one-to-one. Rather it is two-to-one, which means that for every $A \in SO(3)$ there are exactly two elements in $SU(2)$ whose image under p is A. In other words, $p^{-1}(A) := \{U \in SU(2) \mid p(U) = A\}$ is a set with two elements. For example, $\ker p = p^{-1}(I_3) = \{-I_2, I_2\}$, the two element abelian group. Here I_3 is the 3×3 identity matrix and I_2 is the 2×2 identity matrix. It turns out that $U \in SU(2)$ implies $-U \in SU(2)$ and $U \neq -U$. Moreover, $p(-U) = p(U)$.

The morphism p is the *universal covering map* of the Lie group $SO(3)$, and $SU(2)$ is its *universal covering space*. These topics are covered in most introductory topology texts. The definition of p as well as the proofs of its properties are given in the next section.

Using the identification $SU(2) \cong S^3$ and the morphism p one can prove that $SO(3) \cong \mathbb{R}P^3$, the 3-dimensional real projective space. Some standard results from topology identify the *fundamental homotopy groups*, denoted by π_1, of these spaces. It turns out that $\pi_1(SU(2)) = \pi_1(S^3) = 0$, that is $SU(2)$ is *simply connected*, and $\pi_1(SO(3)) = \ker p \cong \mathbb{Z}_2$, the abelian, *cyclic* group with two elements. See a topology text for more details.

For a Lie algebra $\mathcal{G}$ whose Lie bracket is $[\cdot, \cdot]$ and an integer $n \geq 1$ we say that a Lie algebra morphism $\rho : \mathcal{G} \to Mat(n \times n; \mathbb{C})$ is a *(Lie algebra) representation* of $\mathcal{G}$ of dimension n. A *Lie algebra morphism* is a linear map that preserves the Lie bracket operation. So this means that ρ is linear and that $\rho([A, B]) = [\rho(A), \rho(B)]$ for all $A, B \in \mathcal{G}$. (Recall Exercise 13.4.7, which defines the Lie bracket on the right side.)

It turns out that the *representation theory as Lie algebra* for the the Lie algebra $su(2)$ is equivalent to the *representation theory as Lie group* for the Lie group $SU(2)$. This is a theorem that holds for all simply connected Lie groups. The representation theory of $su(2)$ is well known. It turns out that for every integer $n \geq 1$ there is a unique *irreducible* representation of $su(2)$ denoted by

$$\rho_n : su(2) \to Mat(n \times n; \mathbb{C}).$$

Here *unique* means up to *isomorphism* (of representations of Lie algebras). Also, *irreducible* means that any subspace $V \subset \mathbb{C}^n$ *invariant* under ρ_n (meaning that $\rho_n(A)(V) \subset V$ for all $A \in su(2)$) must be either $V = 0$ or $V = \mathbb{C}^n$. For physical reasons, we write $n = 2s + 1$, where $s \in \{0, 1/2, 1, 3/2, 2, \dots\}$. This is because s is the *spin* of the irreducible representation ρ_n in units where $\hbar = 1$. Every irreducible representation of $su(2)$ is interpreted physically as corresponding to angular momentum. For $s \in \mathbb{N}$ we saw this in the solution of the eigenvalue problem for the hydrogen atom, in which case we had what is known as *orbital angular momentum* with the notation $l \in \mathbb{N}$ instead of s. So, we are led to the idea that spin is a type of angular momentum, even when $s \in \{1/2, 3/2, 5/2, \dots\}$. We will come back to this idea.

The Lie algebra representation ρ_n always corresponds infinitesimally to a Lie group representation (meaning a C^∞ group morphism)

$$\tau_n : SU(2) \to Gl(n \times n; \mathbb{C}),$$

where $Gl(n \times n; \mathbb{C})$ is the Lie group of all invertible $n \times n$ matrices with complex entries. One can 'factor τ_n through $SO(3)$' if and only if $n = 2s + 1$ with $s \in \mathbb{N}$. To *factor τ_n through $SO(3)$* means that there exists a Lie group representation $\tilde{\tau}_n : SO(3) \to Gl(n \times n; \mathbb{C})$ such that τ_n is equal to the factorization (i.e., composition) $\tau_n = \tilde{\tau}_n \circ p$, namely

$$SU(2) \xrightarrow{p} SO(3) \xrightarrow{\tilde{\tau}_n} Gl(n \times n; \mathbb{C}) \quad \text{for odd integers } n \geq 1.$$

Thus the representation τ_n of $SU(2)$ has an associated representation of $SO(3)$ if and only if n is an odd integer. At the level of Lie algebras we have an isomorphism $su(2) \cong so(3)$, so there is no difference between Lie algebra representations of $su(2)$ and $so(3)$. But at the level of Lie groups there is a difference between the Lie group representations of $SU(2)$ and $SO(3)$. Actually, $Gl(n \times n; \mathbb{C})$ can be replaced in this discussion by $U(n)$, the Lie group of all unitary $n \times n$ matrices with complex entries, thereby making τ_n a *unitary representation* for every integer $n \geq 1$.

A *state with spin s* is a unit vector $\psi \in \mathbb{C}^{2s+1}$. But this ignores any possible spatial dependence. More generally, a *spatially dependent state with spin s* is a unit vector in $L^2(\mathbb{R}^3; \mathbb{C}^{2s+1})$, this being the space of (equivalence classes of) measurable *square integrable* functions $\psi : \mathbb{R}^3 \to \mathbb{C}^{2s+1}$, meaning

$$\int_{\mathbb{R}^3} d^3\mathbf{x} \, \|\psi(\mathbf{x})\|^2 = \int_{\mathbb{R}^3} d^3\mathbf{x} \sum_{j=1}^{2s+1} |\psi_j(\mathbf{x})|^2 < \infty,$$

where $\psi(\mathbf{x}) = (\psi_1(\mathbf{x}), \dots, \psi_{2s+1}(\mathbf{x})) \in \mathbb{C}^{2s+1}$ for all $\mathbf{x} \in \mathbb{R}^3$. The special case $s = 0$, *spin zero*, has the usual Hilbert space $L^2(\mathbb{R}^3; \mathbb{C})$ for one 'particle'.

The Lie algebra $su(2)$ also has an associated representation on the Hilbert space $L^2(\mathbb{R}^3; \mathbb{C}^{2s+1})$ given for each $A \in su(2)$ by

$$\tilde{\rho}_{2s+1}(A)\psi := \rho_{2s+1}(A) \circ \psi : \mathbb{R}^3 \to \mathbb{C}^{2s+1}.$$

Exercise 14.2.1 *The Lie group $SU(2)$ acts on $L^2(\mathbb{R}^3; \mathbb{C}^{2s+1})$ in a way which corresponds to $\tilde{\rho}_{2s+1}$. Find this action. (This is a rather advanced exercise.)*

In atomic, nuclear and particle physics there many observed states with integer spin and with half-integer spin. Spin is also a particle property. For example, the spin of a pion is 0; the spin of an electron is 1/2; the spin of a photon is 1; the spin of a Δ^{++} is 3/2. (All of these in units with $\hbar = 1$.)

For other physical reasons the states with spin $s \in \{0, 1, 2, \dots\} = \mathbb{N}$ are called *bosons* in honor of the physicist S.N. Bose. On the other hand the states with spin $s \in \{1/2, 3/2, 5/2, \dots\}$ are called *fermions* in honor of the physicist E. Fermi. These two types of states have quite different properties as we shall discuss later in Chapter 15.

14.3   The Map p

In this section we present the details of the C^∞ group morphism

$$p : SU(2) \to SO(3).$$

First off, we have to define it. To do this we take an arbitrary self-adjoint matrix $X = X^* \in Herm(2 \times 2; \mathbb{C})$, the *real* vector space of 2×2 *hermitian* (i.e., *self-adjoint*) matrices with entries in $\mathbb{C}$. Then we write X uniquely as

$$X = x_0 I + x_1 \sigma_1 + x_2 \sigma_2 + x_3 \sigma_3 = \begin{pmatrix} x_0 + x_3 & x_1 - ix_2 \\ x_1 + ix_2 & x_0 - x_3 \end{pmatrix}, \quad (14.3.1)$$

where $(x_0, x_1, x_2, x_3) \in \mathbb{R}^4$ and $\sigma_1, \sigma_2, \sigma_3$ are the Pauli matrices (see (14.1.2)). This shows that $Herm(2 \times 2; \mathbb{C}) \cong \mathbb{R}^4$ as real vector spaces. Next, we take $A \in Mat(2 \times 2; \mathbb{C})$ and define $q(A)$ for all $X \in Herm(2 \times 2; \mathbb{C})$ by

$$q(A)(X) := AXA^*,$$

a product of three 2×2 matrices. Then $q(A)(X)$ is linear over $\mathbb{R}$ in the vector X. Also, $q(A)(X)$ is self-adjoint (that is, $q(A)(X) \in Herm(2 \times 2; \mathbb{C})$), since

$$(q(A)(X))^* = (AXA^*)^* = A^{**}X^*A^* = AXA^* = q(A)(X).$$

So, $q: Mat(2 \times 2; \mathbb{C}) \to \mathcal{L}(Herm(2 \times 2; \mathbb{C})) \cong \mathcal{L}(\mathbb{R}^4)$, where for any real Hilbert space $\mathcal{H}$ we define $\mathcal{L}(\mathcal{H}) := \{T : \mathcal{H} \to \mathcal{H} \,|\, T \text{ is real linear and bounded}\}$. We note that $q(A)(X)$ is *not* a linear function in A, although its entries are homogeneous polynomials of degree 1 in the four variables x_0, x_1, x_2, x_3 with coefficients that are *quadratic* expressions in the entries of A and A^*. Consequently, $q(A)$ is a C^∞ function of the entries of A.

Exercise 14.3.1 *This function q is a* unital ring morphism, *meaning:*

- $q(I) = I$ *where each occurrence of I is the appropriate identity.*

- $q(AB) = q(A)q(B)$ *for all $A, B \in Mat(2 \times 2; \mathbb{C})$.*

This map q has another property related to determinants, namely

$$\det\big(q(A)(X)\big) = \det AXA^* = \det A \det X \det A^* = |\det A|^2 \det X. \quad (14.3.2)$$

Exercise 14.3.2 *This exercise is a fun interlude for those who have seen a little of special relativity. Using the matrix representation (14.3.1) of X prove*

$$\det X = x_0^2 - x_1^2 - x_2^2 - x_3^2. \quad (14.3.3)$$

This is a familiar expression from special relativity known as the Minkowski metric *on the* spacetime $\mathbb{R}^4$.

We next define $r = q|_{SU(2)}$, the restriction of q to $SU(2)$. Since $SU(2)$ is a group and r preserves multiplication, the range of r will be contained in the group $Gl(4;\mathbb{R})$ of all invertible maps acting on $Herm(2 \times 2;\mathbb{C}) \cong \mathbb{R}^4$. Now the three Pauli matrices $\sigma_1, \sigma_2, \sigma_3$ span a 3-dimensional real subspace V of the real vector space $Herm(2 \times 2;\mathbb{C})$.

Exercise 14.3.3 *V consists exactly of the matrices $A \in Herm(2 \times 2;\mathbb{C})$ with trace zero, that is $Tr\, A = 0$. Therefore, $V = su(2)$.*
For convenience we will use the shorter notation V in the following.

For each Pauli matrix σ_j we see that

$$Tr(r(U)(\sigma_j)) = Tr(U\sigma_j U^*) = Tr(U^*U\sigma_j) = Tr(\sigma_j) = 0$$

holds for all $U \in SU(2)$, where $Tr(M)$ is the trace of the matrix M. Hence, $r(U)(\sigma_j) \in V$ and then by linearity it follows that $r(U)V \subset V$ for every $U \in SU(2)$. So we define $p(U) := r(U)|_V$, the restriction of $r(U)$ to V. Therefore, $p(U) \in Gl(V) \cong Gl(\mathbb{R}^3)$, the group of invertible, real linear maps of $\mathbb{R}^3$ to itself. Here, we are identifying V with $\mathbb{R}^3$ by using the basis $\sigma_1, \sigma_2, \sigma_3$ of V. We take the inner product of V that makes $\sigma_1, \sigma_2, \sigma_3$ an orthonormal basis; this then corresponds to the standard orthonormal basis on $\mathbb{R}^3$.

Exercise 14.3.4 *Show that this inner product on V is given for $A, B \in V$ by*

$$\langle A, B \rangle = \frac{1}{2}Tr(AB).$$

Using this formula, prove that

$$\langle p(U)A, p(U)B \rangle = \langle A, B \rangle$$

for all $A, B \in V$ and $U \in SU(2)$.
In other words, the linear map $p(U)$ is an orthogonal map on $V \cong \mathbb{R}^3$, that is $p(U) \in O(3)$.

So at this point in the argument we have that

$$p : SU(2) \to O(3)$$

is a C^∞ group morphism. It is time for a little topology. (The rest of this paragraph might be too much for some of you. Not to worry. Just skip ahead.) We use the homeomorphism of the topological space $SU(2)$ with the unit sphere $S^3 \subset \mathbb{R}^4$ (see Exercise 14.1.3) in order to see that $SU(2)$ is a *connected* topological space. But the image of a connected space under a continuous function is again connected. Now the composition

$$SU(2) \xrightarrow{p} O(3) \xrightarrow{\det} \{-1, +1\}$$

of two continuous functions is itself continuous. So its range is connected. But $\{-1, +1\}$ is not a connected space. Since $(\det \circ p)(I_2) = \det I_3 = 1$ and

$I_2 \in SU(2)$, it follows that $\det \circ p$ has its range in the singleton set $\{+1\}$, which means that p has its range in $SO(3)$. Expressing this as a diagram, we have

$$p : SU(2) \to SO(3).$$

Beware that we have not yet proved that p is *onto* $SO(3)$. An explicit way of seeing that is to compute the 'images' of various one-parameter subgroups of $SU(2)$ under the map p. For example consider this one-parameter subgroup of $SU(2)$:

$$\mathbb{R} \ni \theta \mapsto U_\theta := e^{-i\theta\sigma_3} = \begin{pmatrix} e^{-i\theta} & 0 \\ 0 & e^{i\theta} \end{pmatrix}.$$

Exercise 14.3.5 *Write $X \in V$ in the form (14.3.1) with $x_0 = 0$. Prove*

$$p(U_\theta)(X) = \begin{pmatrix} x_3 & e^{-2i\theta}(x_1 - ix_2) \\ e^{2i\theta}(x_1 + ix_2) & -x_3 \end{pmatrix}.$$

Show that this is the rotation in $V \cong \mathbb{R}^3$ with coordinates x_1, x_2, x_3 given by

$$R_3(2\theta) = \begin{pmatrix} \cos(2\theta) & -\sin(2\theta) & 0 \\ \sin(2\theta) & \cos(2\theta) & 0 \\ 0 & 0 & 1 \end{pmatrix} \in SO(3),$$

the rotation around the x_3-axis by the angle 2θ.

Note that U_θ is a periodic function in θ with (smallest, positive) period 2π, while $p(U_\theta)$ is a periodic function in θ with (smallest, positive) period π. So when θ goes through one period of U_θ and thereby goes around a circle once, its image $p(U_\theta)$ goes through two periods and thereby goes around a circle twice. This is a reflection of the fact that p is two-to-one. Similarly, the one-parameter subgroups $e^{-i\theta\sigma_1}$ and $e^{-i\theta\sigma_2}$ of $SU(2)$ have images under p that are one-parameter subgroups of $SO(3)$ of rotations around the x_1-axis and the x_2-axis, respectively.

Exercise 14.3.6 *Prove the previous sentence and then understand why these three one-parameter subgroups generate $SO(3)$, thereby proving that p is onto.*

Alternatively, compute the one-parameter subgroup $p(e^{-i\theta\, u\cdot\sigma})$ in $SO(3)$, where $u = (u_1, u_2, u_3)$ is a unit vector in $\mathbb{R}^3$ and $\sigma = (\sigma_1, \sigma_2, \sigma_3)$, the 'vector' of the three Pauli matrices. Also, we put $u \cdot \sigma := u_1\sigma_1 + u_2\sigma_2 + u_3\sigma_3$. Then use this result to prove that p is onto.

The last remaining property to prove about p is that it is two-to-one.

Exercise 14.3.7 $\ker p = \{-I, I\}$ *implies that p is two-to-one. (Recall that $-I \in SU(2)$ so this statement does make sense.)*

So we have to show that $\ker p = \{-I, I\}$. It is clear that $\{-I, I\} \subset \ker p$. So take an arbitrary element $U \in \ker p$, that is, $p(U)X = UXU^* = X$ for all

$X \in V$. We have to show that $U = \pm I$. Well, by taking $X = \sigma_3$ we have that $U\sigma_3 U^* = \sigma_3$ must hold. We write U in the form

$$U = \begin{pmatrix} \alpha & \beta \\ -\beta^* & \alpha^* \end{pmatrix},$$

where $\alpha, \beta \in \mathbb{C}$ and $|\alpha|^2 + |\beta|^2 = 1$. So we calculate

$$\begin{aligned} U\sigma_3 U^* &= \begin{pmatrix} \alpha & \beta \\ -\beta^* & \alpha^* \end{pmatrix} \begin{pmatrix} 1 & 0 \\ 0 & -1 \end{pmatrix} \begin{pmatrix} \alpha^* & -\beta \\ \beta^* & \alpha \end{pmatrix} \\ &= \begin{pmatrix} \alpha & -\beta \\ -\beta^* & -\alpha^* \end{pmatrix} \begin{pmatrix} \alpha^* & -\beta \\ \beta^* & \alpha \end{pmatrix} \\ &= \begin{pmatrix} |\alpha|^2 - |\beta|^2 & -2\alpha\beta \\ -2\alpha^*\beta^* & -|\alpha|^2 + |\beta|^2 \end{pmatrix}. \end{aligned}$$

This matrix is equal to σ_3 if and only if $|\alpha|^2 - |\beta|^2 = 1$ and $\alpha\beta = 0$. Now the second equality implies that either $\alpha = 0$ or $\beta = 0$. But $\alpha = 0$ together with the first equality implies $-|\beta|^2 = 1$, which has no solution. So we must have $\beta = 0$ and $|\alpha|^2 = 1$. This means that $U \in \ker p$ must have the form

$$U = \begin{pmatrix} \alpha & 0 \\ 0 & \alpha^* \end{pmatrix},$$

where $\alpha \in \mathbb{C}$ and $|\alpha|^2 = 1$. To learn more about α we use that $U \in \ker p$ implies the equation $U\sigma_1 U^* = \sigma_1$. So we calculate

$$\begin{aligned} U\sigma_1 U^* &= \begin{pmatrix} \alpha & 0 \\ 0 & \alpha^* \end{pmatrix} \begin{pmatrix} 0 & 1 \\ 1 & 0 \end{pmatrix} \begin{pmatrix} \alpha^* & 0 \\ 0 & \alpha \end{pmatrix} \\ &= \begin{pmatrix} 0 & \alpha \\ \alpha^* & 0 \end{pmatrix} \begin{pmatrix} \alpha^* & 0 \\ 0 & \alpha \end{pmatrix} \\ &= \begin{pmatrix} 0 & \alpha^2 \\ (\alpha^*)^2 & 0 \end{pmatrix}. \end{aligned}$$

So this matrix is equal to σ_1 if and only if $\alpha^2 = 1$, which has the two solutions $\alpha = \pm 1$. And this implies $U = \pm I$, which concludes the proof that $\ker p = \{-I, I\}$. ■

14.4   The Representations ρ_s

In this section we define the irreducible representations

$$\rho_s : SU(2) \to Gl(V_s),$$

where the dimension of V_s is $2s + 1$ for each $s \in \{0, 1/2, 1, 3/2, 2, \dots\}$. To achieve this we define V_s to be the complex vector space of all homogeneous polynomials of degree $2s$ in the two indeterminants x and y, namely

$$V_s := \operatorname{span}_{\mathbb{C}} \{x^j y^{2s-j} \mid j = 0, 1, \ldots, 2s\} = \Big\{ \sum_{j=0}^{2s} a_j \, x^j y^{2s-j} \mid a_j \in \mathbb{C} \Big\}.$$

For our choice of s we have that $2s \geq 0$ is an integer and $\dim_{\mathbb{C}} V_s = 2s + 1 \geq 1$. Conversely, any integer $n \geq 1$ can be written as $n = 2s + 1$ for a unique $s \in (1/2)\,\mathbb{N}$.

Let $f = \sum_{j=0}^{2s} c_j x^j y^{2s-j} \in V_s$ be given, where each $c_j \in \mathbb{C}$. This gives a *polynomial function* $f : \mathbb{C}^2 \to \mathbb{C}$. Next for any $U \in SU(2)$ consider the composition

$$\mathbb{C}^2 \xrightarrow{U^{-1}} \mathbb{C}^2 \xrightarrow{f} \mathbb{C}.$$

Then we define $\rho_s(U)(f)$ by $\rho_s(U)(f) := f \circ U^{-1}$. The reason for using U^{-1} here instead of U is given in the next exercise.

(Parenthetically, let me note for the *cognescetti* that $\rho_s(U)(f)$ in category theory is called the *pull-back* of f by U^{-1}. The standard categorical notation is $\rho_s(U)(f) = f \circ U^{-1} = (U^{-1})^* f$, which clashes fatally with our notation $*$ for the adjoint. However, this viewpoint shows how the *contravariant functor*, pull-back, is compensated by using U^{-1} instead of U.)

Exercise 14.4.1 *Establish the following basic properties:*

- *Prove that $f \in V_s$ implies that $\rho_s(U)(f) \in V_s$, that is, $\rho_s(U)$ maps each homogeneous polynomial of degree $2s$ to a homogeneous polynomial of degree $2s$.*

- *Prove that $\rho_s(U) : V_s \to V_s$ is linear.*

- *Prove that $\rho_s : SU(2) \to \mathcal{L}(V_s)$ is a representation, that is, $\rho_s(I) = I$ and $\rho_s(U_1 U_2) = \rho_s(U_1)\rho_s(U_2)$.*
  Remark: *The very last identity would not have worked out if we had chosen U rather than U^{-1} in the definition of $\rho_s(U)$.*
  Prove that $\rho_s : SU(2) \to Gl(V_s)$, the group of invertible linear maps $V_s \to V_s$ and that ρ_s is a group morphism.

It remains to prove that ρ_s is irreducible. This result is somewhat more complicated. The proof can be found in many basic texts on representation theory of Lie groups. (One of my favorites is [22].) Or the reader can take it to be a very challenging exercise. Moreover, it is also true that every irreducible representation of $su(2)$ "is equal to" ρ_s for some half-integer s. Of course, the expression in quotes really means: *isomorphic as a representation*, which is a something that must be defined. This is also quite difficult, but very important since it says we have obtained a complete classification of all possible spin states. Here is a more accessible exercise.

Exercise 14.4.2 *Prove that $\rho_s(-I) = I$ if and only if s is an integer. Prove that $\rho_s(-I) = -I$ if and only if $s \in \{1/2, 3/2, 5/2, \ldots\}$, in which case we say* *s is a* half-integer. *Hence, ρ_s factors through $p : SU(2) \to SO(3)$, and thus gives an irreducible representation of $SO(3)$, if and only if s is an integer.*

14.5 ρ_s and Angular Momentum

While spin can be given a classical counterpart, it truly is a child of quantum theory. The fact that the physically motivated $SO(3)$ symmetries of $\mathbb{R}^3$ do not suffice for physics is rather amazing. The mathematical facts are that $SO(3)$ has $SU(2)$ as its universal covering space and that $SU(2)$ has representations that do not come from $SO(3)$. This could have all been a very pretty mathematical theory with no relation to physics. But the basic building blocks of everyday matter are electrons, protons, and neutrons, all of which have spin 1/2. And the stability of that matter depends on the Fermi-Dirac statistics of such states. Dark matter could be a totally new ball game. We just don't know. But to understand the fraction of the universe that interacts strongly with us, we have to include representations of $SU(2)$ that do not come from $SO(3)$. So all the heavy, abstract mathematics has a big physics payoff.

What does the representation ρ_s have to do with angular momentum? First, for the case when $s \geq 0$ is an integer, ρ_s is equal (up to isomorphism) to the irreducible representation of the real Lie algebra spanned by the three operators $(i\hbar)^{-1}L_x$, $(i\hbar)^{-1}L_y$, $(i\hbar)^{-1}L_z$ (the angular momentum operators up to a multiplicative constant) restricted to the complex vector space spanned by the spherical harmonics $\{Y_m^s \mid -s \leq m \leq s\}$ of odd dimension $2s + 1$, as introduced in the spherically symmetric 2-body problem which includes the important special case of the hydrogen atom. Since the mathematics of ρ_s for integer $s \geq 0$ is identical (that is, isomorphic) with the mathematics of angular momentum, it seems reasonable to suppose that the physics is identical as well in this case. Second, for the case when s in not an integer, we assume that this too is a type of angular momentum, called *spin* in this case. To see that this is a reasonable assumption, we define $S^2 := S_1^2 + S_2^2 + S_3^2$ in analogy with L^2 for orbital angular momentum.

Exercise 14.5.1 *Prove that $[S^2, S_j] = 0$ for $j = 1, 2, 3$.*

Now we want to see how S^2 looks like in the representation ρ_s. Up to this point $\rho_s(S^2)$ has not been defined. We remedy this as follows:

$$\rho_s(S^2) := \big(\rho_s(S_1)\big)^2 + \big(\rho_s(S_2)\big)^2 + \big(\rho_s(S_3)\big)^2.$$

Then it can be proved that $\rho_s(S^2) = s(s + 1)I$, where I here is the identity operator acting on V_s, though the proof might be beyond you. This formula should remind you of how L^2 acts on the spherical harmonics. In fact, all of the formulas for the angular momentum operators in Chapter 12 have exact analogues for all allowed values of s, if we put $\hbar = 1$.

But what about the physically important case $s = 1/2$? It turns out that a system with two spin 1/2 *distinguishable* particles (such as an electron and a proton, but not two neutrons) has the representation $\rho_{1/2} \otimes \rho_{1/2}$ of $SU(2)$, where $\otimes$ denotes the tensor product of representations. And $\rho_{1/2} \otimes \rho_{1/2}$ acts on $\mathbb{C}^2 \otimes \mathbb{C}^2 \cong \mathbb{C}^4$, where now $\otimes$ denotes the tensor product of vector spaces. (So this is unintelligible for those without some inkling of what tensor products are about. Anyway, this physically important situation motivates the study of tensor products.) But something a bit untoward happens. Even though $\rho_{1/2}$ is irreducible, the tensor product $\rho_{1/2} \otimes \rho_{1/2}$ is *not* irreducible. In fact, there is a direct sum decomposition $\mathbb{C}^4 = \mathbb{C}^1 \oplus \mathbb{C}^3$ such that

$$\rho_{1/2} \otimes \rho_{1/2} = \rho_0 \oplus \rho_1,$$

where ρ_0 with $s = 0$ acts on the first summand $\mathbb{C}^1 \cong \mathbb{C}$ of dimension $2s+1 = 1$ while ρ_1 with $s = 1$ acts on the second summand $\mathbb{C}^3$ of dimension $2s+1 = 3$. We say that we have a *singlet state* with spin 0 and a *triplet of states* with spin 1. The idea is that since we get angular momentum (according to the previous paragraph) by combining two spin 1/2 states, then it is plausible to think that spin 1/2 itself is a type of angular momentum. So a mathematical formalism leads to a physical interpretation, which then in practice works quite well.

14.6   Magnetism

How do we see spin in experiments? One technique is to measure the angular correlations among the tracks of products of radioactive decays, a fascinating story in its own right. But another way is via interactions with magnetic fields. It turns out that some spin 1/2 particles not only have electric charge, but also a *magnetic moment*. This can be thought of as a miniature compass needle that tends to align with any ambient magnetic field. Even the neutron with zero electric charge has a non-zero magnetic moment. And any nucleus with non-zero spin has a non-zero magnetic moment. One example of this is the most common nucleus of a hydrogen atom: a proton.

The way this works in classical physics is that the magnetic moment is a vector $\vec{\mu} = (\mu_1, \mu_2, \mu_3)$ that interacts with a magnetic field $\vec{B} = (B_1, B_2, B_3)$ to give an interaction *magnetic energy* $E = -\vec{B} \cdot \vec{\mu} = -(B_1\mu_1 + B_2\mu_2 + B_3\mu_3)$. (The negative sign is a convention made so that for a given magnetic field the energy is minimized when the magnetic moment points in the same direction as the magnetic field.) The non-trivial step is to quantize this. The way to do this for an electron is to replace $\vec{\mu}$ with $\hat{\mu}$ defined by

$$\hat{\mu} := g \frac{q_e \hbar}{2m_e} \hat{S},$$

where $q_e < 0$ is the electric charge of an electron, m_e is the mass of an electron, g is a dimension-less, real quantity known as the *Landé g-factor* of an electron and

$$\hat{S} := (S_1, S_2, S_3)$$

is the vector whose entries are the three 2×2 spin matrices. (The expression $\mu_B := -q_e\hbar/2m_e$ is known as the *Bohr magneton*. So, $\hat{\mu} = -g\,\mu_B\,\hat{S}$.) Then the quantum Hamiltonian is defined by

$$H := -\overrightarrow{B} \cdot \hat{\mu} = -g\frac{q_e\hbar}{2m_e}(B_1 S_1 + B_2 S_2 + B_3 S_3),$$

which is a 2×2 self-adjoint matrix since the components of $\overrightarrow{B}$ are real. All of this is valid for a magnetic field that varies in space and in time. If all this seems very implausible to you as it originally did to me, just remember that this is the sort of imaginative theorizing that led to a Nobel prize. Because it all works out!

A simple case is a time independent, constant (in both magnitude and direction) magnetic field, say $\overrightarrow{B} = (0, 0, B)$. Then the dynamics is obtained from the unitary group

$$e^{-itH/\hbar} = e^{itgq_e BS_3/2m_e} = \begin{pmatrix} e^{itgq_e B/4m_e} & 0 \\ 0 & e^{-itgq_e B/4m_e} \end{pmatrix}.$$

(Recall that $S_3 = \sigma_3/2$.) The time evolution of the state

$$\psi_3^\uparrow = \begin{pmatrix} 1 \\ 0 \end{pmatrix}$$

is $e^{-itH/\hbar}\psi_3^\uparrow = e^{itgq_e B/4m_e}\psi_3^\uparrow$, which gives just a trivial phase factor and so is the same state $\psi_3^\uparrow$. But the time evolution of the state

$$\psi_1^\uparrow = 2^{-1/2} \begin{pmatrix} 1 \\ 1 \end{pmatrix},$$

an eigenvector of S_1 with eigenvalue $1/2$, is not trivial, but rather is

$$\psi_t := e^{-itH/\hbar}\psi_1^\uparrow = 2^{-1/2} \begin{pmatrix} e^{itgq_e B/4m_e} \\ e^{-itgq_e B/4m_e} \end{pmatrix} = 2^{-1/2} \begin{pmatrix} e^{i\omega_B t} \\ e^{-i\omega_B t} \end{pmatrix},$$

where $\omega_B := gq_e B/4m_e$. So a measurement of S_1 in this time dependent state ψ_t yields $+1/2$ with probability

$$\begin{aligned} P(S_1 = 1/2 \mid \psi_t) &= \langle \psi_t, P_{S_1}(\{1/2\})\psi_t \rangle \\ &= |\langle \psi_1^\uparrow, \psi_t \rangle|^2 \\ &= |2^{-1}(e^{i\omega_B t} + e^{-i\omega_B t})|^2 \\ &= \cos^2(\omega_B t) \\ &= \frac{1}{2}(1 + \cos(2\omega_B t)), \end{aligned}$$

a time dependent probability which oscillates between 0 and 1 with angular frequency $2\omega_B$. The first two equalities in this calculation are fully justified in Chapter 16. (Or the reader may be more quickly pleased by the remark that $\langle \psi_1^\uparrow, \psi_t \rangle$ is the probability amplitude for finding the state ψ_t in the state ψ_1, which is the eigenvector for S_1 with eigenvalue $1/2$, and thus $|\langle \psi_1^\uparrow, \psi_t \rangle|^2$ is the corresponding probability. This way the first step in this calculation can be skipped.) Clearly,

$$P(S_1 = -1/2 \,|\, \psi_t) = 1 - P(S_1 = 1/2 \,|\, \psi_t) = \sin^2(\omega_B t) = \frac{1}{2}(1 - \cos(2\omega_B t)),$$

which also oscillates between 0 and 1 with angular frequency $2\,\omega_B$. The other eigenvector of S_1 as well as the eigenvectors of S_2 have the same oscillatory time dependence with angular frequency $2\omega_B$.

The Landé g-factor of an electron is bit larger than 2, closer to 2.002319, while the muon g-factor is around 2.002332 and is definitely not equal to that of an electron. (A *muon* is a particle that is rather like an electron though with much larger mass.) The analysis we have done can readily be adapted to other particles with a magnetic moment, provided that the appropriate Landé g-factor, mass and charge are used.

This mathematics of spin is what is behind *nuclear magnetic resonance (NMR)* and its technological offshoot, *magnetic resonance imaging (MRI)*, which has many applications in medicine among other sciences. It is also the theory that accounts for the *Zeeman effect* and the *Stern-Gerlach experiment*. Quantum computation based on spin systems is also understood in terms of this mathematical language, which is found in the very nice text [21].

14.7   Notes

H. Weyl published the book *Gruppentheorie und Quantenmechanik* [35] in 1928. This remarkable achievement either went unnoticed or was disparaged as 'Gruppenpest', a plague of groups. E. Noether's famous theorem relating symmetries in classical physics to conservation laws appeared in 1918, well before modern quantum theory. It seems hard to believe that so little interest and so much antagonism was paid to these ideas for so long. Of course, W. Pauli was not shy about introducing his eponymous matrices in order to explain spin.

But a most amazing upshot of this emphasis on using Lie groups in physics is that it provides a new mathematical structure for describing what are otherwise ambiguously defined, basic constituent states of matter. These basic states are identified with the unit vectors in a Hilbert space on which an irreducible unitary representation of a Lie group acts. The mathematics of unitary representations acting on a Hilbert space is far beyond the scope of this introductory book, but it must be remarked that it is a central aspect of quantum theory. This seminal breakthrough is due to E.P. Wigner in [37] for a particular Lie group, the Poincaré group of special relativity theory.

For an up-to-date reference on the relation of groups to quantum theory I highly recommend the recent, voluminous tome [38] by P. Woit. Also, the presentation in [22] is quite nice, but it only takes one up to the early 70's. A definite, encyclopedic text of the pure mathematics of Lie theory is the classic [31].

Folk wisdom says that one should not do quantum theory without having a quantum Hamiltonian in hand. At my own peril I have not paid much heed to that advice in this book. Besides the Hamiltonians for the one-dimensional harmonic oscillator and the hydrogen atom (plus some simple Schrödinger operators in the exercises), we have now added that for a particle with magnetic moment in a magnetic field. And there will be no others. As I remarked earlier, a lot of understanding comes from a hand-on approach to solving or approximating the Schrödinger equation in many concrete examples. It is how we understand crystals, molecules, semi-conductors, lasers, superconductivity, and the Periodic Table. An excellent reference for some of this is [3] on solid state physics. I encourage my readers to pursue the paths of their interests beyond the limitations of this book. It is not an exaggeration to say that the variety of applications of quantum theory is due to the variety of quantum Hamiltonians.

Chapter 15

Bosons and Fermions

Yin and Yang.
Ancient Chinese concepts

The rest of this book is dedicated to the quantum theory of a one-body (or one-particle) system. Even the hydrogen atom, which is called the quantum two-body problem, reduces to two one-body problems: one for the center of mass of the two bodies and one for an electron (with a slightly adjusted mass) interacting with the external potential energy generated by the electric charge of a proton located at a fixed position. However, we must make some mention in this short chapter of how multi-body problems are handled.

15.1 Multi-particle Statistics

The general theory of multi-particle systems depends on the types of the particles involved. It turns out that there are two basic types of particles: bosons and fermions. We have already seen this dichotomy when we discussed spin. Recall that in units with $\hbar = 1$ bosons have spin that is an integer $n \geq 0$, while a fermion has half-integer spin $n + 1/2$, where $n \geq 0$ is an integer. But the important difference for multi-particle systems is that each of these two particle types has its own *statistics*. This is an unfortunate, and unchangeable, choice of terminology.

The bosons satisfy *Bose-Einstein statistics*. This starts with a Hilbert space $\mathcal{H}$ that describes the one-body system of a boson. Then to describe the n-body system with $n \geq 2$ identical bosons one forms states $\psi_1 \circ \cdots \circ \psi_n$ where each $\psi_j \in \mathcal{H}$. The 'tricky bit' is to understand what the product $\circ$

S. B. Sontz, *An Introductory Path to Quantum Theory*,
https://doi.org/10.1007/978-3-030-40767-4_15

is and then how these states combine to form an associated Hilbert space, denoted as $\mathcal{H} \circ \cdots \circ \mathcal{H}$ with n factors. Sweeping these important technical details to a side, let's simply note that the *symmetric product* $\circ$ satisfies

$$\psi_1 \circ \cdots \circ \psi_i \circ \cdots \circ \psi_j \circ \cdots \circ \psi_n = \psi_1 \circ \cdots \circ \psi_j \circ \cdots \circ \psi_i \circ \cdots \circ \psi_n$$

for all pairs of integers i, j for which $1 \leq i < j \leq n$. We describe this by saying that the state does not change under the interchange of any pair of identical bosons. This interchange property is the essential ingredient of Bose-Einstein statistics.

Fermions satisfy *Fermi-Dirac statistics*. This also starts with a Hilbert space $\mathcal{H}$ that describes the one-body system of a fermion. Then to describe the n-body system with $n \geq 2$ identical fermions one forms states $\psi_1 \wedge \cdots \wedge \psi_n$ where each $\psi_j \in \mathcal{H}$. As before the 'tricky bit' is to understand what this new product $\wedge$ is and then how these states combine to form an associated Hilbert space, denoted as $\mathcal{H} \wedge \cdots \wedge \mathcal{H}$ with n factors. Continuing to sweep important technical details to a side, we note that the *wedge product* $\wedge$ satisfies

$$\psi_1 \wedge \cdots \wedge \psi_i \wedge \cdots \wedge \psi_j \wedge \cdots \wedge \psi_n = -\psi_1 \wedge \cdots \wedge \psi_j \wedge \cdots \wedge \psi_i \wedge \cdots \wedge \psi_n$$

for all pairs of integers i, j for which $1 \leq i < j \leq n$. We describe this by saying that the state changes sign under the interchange of any pair of identical fermions. This interchange property is the essential ingredient of Fermi-Dirac statistics. You may have already seen the wedge product in the study of *differential forms* in an advanced calculus course or in a differential geometry course.

Of course, at a mathematical level inner products must be defined for these new Hilbert spaces, and they must be proved to be complete. The technical details require a clear understanding of how to construct the *tensor product* of Hilbert spaces even though neither of these new Hilbert spaces is a tensor product. We leave those details to other texts.

More importantly, at a physical level one must grapple with the idea that identical 'particles' are really and truly (that is, fundamentally) identical. Such a concept is not found in classical physics. For example, a dust particle in a dust storm is distinguishable from all the other nearby dust particles by its shape, mass, velocity, angular momentum, and (if none of the previous apply) its position. But electrons in a cloud, just as one example, are *not* so distinguishable. As a concrete case, the electrons in a solid crystal form such a cloud of identical fermions. With all due respect to those ancient Greeks who advocated an atomic theory of matter, I do not think they were talking about bosons and fermions.

There is also a curious duality going on here. Each of the fundamental particles of the *standard model* of particle physics is either a boson or a fermion. Of course. But something else is also true. It turns out that all of the building blocks of ordinary matter are fermions, and all the mediators of interactions are bosons. So the two standard modes of scientific explanation, the material cause, and the effective cause of Bacon are described by two very distinct, mathematically precise Hilbert spaces, each with its own statistics. I am led to think that somehow this did not have to be so. Certainly, Bacon could not have had this in mind when he advocated in favor of his new logic. And these are complementary aspects of physical reality. After all, matter without interactions is undetectable. And interactions without matter is almost inconceivable. This neat duality may not hold for the so-called *dark matter*, about which we know almost nothing. For example, we do not know whether quantum theory applies to dark matter. We do not know how that will play out.

15.2 Notes

The words *boson* and *fermion* were coined by Paul Dirac. The classification of all matter and interactions in terms of just these two concepts is one of the greatest discoveries ever in physics, although dark matter might not be so classified. Three of the physicists whose names appear in the expressions "Bose-Einstein statistics" and "Fermi-Dirac statistics" received the Nobel prize. The one who did not, although he was nominated several times, was S.N. Bose, who was also the only one not born in Europe.

To understand why spin and statistics are so intimately linked is beyond the scope of this book. It is a consequence of *quantum field theory*, where the result is known as the *Spin-Statistics theorem*.

Another aspect unique to the quantum theory of multi-particle systems is *entanglement*. There is nothing like this in classical theory. But this is one among many other advanced topics in quantum theory which we will not discuss, except to say that its experimental success is another reason why quantum theory has replaced classical theory. The curious reader is invited to consult the extensive literature on this.

Chapter 16

Classical and Quantum Probability

> Curiouser and curiouser.
>
> Lewis Carroll

This chapter requires much more mathematical sophistication. It is crucial, however, for the understanding of the profound difference between classical probability and quantum probability. In my opinion it is quantum probability and nothing else that makes quantum theory so strange and difficult to grasp. Whatever else it is, quantum probability is *not* classical probability. Hence, Einstein's oft-quoted critique of quantum theory, namely that "the Old One does not play dice", is quite true, though not in the way he meant, since playing dice is described purely by classical probability. So I start this chapter with a review of the basics of classical probability in order to contrast it with quantum probability. As I said in the Preface, this is the most important chapter in this book. Undoubtedly, it is the most difficult.

16.1 Classical Kolmogorov Probability

The perfectly respectable scientific theory of classical probability was born in 1654 in an exchange of letters between P. Fermat and B. Pascal about games of chance, especially dice. In 1657 C. Huygens published the first book on the topic, where he explained expected value. It took almost 280 years to arrive at the modern theory of *classical probability* as axiomatized by Kolmogorov in 1933. (See [18].) At a mathematical level, this is a special topic within *measure theory*. However, the actual details of this specific topic give it a sweet, wonderful flavor that is quite distinct from the unpalatable sawdust of pure measure theory.

© Springer Nature Switzerland AG 2020   169
S. B. Sontz, *An Introductory Path to Quantum Theory*,
https://doi.org/10.1007/978-3-030-40767-4_16

Here is a crash course on the Kolmogorov axiomatization. We start with the first case where we have a finite set Ω whose elements are interpreted to be the elementary results of seemingly identical experiments. The whole point of probability theory is that identical situations can produce a variety of different results, or as they are commonly called: *outcomes*. Why this is so, and even how this might be so, is not the business of probability theory, but rather its starting point. So we suppose that the set of outcomes contains at least $n \geq 2$ elements, say $\Omega = \{\omega_1, \ldots, \omega_n\}$. We further suppose that each of the possible outcomes ω_k has an experimentally measurable *frequency* f_k of occurring, that is, in N measurements of the identical experiments which give n_k occurrences of the outcome ω_k the ratio n_k/N approaches the number f_k as N becomes large. Clearly, $0 \leq f_k \leq 1$ and $f_1 + \cdots + f_n = 1$ hold. That's on the experimental side.

On the theoretical side we suppose that there are real numbers p_k, for $1 \leq k \leq n$, associated with each possible outcome ω_k. We want these numbers p_k to correspond to the experimental quantities f_k. So we require that these numbers satisfy $0 \leq p_k \leq 1$ and $p_1 + \cdots + p_n = 1$. We include the limiting cases $p_k = 0$ and $p_k = 1$ for theoretical completeness. If ω_k never occurs (and so should not have been included in Ω, but was anyway), then we put $p_k = 0$. If ω_k always occurs (and so none of the other outcomes in Ω ever occurs), then we put $p_k = 1$. We also think of every subset $E \subset \Omega$ as an *event* and we define the *probability of the event E* to be

$$P(E) := \sum_{k \,:\, \omega_k \in E} p_k. \tag{16.1.1}$$

So we have the *probability measure* $P : \mathrm{Pow}(\Omega) \to [0,1]$, where $\mathrm{Pow}(\Omega)$ is the set whose elements are all of the subsets of Ω. The probability measure P satisfies $P(\emptyset) = 0$, where $\emptyset$ denotes the *empty set* (since by convention the sum over an empty index set is 0), and it satisfies

$$P(E_1 \cup E_2) = P(E_1) + P(E_2) \tag{16.1.2}$$

for any pair of events E_1, E_2 such that $E_1 \cap E_2 = \emptyset$. (In such a case one says that the events E_1 and E_2 are *disjoint*.)

Exercise 16.1.1 *Prove that $0 \leq P(E) \leq 1$, $P(\Omega) = 1$ and that* (16.1.2) *holds.*

The classical example of this case is the tossing of two *dice*. Each die (singular of 'dice') is a cube with each of its six sides labeled with a number of dots that goes from 1 to 6 dots. The 'experiment' is to toss the dice onto a horizontal plane surface and wait until they come to rest. The pair of numbers that are on the top side of the dice is the outcome. This means we can put $\Omega = S \times S$ where $S = \{1, 2, 3, 4, 5, 6\}$, giving $6^2 = 36$ possible outcomes. We can now assign probabilities p_k to these 36 outcomes in many, many ways. However, if we think that the dice are fair and neither affects the other, then we put $p_k = 1/36$ for each outcome. We can check whether actual physical dice conform to this probabilistic model by tossing them many times to see whether the measured frequency of each outcome is reasonably close to the theoretical value of $1/36$. The science for making this sort of assessment is *statistics* about which I will have nothing further to say. Of course, the manner in which the dice are tossed may influence the validity of this model. I was once told by a rather street smart person to never trust someone who just tosses dice down onto the sidewalk, but that I should insist that they bounce the dice off a wall! Behind this advice is the idea that the motion of the dice is described by a deterministic, classical mechanics model which must include friction forces. Since the initial conditions of the dice at the moment of release are not usually controlled, a probabilistic model is used in such circumstances. But with practice one might have enough control over the initial conditions to get any desired outcome if the dice are tossed onto a horizontal uniform surface. However, when tossed against a wall, one can not expect to control the angle of impact against it, and so a probabilistic model is adequate. I suppose that this sort of classical probability model is what Einstein was referring to in his famous quote.

The second case is to allow the set of outcomes Ω to be an infinite, discrete set, which is almost the same notation: $\Omega = \{\omega_1, \ldots, \omega_n, \ldots\}$. Again, we suppose that there is a real number $p_k \in [0, 1]$ associated with each possible outcome ω_k and that $\sum_{k=1}^{\infty} p_k = 1$. Again, we define the *probability measure* $P : \mathrm{Pow}(\Omega) \to [0, 1]$ by (16.1.1), but now the index set of the summation can be infinite. Then we have $P(\emptyset) = 0$ and

$$P(\cup_{j \subset J} F_j) = \sum_{j \in J} P(E_j). \tag{16.1.3}$$

provided that $E_k \cap E_l = \emptyset$ whenever $k \neq l$. And now the index set J can be finite or countably infinite.

Exercise 16.1.2 *Prove for this second case that* $0 \leq P(E) \leq 1$, $P(\Omega) = 1$, $P(\emptyset) = 0$ *and* (16.1.3).

One famous example of this second case is the *Poisson distribution*. In this case $\Omega = \mathbb{N} = \{0, 1, 2, \ldots\}$, the non-negative integers. The experiment typically consists in counting the number of 'random' events that occur in a given time period, such as the number of cosmic rays that arrive at a fixed detector per hour or the number of beta decays per hour of a radioactive substance. Suppose that the average value of such a count is the real number $m > 0$. (Recall that the average value of an integer valued quantity can be any positive real number as, for example, the average number of children per family can be 2.4.) For the Poisson distribution with parameter $m > 0$, the theoretic probability model is

$$p_k := e^{-m} \frac{m^k}{k!} \tag{16.1.4}$$

for each outcome $k \in \Omega = \mathbb{N}$. Justification for this formula, given certain hypotheses, can be found in elementary texts on probability theory.

Exercise 16.1.3 *For the Poisson distribution as given in* (16.1.4) *verify that this is actually a probability model, namely that $p_k \in [0,1]$ and $\sum_{k \in \mathbb{N}} p_k = 1$. As a challenge try proving that the* mean value, *defined to be $\sum_{k \in \mathbb{N}} k \, p_k$, is equal to m. This says the parameter m of the probability model corresponds to the experimentally determined average (or mean) value of an ensemble of outcomes.*

For the general case of classical probability, which includes as special cases the previous two cases, one starts with an arbitrary non-empty set Ω, which is called the *sample space*. One lets $\mathcal{F}$ be a collection of subsets of Ω such that

- $\emptyset, \Omega \in \mathcal{F}$. We have denoted the empty set by $\emptyset$.

- If $E \in \mathcal{F}$, then the *complementary set* satisfies $\Omega \setminus E \in \mathcal{F}$, where $\Omega \setminus E := \{\omega \in \Omega \,|\, \omega \notin E\}$.

- Whenever $E_j \in \mathcal{F}$ for all $j \in J$, any finite or countably infinite index set, then their union satisfies $\cup_{j \in J} E_j \in \mathcal{F}$.

One says that $\mathcal{F}$ is a *σ-algebra in* Ω and that an element $E \in \mathcal{F}$ is an *event*. In this most general case, we do not assign probabilities to the elements of Ω, but rather directly to the events. In other words, we suppose there is a function $P : \mathcal{F} \to \mathbb{R}$, called the *probability measure*, such that

- $0 \leq P(E) \leq 1$ for all $E \in \mathcal{F}$.

- $P(\emptyset) = 0$ and $P(\Omega) = 1$.

- Whenever $E_j \in \mathcal{F}$ for all $j \in J$, any finite or countably infinite index set, is a *disjoint family* (that is, $E_j \cap E_k = \emptyset$ for all $j \neq k$), then

$$P(\cup_{j \in J} E_j) = \sum_{j \in J} P(E_j).$$

We read $P(E)$ as "the probability that the event E occurred" or more simply as "the probability of E". The triple $(\Omega, \mathcal{F}, P)$ is called a *(classical) probability space*. In analogy to quantum theory P is also called a *state*.

If we take $\Omega = \mathbb{R}$ and no σ-algebra is indicated, then implicitly it is understood that we are taking $\mathcal{F}$ to be the *Borel σ-algebra* $\mathcal{B}(\mathbb{R})$, which by definition is the smallest σ-algebra of subsets of $\mathbb{R}$ that contains all of the open intervals in $\mathbb{R}$. We say that $B \subset \mathbb{R}$ is a *Borel set* if $B \in \mathcal{B}(\mathbb{R})$. If you have never heard of Borel sets, then do not worry, since every set you will ever see is a Borel set.

A famous example, which is not covered by the first two cases, is the *Gaussian probability measure* defined on $\Omega = \mathbb{R}$. For this there are two parameters, a real number $\mu \in \mathbb{R}$ and a positive real number $\sigma > 0$. The Gaussian probability of the open interval (a, b), where $a < b$ are extended real numbers (that is, $a, b \in [-\infty, \infty]$), is given by

$$P\big((a, b)\big) = \frac{1}{\sqrt{2\pi\sigma^2}} \int_a^b e^{(x-\mu)^2/(2\sigma^2)} \, dx. \tag{16.1.5}$$

Those who understand a lot of measure theory understand that this extends in a unique way to a definition of a probability measure $P(B)$ defined for every Borel subset B of $\mathbb{R}$. But in practice (16.1.5) is all you need to know.

Exercise 16.1.4 *For the Gaussian probability measure defined in* (16.1.5) *prove that* $P(\mathbb{R}) = 1$. *(**Hint:** $\mathbb{R} = (-\infty, \infty)$.)*
    For those who know measure theory, try showing that $P(\{a\}) = 0$ *for any* $a \in \mathbb{R}$, *where the* singleton set *$\{a\} \subset \mathbb{R}$ is the (Borel, of course) subset of $\mathbb{R}$ that contains only one element, namely a.*

Exercise 16.1.5 *Prove there exists a subset of $\mathbb{R}$ which is not a Borel set.* **Warning:** *This is an extremely non-trivial problem that requires some rather advanced mathematical tools. Don't be disappointed if it is beyond you. And besides you will never be needing this bizarre fact.*

Exercise 16.1.6 *Let $(\Omega, \mathcal{F}, P)$ be a probability space. Prove that for all $E \in \mathcal{F}$ we have that $P(\Omega \setminus E) = 1 - P(E)$.*

A measured real-valued quantity of a system, which is being described probabilistically, is modeled by a *random variable* which is defined to be a *measurable function* $X : \Omega \to \mathbb{R}$ meaning that $X^{-1}(B) \in \mathcal{F}$ for all Borel subsets $B \subset \mathbb{R}$. Recall from set theory that the *inverse image of B by X* is defined to be $X^{-1}(B) := \{\omega \in \Omega \,|\, X(\omega) \in B\}$. In this context $X^{-1}(B)$ is called the *event that X takes a value in B*. We recall something we have already seen in Section 8.2.

Definition 16.1.1 *The* expected value *of a random variable $X : \Omega \to \mathbb{R}$ is $\mathcal{E}(X) = \langle X \rangle := \int_\Omega X(\omega) \, P(d\omega)$ provided that this integral exists.*

Exercise 16.1.7 *Find a formula for $\mathcal{E}(X)$ in the context of Exercise 16.1.2. This should convince you that the definition of $\mathcal{E}(X)$ is reasonable.*

Exercise 16.1.8 *Let $X : \Omega \to \mathbb{R}$ be a random variable for the probability space $(\Omega, \mathcal{F}, P)$. Show that $\mathcal{B}(\mathbb{R}) \ni B \mapsto P(X^{-1}(B)) =: (X_* P)(B)$ is a probability measure on $\mathbb{R}$.*
    For those who know measure theory show that $\mathcal{E}(X) = \int_{\mathbb{R}} \lambda \, (X_* P)(d\lambda)$ and, more generally, that the nth moment $\mathcal{E}(X^n) = \int_{\mathbb{R}} \lambda^n \, (X_* P)(d\lambda)$ for every integer $n \geq 0$ provided the relevant integrals exist.

We say that $X_* P$ is the *(probability) distribution* induced by X on $\mathbb{R}$. (Or in the language of category theory, that $X_* P$ is the *push-forward* of P by the random variable X.) But more importantly, $(X_*P)(B)$ is interpreted as the probability of the event that X takes a value in $B \subset \mathbb{R}$.

The overwhelming success of probability theory is due in no small measure to its amazing range of applications in all areas of modern science. And this success arises from a nearly universal aspect of these applications. And what is that aspect? It is *incomplete knowledge*. The ideal of having complete knowledge at some moment of time about a physical, chemical, biological or economic system is rarely realized. So even though a deterministic approach may seem to be appropriate, in practice this just can not be carried out. This is because determinism requires complete knowledge at some moment in time in order to be able to predict what will happen at later times.

Now incomplete knowledge can be described more positively as *partial information*. And partial information about what is happening at some time should help us to understand *something*, if not *everything*, about what will happen at some later time. And this is what classical probability theory sets out to do. Moreover, this admirable goal quite often yields rather good, though less than ideal, results.

But it also gives us a way of thinking about probability. For example, if we can increase our information about a particular situation, we can change our probabilistic analysis to yield even better results. As a specific example, if horse racing is not being fixed behind the scenes (which in general is true, I believe), then getting ever more information about the horses allows you to improve your chances of winning at the race track without ever arriving at complete certainty of winning. This is consistent with the notion that increasing our information to the maximum possible should give us complete information about what happens at future times. Of course, this can be checked experimentally to some extent, namely it can be falsified. Continuing with the example above, the fact that some people are able to win consistently at the horse track, by digging up more and more information on the horses, leads me to my belief that horse racing in general is not fixed.

There are two types of events of special interest in classical probability. First, if $P(E) = 0$, then E is interpreted as an event which never occurs. On the other hand if $P(E) = 1$, then E is interpreted as an event which always occurs. In either case we have complete knowledge about the event E. Such an event is called a *deterministic event*. We know all of its properties, whatever reasonable meaning we care to give to 'property'. Of course, we do not introduce the machinery of classical probability theory just to study these deterministic events. But they are there. And they are the limiting, ideal cases as knowledge is increased to be used to construct ever more accurate probability theories that approximate some underlying deterministic theory.

However, verifying determinism directly seems to be out of the question. After all, how can one know that there exists some missing information, if we do not have that information? Nonetheless, so long as a deterministic basis yields good results, there is no reason to use any other approach.

And there's the rub! Quantum theory involves a probability theory that does not admit improvements with ever more available information, let alone with complete knowledge. The maximum possible information of a quantum system is encoded in the quantum state of the system. And that's it, no more. (And sometimes not even all of this information is available, but more on that later.) So, even with full knowledge of the quantum state, the theory remains probabilistic. No deterministic limiting case is there to be achieved, not even in principle. It can be difficult to wrap one's mind around this.

Therefore this is not consistent with the approach of classical probability theory. Then it should be no surprise to learn that a new type of probability comes into play. This is called *quantum probability*. And it is quite different from classical probability. An ongoing problem is that valid notions formed while learning classical probability do not always remain so when introduced into quantum probability. Here is what I consider to be the most important example of that. In classical probability *a la* Kolmogorov any two distinct measurements, corresponding to the random variables X and Y respectively, have a *joint probability distribution* $P(X^{-1}(B_1) \cap Y^{-1}(B_2))$, which does not depend on the time order of the two measurements. Here B_1 and B_2 are Borel subsets of $\mathbb{R}$. So, $X^{-1}(B_1) \cap Y^{-1}(B_2)$ is the event that X takes a value in B_1 and Y takes a value in B_2. As we shall see in detail in Section 16.10, there is no construction like this in general in the quantum theory of two observables that do not commute. For non-commuting quantum observables there is no joint probability distribution. Rather in quantum theory the order in which two such measurements are performed in time will non-trivially impact their individual probability distributions.

16.2 Quantum Probability

Now one major difference in quantum probability is that to get numbers in $[0, 1]$, which are to be compared with experiment, we need a *quantum observable* A (that is, a *self-adjoint operator* acting in some Hilbert space), a *quantum state* ψ of the system and a (Borel) subset B of the real line $\mathbb{R}$. (For measure theory challenged readers, just think of B as an open interval $(a, b) = \{r \in \mathbb{R} \mid a < r < b\}$ with $a < b$ being real numbers.) For a system in the state ψ, we want to know what is the probability that a measurement of A lies in the set B. We denote that as

$$P(A \in B \mid \psi) \in [0, 1]. \tag{16.2.1}$$

This notation looks like the notation for a *conditional probability* in classical probability theory, that is, the probability of the *quantum event $A \in B$* given that the system is in a quantum state ψ. Our goal is to define mathematically this new type of probability which for now appears in (16.2.1) as a new notation only. However, (16.2.1) will include (10.1.2) as a special case.

We already know what A, B, and ψ are. But what about a quantum event? Well, a classical event $E \in \mathcal{F}$ corresponds to its characteristic function

$\chi \equiv \chi_E$, which satisfies $\chi = \chi^* = \chi^2$. Conversely, every such real-valued (Borel) function χ equals χ_E for a unique $E \in \mathcal{F}$. This helps to motivate the following.

Definition 16.2.1 *A quantum event for a quantum theory that is based on a Hilbert space $\mathcal{H}$ is a projection operator E acting on $\mathcal{H}$. We recall that a projection operator E is a linear operator $E : \mathcal{H} \to \mathcal{H}$ which is self-adjoint (that is, $E^* = E$) and idempotent (that is, $E^2 = E$).*

As we have seen in Chapter 9 each quantum event E determines a closed subspace of $\mathcal{H}$, namely its range $\operatorname{Ran} E$. Conversely, to every closed subspace S of $\mathcal{H}$ there is a unique quantum event E whose range is S, that is to say, $S = \operatorname{Ran} E$. So, an alternative definition is that a quantum event is any closed subspace of $\mathcal{H}$. Definition 16.2.1 actually looks more natural after one has studied the spectral theorem for self-adjoint operators. Here is one version of the spectral theorem. It relates self-adjoint operators to quantum probability. For the rest of this Chapter we assume the Hilbert space $\mathcal{H} \neq 0$.

Theorem 16.2.1 (Spectral Theorem) *Suppose A is a densely defined self-adjoint operator defined in a Hilbert space $\mathcal{H}$. Then there exists a unique projection valued measure (or pvm), denoted by P_A, such that $A = \int_{\mathbb{R}} \lambda \, dP_A(\lambda)$.*

It may seem preposterous, but this theorem is saying that A is *diagonalizable*. But to understand all this, we must attend to the mundane business of giving more definitions.

Definition 16.2.2 *A projection valued measure (pvm) in a Hilbert space $\mathcal{H}$ is a function $P : \mathcal{B}(\mathbb{R}) \to \mathcal{L}(\mathcal{H})$, where $\mathcal{L}(\mathcal{H})$ is the space of all (bounded!) linear maps $T : \mathcal{H} \to \mathcal{H}$, that satisfies the following:*

- *$P(B)$ is a projection for all $B \in \mathcal{B}(\mathbb{R})$, that is, for all Borel subsets B of $\mathbb{R}$. Equivalently, $P(B)$ is a quantum event for all Borel subsets B.*

- *$P(\emptyset) = 0$ and $P(\mathbb{R}) = I$, the zero and identity map of $\mathcal{H}$, respectively.*

- *If B_j for $j \in J$, a finite or countably infinite index set, is a disjoint family of Borel subsets (that is, $B_j \cap B_k = \emptyset$ for $j \neq k$), then*

$$P(\cup_{j \in J} B_j) = \sum_{j \in J} P(B_j)$$

  in the sense that

$$P(\cup_{j \in J} B_j)\phi = \sum_{j \in J} P(B_j)\phi$$

  for all $\phi \in \mathcal{H}$ with the convergence of the sum on the right side of the last equation (when J is infinite) being with respect to the norm of the Hilbert space $\mathcal{H}$.

This is very analogous to a probability measure, except now the values of the 'probabilities' are quantum events ($\equiv$ projections) instead of being real numbers between 0 and 1. For example, suppose that we are considering a quantum observable A and a Borel subset B of the real line $\mathbb{R}$. Then the expression $P_A(B)$ has a physical interpretation, namely it is the quantum event that says a measurement of A has resulted in a measured value in the subset B. We say that the quantum event $P_A(B)$ is the *quantum probability* of A being in B. Despite this (more or less?) intuitive physical interpretation, the mathematical object $P_A(B)$ is not a number in $[0, 1]$, which we could then compare with a relative frequency measured in experiment.

The point of the first part of Exercise 9.3.20 is that the projections lie between the extreme values 0 and I (which are projections themselves) in analogy with the values of a classical probability measure, which lie between the real numbers 0 and 1. In particular, we have that $0 \leq P(B) \leq I$ for any pvm P, where $\leq$ is the partial order for self-adjoint operators. (See Exercise 9.3.16.) Moreover, the limiting cases $P(\emptyset) = 0$ and $P(\mathbb{R}) = I$ are the quantum events that never occur and that always occur, respectively. This is analogous to a classical probability measure P for which $P(\emptyset) = 0$ and $P(\Omega) = 1$ are the limiting values corresponding to the classical events that never occur and that always occur, respectively.

Since this is a sort of measure theory, there should be integrals lurking around. In the first instance the goal of measure theory is to define integrals and establish their properties. This simple fact is usually lost in a blizzard of technical details that is the standard mathematics course in measure theory. Yet another fact often omitted in such courses is that the *full* generalization in analysis of the purely algebraic concept of finite sum is not infinite sum, but rather *integral*. Small wonder that integrals are so important. Infinite sums as well as finite sums are just special cases of integrals. But I digress.

So, the expression $\int_{\mathbb{R}} \lambda \, dP_A(\lambda)$ in Theorem 16.2.1 is just the integral of the real-valued function $\mathbb{R} \to \mathbb{R}$ that maps λ to λ (the identity function) but integrated with respect to the pvm P_A. The generalized measure theory gives this integral a meaning (actually as a densely defined self-adjoint operator), and then the Spectral Theorem 16.2.1 asserts that this integral is exactly the densely defined self-adjoint operator A itself.

The Spectral Theorem 16.2.1 has a converse, which says that for any pvm P on $\mathbb{R}$ we can define an operator A via the formula $A := \int_{\mathbb{R}} \lambda \, dP(\lambda)$. Then A is a densely defined, self-adjoint operator whose uniquely associated pvm P_A satisfies $P_A = P$. In short, there is a bijective (i.e., one-to-one and onto) correspondence between densely defined, self-adjoint operators and pvm's defined on $\mathbb{R}$. So, we can pass freely in a discussion back and forth between the analytic object A and the quantum probabilistic object P_A.

We have yet to produce real numbers in the interval $[0, 1]$ that can then be compared with experimentally measured frequencies. It is about time to do this. The idea behind this was first published by M. Born.

Definition 16.2.3 (Born's rule) *Let E be a quantum event and ψ be a quantum state. Then the probability that the event E occurs when the system is in the state ψ is defined to be*

$$P(E \,|\, \psi) := \langle \psi, E\psi \rangle. \tag{16.2.2}$$

This definition had better give us a real number in $[0, 1]$.

Exercise 16.2.1 *Prove that $P(E \,|\, \psi) = ||E\psi||^2$. Using this identity show that $0 \leq P(E \,|\, \psi) \leq 1$.*

A minor technical point should be dealt with as well. In the next exercise you are asked to prove that $P(E \,|\, \psi)$ depends only on the state represented by the equivalence class of norm 1 vectors determined by ψ.

Exercise 16.2.2 *Suppose that ψ_1 and ψ_2 with $||\psi_1|| = ||\psi_2|| = 1$ represent the same state, that is $\psi_1 = \lambda \psi_2$ for some complex number λ that satisfies $|\lambda| = 1$. Prove that $P(E \,|\, \psi_1) = P(E \,|\, \psi_2)$.*

Quantum theory is mostly a linear theory. For example, it is a theory of linear operators acting in a Hilbert space, and the time evolution is given by the linear Schrödinger equation. But Born's rule (16.2.2) for computing probabilities of an event E is not linear in ψ, the solution of the Schrödinger equation. This non-linearity of quantum theory seems to be basic.

Note that a quantum event E is a self-adjoint operator by definition, that is, it itself is a quantum observable. (Strictly speaking, *super-selection rules* exclude certain self-adjoint operators, and in particular certain projections, from being associated with experimental observables. But super-selection rules do not occur in the quantum theory of many systems, including all those considered in this book.)

Exercise 16.2.3 *Let E be a quantum event in a Hilbert space $\mathcal{H}$. Prove that the spectrum $\mathrm{Spec}(E)$ is a subset of the two-point subset $\{0, 1\}$ of $\mathbb{R}$. There are exactly four subsets of $\{0, 1\}$, and the reader may wish to accept the challenge to prove that all four of these possibilities occur.*
Hint: *By functional analysis the spectrum of E consists only of eigenvalues. If you can not prove that statement, then simply assume it is so and then find all possible eigenvalues.*

So the measured values of E are either 0 or 1. We interpret 0 as saying that the event E did not occur and 1 as saying that E did occur. Notice that $P(E \,|\, \psi) = \langle \psi, E\psi \rangle$ can also be interpreted as the *expected value* of the observable E in the state ψ. In some formulations of quantum theory, there are self-adjoint operators that do not correspond to physical observables.

Nonetheless a quantum event is usually understood to represent a special type of physical observable known as a *Yes/No experiment* which answers a question with Yes meaning that 1 was measured and No meaning that 0 was measured. The exact question being asked is: "Is the system in a state that is in the range of E?" Note that the answer does not concern the initial state of the system (which is often unknowable) but rather is about the final state after the measurement has been made. The answer Yes means that the final state is in the range of the quantum event E, while the answer No means that the final state is in the range of the quantum event $I - E$.

For the case $E = 0$, the zero operator, (which *is* a projection) we have $\text{Spec}(0) = \{0\}$ (recalling that the Hilbert space $\mathcal{H} \neq 0$) and $P(0 \,|\, \psi) = 0$ for all states ψ. Therefore $E = 0$ is an event which never occurs, no matter what the state is.

For the case $E = I$, the identity operator, (which *is* a projection) we have that $\text{Spec}(I) = \{1\}$ and $P(I \,|\, \psi) = 1$ for all states ψ. So, $E = I$ is an event which always occurs, no matter what the state is.

Exercise 16.2.4 *Let E be a quantum event with $E \neq 0$ and $E \neq I$. Prove that* $\text{Spec}(E) = \{0, 1\}$.

So all the intermediate events $0 < E < I$ will occur for some states and will not occur for some (other) states. There are three mutually exclusive cases here for the state ψ:

- $\psi \in \text{Ran}\, E$, in which case ψ is an eigenvector of E with eigenvalue 1 and so $P(E \,|\, \psi) = \langle \psi, E\psi \rangle = \langle \psi, \psi \rangle = 1$.

- $\psi \in (\text{Ran}\, E)^\perp = \ker E$, in which case ψ is an eigenvector of E with eigenvalue 0 and so $P(E \,|\, \psi) = \langle \psi, E\psi \rangle = \langle \psi, 0 \rangle = 0$.

- $\psi \notin \text{Ran}\, E \cup (\text{Ran}\, E)^\perp$, in which case ψ is not an eigenvector of E and moreover $0 < P(E \,|\, \psi) < 1$.

Let's underline that for $0 < E < I$, the third case does occur, that is, $\text{Ran}\, E \cup (\text{Ran}\, E)^\perp \neq \mathcal{H}$. Of course, $\text{Ran}\, E \oplus (\text{Ran}\, E)^\perp = \mathcal{H}$ by the Projection Theorem 9.3.3, since $\text{Ran}\, E$ is closed whenever E is a quantum event.

Exercise 16.2.5 *Prove* $(\text{Ran}\, E)^\perp = \ker E = \text{Ran}(I - E) = (\ker(I - E))^\perp$ *provided that E is a quantum event.*

At this point we can see quite clearly how probability 1 events in quantum probability are different from probability 1 events in classical probability. In the first place a quantum event E has *probability 1 in a given state ψ* means that $P(E \,|\, \psi) = 1$. The mapping $E \mapsto P(E \,|\, \psi)$ for fixed ψ is not a classical probability measure, since its domain is not a σ-algebra if $\dim \mathcal{H} \geq 2$. The only quantum event that has probability 1 for *all* states ψ is $E = I = I_\mathcal{H}$, the identity operator on the Hilbert space $\mathcal{H} \neq 0$. But more profoundly consider the observable $A = S_1$, one of the spin matrices acting in the Hilbert space $\mathbb{C}^2$, that takes exactly the two values $-1/2$ and $1/2$. (Recall Exercise 14.1.5.) Let $\psi \in \mathbb{C}^2$ be a fixed state. Then consider these two statements:

- The observable A has a value in the set $\{-1/2, 1/2\}$.

- Either the observable A has the value $1/2$ or the observable A has the value $-1/2$.

Recall that "the observable A has value in the Borel subset $B \subset \mathbb{R}$" is the quantum event $E_A(B)$ where E_A is the pvm associated with A. The first statement refers to the quantum event $E_A(\{-1/2, 1/2\}) = I$, the quantum event which always occurs. So the first statement is true.

The second statement refers to the quantum event $E_A(\{1/2\})$ in the first half of that statement and to the quantum event $E_A(\{-1/2\})$ in the second half of that statement. Each subspace $\operatorname{Ran} E_A(\{1/2\})$ and $\operatorname{Ran} E_A(\{-1/2\})$ has dimension 1 and so $\operatorname{Ran} E_A(\{1/2\}) \cup \operatorname{Ran} E_A(\{-1/2\}) \neq \mathbb{C}^2$.

Now we see that the word 'has' is playing an important role. Classically, A has the value $1/2$ means that the system is in a state in which a measurement of A always gives the value $1/2$. Taking this interpretation directly over into quantum theory, A has the value $1/2$ means that the system is in a given quantum state $\psi \in \operatorname{Ran} E_A(\{1/2\})$. Similarly, A has the value $-1/2$ means that the system is in a given quantum state $\psi \in \operatorname{Ran} E_A(\{-1/2\})$.

If we now interpret the disjunctive 'or' classically (in the manner going back at least to Aristotle and continuing with Boole), then in the second statement we are speaking of $\operatorname{Ran} E_A(\{1/2\}) \cup \operatorname{Ran} E_A(\{-1/2\})$. Therefore, classically speaking, for any state $\psi \notin \operatorname{Ran} E_A(\{1/2\}) \cup \operatorname{Ran} E_A(\{-1/2\})$ the second statement is false. And there are plenty of such states! But if we interpret 'or' as the *quantum disjunction* (or *lattice meet* operation, that is, direct sum in this finite dimensional setting), then

$$\operatorname{Ran} E_A(\{1/2\}) \oplus \operatorname{Ran} E_A(\{-1/2\}) = \mathbb{C}^2,$$

which corresponds to the quantum event I, which always occurs, then the second statement is true. So the meaning of little words like 'or' can be crucially important for thinking correctly about quantum theory.

The upshot is that quantum probability leads to a new way to think about the logical connective 'or', which is studied in *quantum logic*. For whatever reasons, quantum logic has not become a very active area of current research, while *quantum probability* has. Nonetheless, it is important to understand some of these basics of quantum logic.

Exercise 16.2.6 *Review the details in the analysis of these two statements. Make sure you clearly understand this, especially the properties of the pvm E_A.* (**Hint:** *Find a basis that diagonalizes $A = S_1$.*)

Also, while this is an exercise about the specific operator S_1, do understand why the same sort of situation holds for any observable that takes two or more values. And try to keep in mind what your 'classical intuition' tells you about all this!

One can also see why a one-dimensional Hilbert space is trivial in quantum theory, because in that case there are no intermediate quantum events, but only the two quantum events, $E = 0$ and $E = I$. And these are somewhat trivial events in the sense that they are deterministic, and they also 'form' a σ-algebra, that is, profoundly these two quantum events have the same lattice structure (see discussion below) as the σ-algebra $\mathcal{F} = \{\emptyset, \Omega\}$ where Ω is any non-empty set. Of course, as we have already seen, in the setting of the Hilbert space $\mathcal{H} = \mathbb{C}^2$ there is a non-trivial quantum theory, namely spin $1/2$. In fact, we have the following.

Exercise 16.2.7 *Suppose that $\mathcal{H}$ is a Hilbert space with dimension ≥ 2. Prove that there are quantum events E in $\mathcal{H}$ with $0 < E < I$.*

The next exercise is a long excursion into basic set theory. It is meant to provide insight since it contains some of the language which we will be seeing in quantum theory.

Exercise 16.2.8 *If A, B are sets that satisfy $a \in B$ for all $a \in A$, then we say that A is a* subset *of B. Notation: $A \subset B$.*

Let S be a set. We define its power set *to be the set of all of its subsets, that is $\mathrm{Pow}(S) := \{A \mid A \subset S\}$. Prove that the relation $\subset$ defines a partial order on $\mathrm{Pow}(S)$. (See Exercise 9.3.16 for the definition of 'partial order'.)*

Notice that $\emptyset$, the empty set, and S itself are elements in $\mathrm{Pow}(S)$. Prove that $\emptyset \subset A \subset S$ for all $A \in \mathrm{Pow}(S)$. (We say that $\mathrm{Pow}(S)$ has a minimal *element $\emptyset$ and a* maximal *element S.)*

Prove that for any $A, B \in \mathrm{Pow}(S)$ there is a unique minimal upper bound C for A and B. To say that C is an upper bound means that $A \subset C$ and $B \subset C$. To say that C is the minimal *upper bound means that $C \subset D$ whenever D is an upper bound of A and B. We say that C is the* union *of A and B. Notation: $A \cup B := C$.*

Prove that for any $A, B \in \mathrm{Pow}(S)$ there is a unique maximal lower bound E for A and B. To say that E is a lower bound *means that $E \subset A$ and $E \subset B$. To say that E is the* maximal *lower bound means that $F \subset E$ whenever F is a lower bound of A and B. We say that E is the* intersection *of A and B. Notation: $A \cap B := E$. Note that this paragraph is dual (in some sense) to the previous one.*

Prove that for any $A \in \mathrm{Pow}(S)$ there is a unique element $A^c \in \mathrm{Pow}(S)$ such that $A \cup A^c = S$ and $A \cap A^c = \emptyset$. One says A^c is the complement *of A.*

Prove the following two identities, which are called de Morgan's laws:

$$A \cap (B \cup C) = (A \cap B) \cup (A \cap B), \quad A \cup (B \cap C) = (A \cup B) \cap (A \cup B)$$

for all $A, B \in \mathrm{Pow}(S)$. These are actually distributive laws *of each of the binary operations $\cap$ and $\cup$ with respect to the other.*

With these operations $\mathrm{Pow}(S)$ is a Boolean algebra. *(You might want to look up the exact definition of a Boolean algebra.)*

For the next exercise recall that projections E are in bijective correspondence with closed subspaces via $E \mapsto \mathrm{Ran}\, E$. (See Exercise 9.3.21.)

Exercise 16.2.9 *Define* $\mathrm{Proj}(\mathcal{H}) := \{E \in \mathcal{L}(\mathcal{H}) \mid E \text{ is a projection}\}$, *where* $\mathcal{H}$ *is a Hilbert space. Prove that with the partial order* $E \leq F$ *of self-adjoint operators, one has the following properties of* $\mathrm{Proj}(\mathcal{H})$:

- *There is a unique maximal element, namely the identity operator* I, *whose corresponding closed subspace is* $\mathcal{H}$.

- *There is a unique minimal element, namely the zero operator* 0, *whose corresponding closed subspace is the zero subspace.*

- *Each pair of elements* E, F *has a unique minimal upper bound, denoted as* $E \vee F$, *with corresponding closed subspace* $(\mathrm{Ran}\, E \cup \mathrm{Ran}\, F)^{--}$, *the closure of the union of the ranges.*

- *Every pair of elements* E, F *has a unique maximal lower bound, denoted as* $E \wedge F$, *with corresponding closed subspace* $\mathrm{Ran}\, E \cap \mathrm{Ran}\, F$.

- *For every* E *there is a unique element* $E^\perp$ *such that* $E \vee E^\perp = I$ *and* $E \wedge E^\perp = 0$. *The closed subspace corresponding to* $E^\perp$ *is* $(\mathrm{Ran}\, E)^\perp$.

The point of this exercise is that a pvm is a function whose co-domain is $\mathrm{Proj}(\mathcal{H})$, which is a partially ordered set. This set has some similarity with the co-domain of a classical probability measure, namely the interval $[0, 1]$, which is a totally ordered set. However, $\mathrm{Proj}(\mathcal{H})$ is also the domain of the functions $P(\cdot \mid \psi)$ for every state ψ. (See Definition 16.2.3.) Even though these functions bear some resemblance to a classical probability measure, their common domain $\mathrm{Proj}(\mathcal{H})$ is *not* a σ-algebra as you are asked to show next. In fact, it is not even a Boolean algebra.

Exercise 16.2.10 *Show that the de Morgan law*

$$E_1 \wedge (E_2 \vee E_3) = (E_1 \wedge E_2) \vee (E_1 \wedge E_3)$$

does not *hold for all* $E_1, E_2, E_3 \in \mathrm{Proj}(\mathcal{H})$, *provided that* $\dim \mathcal{H} \geq 2$.

Prove that in general $\mathrm{Proj}(\mathcal{H})$ *is not a* σ-algebra. *Identify the exceptional cases when it is.*

Remark: *Every* σ-algebra *is an ortho-complemented lattice. So to say that an ortho-complemented lattice is not a* σ-algebra *really means that it is not isomorphic to any* σ-algebra. *Therefore, you are being asked to understand (and maybe even define) both* isomorphism *and* isomorphic *in this context.*

We now define some classical probability measures that do appear in quantum probability theory.

Definition 16.2.4 *Suppose that* $\mathcal{H}$ *is a Hilbert space. Let* A *be a self-adjoint densely defined operator acting in* $\mathcal{H}$, *let* ψ *be a state in* $\mathcal{H}$, *and let* B *be a Borel subset of* $\mathbb{R}$. *Then we put the previously defined structures together, viewing* A *and* ψ *as fixed and thereby defining a function* $\mathcal{B}(\mathbb{R}) \to [0, 1]$ *by*

$$P(A \in B \mid \psi) := P(P_A(B) \mid \psi) = \langle \psi, P_A(B)\psi \rangle \quad \text{for } B \in \mathcal{B}(\mathbb{R}).$$

The left-hand side is read as "the probability that the observable A yields a measured value in the set B given that the system is initially in state ψ" or more briefly as "the probability that A is in B given ψ".

This fulfills the goal of defining the notation introduced in (16.2.1).

Exercise 16.2.11 *Continuing with the notation of the previous definition, for $A = A^*$ and a state $\psi \in \mathcal{H}$ prove that $B \mapsto P(A \in B \mid \psi)$ is a classical probability measure, which is called a* spectral measure *of A.*

We now have all the tools on the table in order to present the complete form of the Measurement and Collapse Condition as given in preliminary form in Chapter 10.

Measurement and Collapse Axiom: Suppose that a quantum system is in the state represented by ψ and that a measurement of the observable A is made.

a) The probability that this measurement yields a value in the Borel subset $B \subset \mathbb{R}$ is $P(A \in B \mid \psi)$.

b) Suppose that this measurement did yield a value in B. In particular, the corresponding classical probability satisfies

$$P(A \in B \mid \psi) = \langle \psi, P_A(B)\psi \rangle = \|P_A(B)\psi\|^2 \neq 0.$$

Then after this measurement the system will be in the state represented by the unit vector

$$\frac{1}{\|P_A(B)\psi\|} P_A(B)\psi.$$

Part a) seems inescapable. After all, we are obliged to give some physical interpretation to the classical probability measures $B \mapsto P(A \in B \mid \psi)$, which arise from the self-adjoint operator A. (Also, the set of all these probability measures, one for each ψ, in turn determine uniquely A.) And Part a) does just that in a 'natural' way. However, due to its essential use of probability, Part a) is controversial.

Part b) is even more controversial, even though it is totally deterministic. However, experimental evidence has never falsified it. Moreover, it is used extensively in *quantum computation* theory, which is a theory as its name implies. So Part b) has applications in theory as well as in experiments. But, as noted earlier, the collapse from the initial to the final state is not linear. Nor could it be, since the space of states is not a vector space. (See Section 9.4.) This is not a paradox, but it remains a curious, puzzling aspect of quantum theory.

Of course, one would like an extension of this axiom to situations that are not experiments. Or some more general understanding of 'measurement' might be adequate. Even so, the initial state may be unknown—or worse yet, unknowable. In classical mechanics the state of a physical system can be known to any level of precision desired. But the state ψ of a quantum mechanical system may be unknowable.

It turns out that Part a) of this axiom is not only the origin of much controversy; it also leads to a rejection of the intuitive notion that quantum states only have properties with *probability* 1. This is a tricky business, because there are some such probability 1 properties in quantum theory. For example, the measured value m_e of the mass of an electron is always the same; it is a *property* that all electrons have. Such properties are cataloged and can be found in *particle properties tables* in the physics literature. (By the way such a table is never called a *wave properties table*.) We fall into language dripping with this classical viewpoint. For example, we say that the "mass of the electron *is* m_e" or "the electron *has* mass m_e". While these usages can be defended, the verbs "to be" and "to have" are linguistic traps when used cavalierly in quantum theory as we already saw in studying the spin components of an electron. The spin itself of an electron is a particle property with value $\hbar/2$, where $\hbar$ is Planck's constant. Thus its spin in units where $\hbar = 1$ is $1/2$ with probability 1.

But there are other aspects of the state of an electron (such as the just mentioned components of spin) that are not properties in this deterministic sense but nonetheless appear to be very similar to 'properties'. And such aspects invariably concern the results of two (or more!) measurements of observables that do not commute. The point here is that the time order of these measurements matters due to the collapse of the state vector with each measurement. And this is a question of an axiom (that is, basic principle) of quantum theory.

In classical physics the time order of measurements does not matter *in principle*, though it might matter in *practice*. Why in practice? Because the first measurement could perturb the state of the system, thereby having a significant effect on the second measurement. Why not in principle? Because the perturbations due to the first measurement could be made so small as to not impact significantly on the second measurement. Of course, also in practice, it might be quite difficult to control the size of these perturbations. However, this results in exactly the sort of partial information that classical probability theory was built to deal with. And besides, measurements never give precise values. All of this is handled, and handled quite well, by classical probability theory.

But quantum systems are quite different, and so our intuitions that come from years of using classical probability theory do not apply. We are obliged to develop something else in order to deal with quantum theory, whether it be quantum probability theory as presented here or something else. The goal here is to present standard quantum theory, so we will only discuss quantum probability theory. Any other new, proposed theory must be better in the scientific sense that, at least in some cases, it makes predictions different than the standard theory *and* that the appropriate experiments support the new theory and therefore also falsify the standard quantum theory. I do not wish to discourage such alternative proposals, but I do wish to describe what the scientific test for them is. Of course, if a proposed new quantum theory makes *exactly* the same predictions as standard quantum theory, then it is neither better nor worse than the standard theory. Rather, it would be an alternative way of formulating the standard theory. By the way, this is more than intellectual chitchat. The *Heisenberg picture* of quantum theory, presented in Chapter 17, gives such an alternative, equivalent reformulation of standard quantum theory.

16.3 States as Quantum Events

The framework of quantum probability leads to an interesting way to view quantum states. Recall that a vector $\psi \neq 0$ represents a state or, equivalently, the one-dimensional subspace $\mathbb{C}\psi$ is the state. And any one-dimensional subspace V of the Hilbert space can be written as $V = \mathbb{C}\psi$ for any non-zero $\psi \in V$. So the states are exactly the one-dimensional subspaces, that is, the states are those quantum events E with $\dim E = 1$. And we are taking the standard point of view that all quantum events correspond to physical observables. Consequently, every quantum state corresponds to a physical observable, namely, a Yes/No experiment.

This is an important point since sometimes even the experts claim that a solution of the Schrödinger equation is not an observable. (For example, see [11].) This is technically correct, since the solution is a normalized vector in the appropriate Hilbert space. However, that normalized solution ψ does uniquely determine a one-dimensional subspace and hence a corresponding quantum event, which is a physical observable and is written as $|\psi\rangle\langle\psi|$ in Dirac notation. (See Section 9.7.) So, a solution of the Schrödinger equation does give us immediately a quantum event, which itself is an observable.

16.4 The Case of Spin $1/2$

Now let's see how the collapse part of the Measurement Axiom works in the specific example of a spin $1/2$ system. (See a description of the *Stern-Gerlach experiment* to see how this was originally done.) First, we establish a Cartesian coordinate system (x_1, x_2, x_3) in an experimental setup. The initial state of the system is unknown. One measures the x_3-component of the spin. This measurement is represented by the self-adjoint spin matrix $S_3 = (1/2)\sigma_3$ with eigenvalues $\{-1/2, 1/2\}$. Say $-1/2$ is measured. So the unknown initial state collapses to the normalized eigenvector of S_3 with eigenvalue $-1/2$, namely

$$\psi_3^\downarrow := \begin{pmatrix} 0 \\ 1 \end{pmatrix}.$$

If we next measure the x_3-component of the spin again, then we get the value $-1/2$ with probability 1, since a quick calculation shows that

$$P(S_3 \in \{-1/2\} \,|\, \psi_3^\downarrow) = 1,$$

where $\{-1/2\}$ is the subset of $\mathbb{R}$ containing exactly the one element $-1/2$.

Exercise 16.4.1 *Do this calculation.*
Hint: *You will have to find the pvm for S_3.*

Classical intuition leads one to think that the state $\psi_3^\downarrow$ 'has' the property that its x_3-component of the spin equals the value $-1/2$. But this is not a property in the usual sense of that word. Or another way to say this is that 'has' is not to be taken in its usual sense. Why is this so? Well, instead of repeating the measurement for the x_3-component of the spin, let's suppose that we measure next the x_1-component of the spin. The corresponding self-adjoint spin matrix is $S_1 = (1/2)\sigma_1$.

Exercise 16.4.2 *Demonstrate that the eigenvalues of S_1 are $\{-1/2, 1/2\}$ with corresponding normalized eigenvectors given uniquely (modulo a complex phase factor) by*

$$\psi_1^\downarrow := \frac{1}{\sqrt{2}} \begin{pmatrix} 1 \\ -1 \end{pmatrix} \qquad \text{and} \qquad \psi_1^\uparrow := \frac{1}{\sqrt{2}} \begin{pmatrix} 1 \\ 1 \end{pmatrix} \qquad \text{respectively.}$$

So the probability of measuring $1/2$ for the x_1-component of the spin, given that the initial state is now known to be $\psi_3^\downarrow$, is

$$P(S_1 \in \{1/2\} \,|\, \psi_3^\downarrow) = \langle \psi_3^\downarrow, P_{S_1}(\{1/2\}) \psi_3^\downarrow \rangle,$$

where P_{S_1} is the pvm for the self-adjoint matrix S_1.

Exercise 16.4.3 *The projection $P_{S_1}(\{1/2\})$ is the projection whose range is $\mathbb{C}\psi_1^\uparrow$, that is, for all $\phi \in \mathcal{H}$ we have*

$$P_{S_1}(\{1/2\})\phi = \langle \psi_1^\uparrow, \phi \rangle \psi_1^\uparrow.$$

In particular, you should show that the formula on the right side does give the projection with range $\mathbb{C}\psi_1^\uparrow$. In Dirac notation (see Section 9.7) we can rewrite this result as $P_{S_1}(\{1/2\}) = |\psi_1^\uparrow\rangle\langle\psi_1^\uparrow|$.

So we can now carry through the calculation as follows:

$$P(S_1 \in \{1/2\} \,|\, \psi_3^\downarrow) = \langle \psi_3^\downarrow, P_{S_1}(\{1/2\}) \psi_3^\downarrow \rangle = \langle \psi_3^\downarrow, \langle \psi_1^\uparrow, \psi_3^\downarrow \rangle \psi_1^\uparrow \rangle$$
$$= |\langle \psi_1^\uparrow, \psi_3^\downarrow \rangle|^2$$
$$= \left| \left\langle \frac{1}{\sqrt{2}} \begin{pmatrix} 1 \\ 1 \end{pmatrix}, \begin{pmatrix} 0 \\ 1 \end{pmatrix} \right\rangle \right|^2$$
$$= \frac{1}{2}.$$

Given that S_1 yields the measured value $1/2$, the state collapses to $\psi_1^\uparrow$. Then further measures of S_1 will yield $1/2$ with probability 1.

Since we are dealing with a classical probability measure on the two-point space $\{-1/2, 1/2\}$, we immediately conclude that

$$P(S_1 \in \{-1/2\} \,|\, \psi_3^\downarrow) = 1 - P(S_1 \in \{1/2\} \,|\, \psi_3^\downarrow) = \frac{1}{2}.$$

Or we could have calculated this directly using the same method. This is known as the *Bernoulli probability measure* for a fair coin. (See any basic probability text for details about this and many other important probability measures.) Consequently, it makes no sense to say that the state $\psi_3^\downarrow$ has a particular, or even a preferred, value of the x_1-component of the spin.

But if we next remeasure S_3 after having measured S_1, we find ourselves in the situation of a Bernoulli probability distribution but now for S_3. The property that the observable S_3 has value $1/2$ with probability 1 has been lost. And so it goes as we alternate back and forth between measurements of S_3 and S_1. Any repeated sequence of measurements of one of these will all give the same value, but continuing with a measurement of the other gives us a Bernoulli distribution. This is quite different from measurements of the mass of an elementary particle, since these give the same value with probability 1 always.

This discussion of spin is easily generalized to take care of measurements of spin components in any pair of directions. The first of these directions can be taken to be the positive x_3-direction. So that is no different. If the second direction is taken to be orthogonal to the first direction, then we are repeating the previous example, and we get a fair coin Bernoulli probability measure. So what happens if we take a general direction, that is not orthogonal to the first direction, for the second direction? Well, we get a skewed Bernoulli probability measure with probabilities p and $1 - p$ for the two possible spin values, where $p \in [0, 1]$. The reader is invited to consider the exact details.

The general case of two quantum observables is discussed in detail in Section 16.10. This section is just a special case of that.

16.5 Expected Value (Revisited)

We will again consider the *expected value* of an observable (see Section 8.2), but now as an aspect of both classical and quantum probability. This is the theoretical structure that corresponds to the empirically determined *average* of a sequence of measurements. As we have seen in classical probability an observable is a (measurable) function $X : \Omega \to \mathbb{R}$, also known as a *random variable*, whose domain Ω is a space with a probability measure P, which is also called a *state*. Then recall from Definition 16.1.1 that

$$\mathcal{E}(X) = \langle X \rangle := \int_\Omega X(\omega) dP(\omega) \qquad (16.5.1)$$

is called the *expected value* of the observable X. An example of this is a finite or countably infinite set $\Omega = \{\omega_1, \omega_2, \dots\}$ where each point ω_j has probability $P(\omega_j) = p_j \in [0,1]$ such that $\sum_j p_j = 1$. In this case any function $X : \Omega \to \mathbb{R}$ is measurable, and by Exercise 16.1.7 its expected value is

$$\mathcal{E}(X) = \langle X \rangle = \sum_j p_j\, X(j). \qquad (16.5.2)$$

As a technical aside you should be aware that the right side (16.5.1) or (16.5.2) may not be well defined in all situations. For that reason we restrict our discussion precisely to the case when those expressions do make sense, namely when the integral in (16.5.1) converges absolutely or when the infinite sum in (16.5.2) converges absolutely.

Now let's see how this is done in quantum probability. We let X denote a quantum observable, which is a self-adjoint operator acting in some Hilbert space, $\mathcal{H}$. For convenience, we assume that X is bounded. In particular, this means that $X\psi \in \mathcal{H}$ is well defined for every $\psi \in \mathcal{H}$. Then we define the *expected value of X in the pure state* $\psi \in \mathcal{H}$ to be

$$\mathcal{E}(X) = \langle X \rangle := \langle \psi, X\psi \rangle. \qquad (16.5.3)$$

Recall that $\|\psi\| = 1$ is the condition on ψ for it to represent a *pure state* and that two such unit vectors ψ, ϕ represent the same pure state if and only if $\phi = \alpha\psi$ for some complex number α satisfying $|\alpha| = 1$. Notice that the value of (16.5.3) does not change if we replace ψ with $\phi = \alpha\psi$ where $|\alpha| = 1$. While the notations $\mathcal{E}(X)$ and especially $\langle X \rangle$ are often used in quantum theory, this is quite unfortunate since the state ψ is often time dependent. In other words, we often are considering a dynamical situation with a changing state of the system. Better notations are $\mathcal{E}_\psi(X)$ and $\langle X \rangle_\psi$.

An important property is that the expected value $\langle X \rangle$ is a real number if $X = X^*$. This follows from elementary Hilbert space calculations:

$$\langle X \rangle = \langle \psi, X\psi \rangle = \langle X^*\psi, \psi \rangle = \langle X\psi, \psi \rangle = \langle \psi, X\psi \rangle^* = \langle X \rangle^*.$$

Even though (16.5.3) looks nothing like an integral, we emphasize that it is the expression in quantum theory that most closely corresponds to an 'integral' of X. This parallelism of mathematical structures between classical and quantum probability is explained more fully in the theory of C^*-algebras, which we commend to the adventurous reader as an advanced mathematical marvel with amazing relations to physics.

The physical intuition behind the mathematical definition (16.5.3) is not trivial. It is a more general formulation of *Born's rule*, which we already saw in (16.2.2). This gives (16.5.3) a lot of authority without actually explaining anything. Before getting into that, let's note that both in classical probability and in quantum probability there are two theoretical ingredients needed for defining the expected value: an observable and a state.

Here's a quite common, but not general, example of Born's rule (16.5.3). Nonetheless we hope this sheds some light on this topic. We suppose that the quantum observable X is *diagonalizable* in the sense that there exists an orthonormal basis $\{\psi_j \mid j \in J\}$ of the Hilbert space such that for every $j \in J$ there exists $\lambda_j \in \mathbb{R}$ (called an *eigenvalue* of X) so that

$$X\psi_j = \lambda_j \psi_j. \tag{16.5.4}$$

(As a technical aside, we also assume that X is a bounded operator.) We take any unit vector $\psi \in \mathcal{H}$ and expand it in the orthonormal basis as

$$\psi = \sum_{j \in J} \alpha_j \psi_j, \tag{16.5.5}$$

where $||\psi|| = 1$ implies that $\sum_{j \in J} |\alpha_j|^2 = 1$. It follows that $\alpha_j = \langle \psi_j, \psi \rangle$. Then we calculate the quantity on the right side of Born's rule to get

$$\langle \psi, X\psi \rangle = \left\langle \sum_j \alpha_j \psi_j, X \sum_k \alpha_k \psi_k \right\rangle = \sum_{j,k} \alpha_j^* \alpha_k \langle \psi_j, X\psi_k \rangle$$

$$= \sum_{j,k} \alpha_j^* \alpha_k \langle \psi_j, \lambda_k \psi_k \rangle = \sum_{j,k} \alpha_j^* \alpha_k \lambda_k \langle \psi_j, \psi_k \rangle = \sum_{j,k} \alpha_j^* \alpha_k \lambda_k \delta_{j,k}$$

$$= \sum_j |\alpha_j|^2 \lambda_j.$$

Here we used (16.5.4) and (16.5.5). The interchange of the summation with X is allowed, since X is bounded. As usual $\delta_{j,k}$ is the Kronecker delta. Now $\langle \psi_j, \psi \rangle = \alpha_j$ is called the *probability amplitude* for the state ψ to transition to the state ψ_j, while $p_j := |\langle \psi_j, \psi \rangle|^2 = |\alpha_j|^2$ is called the *probability* for the state ψ to transition to the state ψ_j. Clearly each $p_j \in [0, 1]$ and $\sum_j p_j = 1$.

So now Born's rule (16.5.3) boils down to simply saying

$$\langle \psi, X\psi \rangle = \sum_j p_j \lambda_j,$$

which is the *classical* expected value of a *classical* observable which takes the value λ_j with probability p_k. But it must be emphasized that these classical probabilities p_j arise in quantum theory from the complex numbers α_j, which need not be real numbers. In short, quantum probability is not classical probability.

While this example may give some of the intuition behind Born's rule, it does not cover the case when X is not diagonalizable in the above sense, for example when X has *continuous spectrum*. And the case when X is unbounded presents the usual technical details. We will see that Born's rule (16.5.3) is a logical consequence of an axiom. (See (21.2.2) and the accompanying discussion.) This shows the power of the axiomatic method to place otherwise *ad hoc* rules into a consistent framework.

16.6   Dispersion

Dispersion is one of the major differences between classical and quantum probability. We now examine this important concept. The definition is the same for either type of probability—or even any more general probability theory with a reasonable definition of expected value! We suppose that X is an observable for which the expected value $\mathcal{E}(X) = \langle X \rangle$ is well defined. Then the *dispersion* (or *variance*) of X is defined to be

$$\Delta(X) := \langle \left(X - \langle X \rangle \right)^2 \rangle$$

provided that the expected value on the right side also is well defined. Notice that the state (whether it be P or ψ) is suppressed from the notation. In classical probability this typically causes no confusion, since there is only one state under consideration. However, in quantum probability one often uses the notation $\Delta_\psi(X)$ to indicate that the state being considered is ψ, and one refers to this as the dispersion of X in the state ψ.

Exercise 16.6.1 *Prove that* $\Delta(X) = \langle X^2 \rangle - \langle X \rangle^2 \geq 0$ *in both classical and quantum probability.*

A related concept, which is important in statistics, is the *standard deviation* of X,

$$\sigma(X) := +\sqrt{\Delta(X)}.$$

We say that X is *dispersion-free* (with respect to some implicitly given state) if $\Delta(X) = 0$. In the case of classical probability the dispersion in the state P is given by

$$\Delta(X) = \int_\Omega \left(X(\omega) - \langle X \rangle \right)^2 dP(\omega),$$

where

$$\langle X \rangle = \int_\Omega X(\omega) \, dP(\omega).$$

So, X is dispersion-free with respect to the (classical) state P simply means that the non-negative function $(X(\omega) - \langle X \rangle)^2$ integrates to 0. In turn, this means that $(X(\omega) - \langle X \rangle)^2 = 0$ for all (well, *almost all* if you speak of measure theory) points $\omega \in \Omega$. So, $X = \langle X \rangle$ (well, *almost everywhere*), which means that X is (well, *almost*) a constant function.

Let's look at this in detail for the discrete case $\Omega = \{\omega_1, \omega_2, \dots\}$ with $P(\omega_j) = p_j \in [0, 1]$ for each j and $\sum_j p_j = 1$. If X is dispersion-free, then

$$\Delta(X) = \sum_j p_j(X(\omega_j) - \langle X \rangle)^2 = 0.$$

So we have a sum of non-negative terms adding up to 0. And that implies that each term must be 0, that is,

$$p_j(X(\omega_j) - \langle X \rangle)^2 = 0$$

for each index j. But this then implies for each j that

$$p_j = 0 \quad \text{or} \quad X(\omega_j) = \langle X \rangle.$$

Of course, the points ω_j for which $p_j = 0$ are not very important; they are the outcomes which never occur! So we give the set of these outcomes a name:

$$N := \{\omega_j \mid p_j = 0\}.$$

Then $X = \langle X \rangle$, a constant, on the set $\Omega \setminus N$. And the values of X on the set N do not matter, since they never happen! So, X is the constant function on the part of Ω that does matter. (N is an example of a *null set*.)

Now here is an extreme case: $p_k = 1$ for some fixed integer $k \geq 1$ and $p_j = 0$ for all $j \neq k$. This is a probability given by the *Kronecker delta*, that is, the probability KD_k of each outcome ω_j is given for all j by

$$KD_k(\omega_j) := \delta_{k,j}.$$

Then, for any $X : \Omega \to \mathbb{R}$ we see that

$$\langle X \rangle = \sum_j KD_k(\omega_j)\, X(\omega_j) = \sum_j \delta_{k,j}\, X(\omega_j) = X(\omega_k). \tag{16.6.1}$$

In particular, $N = \Omega \setminus \{\omega_k\}$ and so X is a constant function on $\Omega \setminus N = \{\omega_k\}$. In other words, for this probability measure every observable is essentially constant and (16.6.1) holds. So, $\langle X^2 \rangle = (X(\omega_k))^2 = \langle X \rangle^2$ and therefore $\Delta(X) = 0$ for every observable X.

The Kronecker delta probability measure KD_k is a *pure state*, meaning that if KD_k is a convex combination, $KD_k = \lambda P_1 + (1 - \lambda) P_2$ with $\lambda \in [0,1]$ and P_1, P_2 being probability measures, then we must have $P_1 = P_2 = KD_k$. Also, if a probability measure P is a pure state, then $P = KD_k$ for some k. While the pure state KD_k is uninteresting as a probability (because it means one outcome occurs 100% of the time and the remaining outcomes never occur), it turns out that every probability measure is a (possibly infinite) convex combination of these pure states.

The situation for dispersions in quantum probability is quite different. Let's consider a quantum observable $X = X^*$. For simplicity we first consider the case $\langle X \rangle = 0$. Next we reintroduce the implicitly given pure state ψ into the notation to obtain

$$\Delta_\psi(X) = \langle X^2 \rangle_\psi = \langle \psi, X^2 \psi \rangle = \langle X\psi, X\psi \rangle = ||X\psi||^2. \tag{16.6.2}$$

(If X is a bounded operator, all this makes sense. If X is unbounded, we must assume that ψ is the domain of X and that $X\psi$ is also in the domain of X.) Therefore in the case $\langle X \rangle_\psi = 0$, X is dispersion-free in the state ψ if and only if $X\psi = 0$ if and only if ψ is an eigenvector of X with eigenvalue 0. For the case of any value of $\langle X \rangle_\psi$ we define a new observable $Y := X - \langle X \rangle_\psi$.

Exercise 16.6.2 *Prove that* $\langle Y \rangle_\psi = 0$ *and* $\Delta_\psi(Y) = \Delta_\psi(X)$.

So we can apply (16.6.2) to Y getting that Y is dispersion-free in the state ψ if and only if $Y\psi = 0$ if and only if $(X - \langle X \rangle)\psi = 0$ if and only if

$$X\psi = \langle X \rangle \psi.$$

By the previous exercise we conclude that if X is dispersion-free in the state ψ, then ψ is an eigenvector of X with eigenvalue $\langle X \rangle_\psi$.

Conversely, if the state ψ is an eigenvector of X, say $X\psi = \lambda\psi$ where $\lambda \in \mathbb{C}$, then $X^2\psi = X(\lambda\psi) = \lambda X\psi = \lambda^2\psi$ and so in the state ψ we get

$$\begin{aligned}
\Delta_\psi(X) &= \langle X^2 \rangle_\psi - \langle X \rangle_\psi^2 \\
&= \langle \psi, X^2\psi \rangle - \langle \psi, X\psi \rangle^2 \\
&= \langle \psi, \lambda^2\psi \rangle - \langle \psi, \lambda\psi \rangle^2 \\
&= \lambda^2 - \lambda^2 = 0.
\end{aligned}$$

We have proved this important result.

Theorem 16.6.1 *Let* $X = X^*$ *be a quantum observable and let* ψ *be a state such that* $\Delta_\psi(X)$ *is defined for the state* ψ. *Then* X *is dispersion-free in the state* ψ *if and only if* ψ *is an eigenvector of* X *with eigenvalue* $\langle X \rangle_\psi$.

The next corollary is one immediate consequence of this theorem.

Corollary 16.6.1 *Let* X *be a quantum observable that acts in a* separable *Hilbert space* $\mathcal{H}$. *Suppose there exists an orthonormal basis of eigenvectors of* X *and that each eigenvalue has multiplicity 1. Define the set of states for which* X *is dispersion-free to be the quotient set:*

$$\mathcal{D} := \{\psi \in \mathcal{H} \mid ||\psi|| = 1, \Delta_\psi(X) = 0\} / \mathbb{S}^1.$$

Then $\mathcal{D}$ *is either a finite set or is countably infinite. Here* $\mathbb{S}^1$ *is the unit circle in the complex plane* $\mathbb{C}$.

Remark: The set of states in $\mathcal{H}$ is uncountably infinite if $\dim \mathcal{H} \geq 2$. For example, if $\mathcal{H} = \mathbb{C}^2$, then the set of states is $\mathbb{S}^3 / \mathbb{S}^1 = \mathbb{S}^2$, where $\mathbb{S}^3$ is the *unit 3-sphere* in $\mathbb{R}^4$. (By the way, this is the *Hopf fibration* of $\mathbb{S}^3$ over $\mathbb{S}^2$.) So in this case (and given the hypotheses on X) there is an abundance of states ψ for which $\Delta_\psi(X) > 0$.

Proof: Let $\mathcal{B} = \{\psi_k\}$ be an orthonormal basis of eigenvectors of X. The map that sends ψ_k to its $\mathbb{S}^1$ equivalence class in $\mathcal{D}$ is an injective function from $\mathcal{B}$ to $\mathcal{D}$. Since each eigenvalue has multiplicity 1, there are no other eigenvectors of X, except for constant multiples of the ψ_k. So this map is surjective as well. Therefore the cardinality of $\mathcal{D}$ is equal to the dimension of $\mathcal{H}$, which by the separability hypothesis is finite or countably infinite. ∎

The previous theorem has another immediate and powerful consequence.

Corollary 16.6.2 *Suppose that $X = X^*$ is a quantum observable that has no eigenvectors. Then $\Delta_\psi(X) > 0$ for any state ψ for which $\Delta_\psi(X)$ is defined.*

This corollary has application to two fundamental observables in physics: position and (linear) momentum. Let's see how that works in the one-dimensional case, where the Hilbert space is $L^2(\mathbb{R})$ and the position operator Q is defined for all $\psi \in D(Q)$ as

$$Q\psi(x) := x\psi(x).$$

(Parenthetically, we remark that

$$D(Q) := \{\psi(x) \in L^2(\mathbb{R}) \mid x\psi(x) \in L^2(\mathbb{R})\}$$

and, by functional analysis, $Q = Q^*$ and $\operatorname{Spec} Q = \mathbb{R}$.) Then the eigenvalue equation for Q is $Q\psi = \lambda\psi$ for some real number λ. But this is equivalent to $(x - \lambda)\psi(x) = 0$ which in turn implies $\psi(x) = 0$ for every real $x \neq \lambda$. No matter what the value of $\psi(\lambda)$ might be chosen to be, we have that $||\psi|| = 0$. (Technically, $\psi = 0 \in L^2(\mathbb{R})$ by measure theory.) So, ψ is not an eigenvector, since an eigenvector by definition is non-zero and has non-zero norm. So the eigenvalue equation for Q has no non-trivial solution, thereby implying that Q has no eigenvalues.

We learn immediately from this that $\Delta_\psi(Q) > 0$ for all pure states ψ for which the left side is defined. This important fact is usually derived in physics texts from the Heisenberg uncertainty principle, which relates Q with the momentum operator P. But we see that this is a 'stand alone' result about the observable Q in its own right.

This contrasts dramatically with the discrete case in classical probability, where there are pure states (namely, the Kronecker delta states KD_j) with respect to which we have $\Delta(X) = 0$ for *every* observable X. Thus Q shows how quantum pure states and quantum observables are very different from their classical analogues. The fact that Q is not a bounded operator is not essential. If we consider a bounded self-adjoint operator with no eigenvectors, we get the same sort of result. For example, we could take $(1 + Q^2)^{-1}$ as such as an operator.

Next, let's see that the momentum operator P, defined by

$$P\psi = \frac{\hbar}{i}\frac{d\psi}{dx},$$

has no eigenvalues. (Here $\psi \in D(P)$ which we do not define, $P = P^*$ and $\operatorname{Spec} P = \mathbb{R}$. However, we do remark that $\psi \in D(P)$ implies that ψ is differentiable and so, in particular, ψ is continuous.) Then the eigenvalue equation says $P\psi = \lambda\psi$ for some real number λ, which is equivalent to

$$\frac{\hbar}{i}\frac{d\psi}{dx} = \lambda\psi(x).$$

The general solution of this first order, linear differential equation is

$$\psi(x) = c\, e^{i\lambda x/\hbar}$$

for all $x \in \mathbb{R}$, where $c \in \mathbb{C}$ is an arbitrary complex constant. Therefore, $|\psi(x)|^2 = |c|^2$ for all $x \in \mathbb{R}$. And so this function is in $L^2(\mathbb{R})$ if and only if $c = 0$, in which case $\psi \equiv 0$ and so ψ is not an eigenvector. So the eigenvalue equation for P has no non-trivial solution, thereby implying that P has no eigenvalues.

We learn immediately from this that $\Delta_\psi(P) > 0$ for all pure states ψ for which the left side is defined. This important fact is usually derived in physics texts from the Heisenberg uncertainty principle, which relates P with the position operator Q. But we see that this is a 'stand-alone' result about the observable P in its own right.

If $X = \lambda I$ for some $\lambda \in \mathbb{R}$ on a Hilbert space $\mathcal{H} \neq 0$, then $X = X^*$ and $\operatorname{Spec} X = \{\lambda\}$. So X is the observable that only has one measured value, namely λ. Since probability theory is about observables with several possible values, this is a trivial situation. Moreover, $\Delta_\psi(X) = 0$ for every state ψ.

Exercise 16.6.3 *Suppose that X is a quantum observable which satisfies $\Delta_\psi(X) = 0$ for every state ψ. Show that $X = \lambda I$ for some real number λ.*

Exercise 16.6.4 *Let $X = X^*$ be a quantum observable with $\operatorname{Spec} X = \{\lambda\}$ for some real number λ. Prove that $X = \lambda I$.*

The previous exercise is an 'eye exam', since the functional analysis needed to solve it can be found somewhere (earlier!) in this book. Also, the industrious reader is invited to find a 2×2 matrix X with $\operatorname{Spec} X = \{\lambda\}$ for some real number λ, but nonetheless $X \neq \lambda I$.

16.7   Probability 1

A limiting case consists of events with probability one. The only quantum event which has probability 1 in all states, as we have seen, is the projection operator I, the identity operator on the Hilbert space, $\mathcal{H}$.

But there are also quantum observables such that $\operatorname{Spec} X = \{\lambda\}$ for some real number λ. Taking the result from the last section, we know that $X = \lambda I$.

Exercise 16.7.1 *Given this situation, show that $P(X = \lambda \,|\, \psi) = 1$ for all states ψ.*

For example, the mass of an elementary particle gives, as far as we know, the same numerical value, say m, in every possible measurement. Then the corresponding quantum observable has to be mI, for which the collapse of the state following a measurement is the identity map since the initial state is an eigenvector of mI. While this is consistent with experiment, it hardly explains the various masses of the elementary particles.

The moral is that probability 1 fits into quantum probability, but there is something lacking that can best be described as *je ne sais quoi*.

16.8   A Feature, Not a Bug

le hasard . . . qui est en réalité
l'ordre de la nature
Anatole France

Some comments are in order about objections to the probabilistic aspect of quantum theory. The probability theory to be improved or replaced is the quantum probability theory as presented in this chapter. And quantum probability I shall argue is an essential feature of quantum theory. That is the issue which must be faced.

Now there are three aspects of quantum theory that are generally accepted by the physics community. The first aspect is that many, though maybe not all, observables are represented by self-adjoint operators. The second is that many, though maybe not all, self-adjoint operators correspond to physical observables. The third aspect is that the expected values of such self-adjoint operators are essential for understanding the associated physical observables. The first two aspects have nothing to do explicitly with probability, while the third aspect does use an expression ('expected value') that comes from classical probability theory. However, as we shall see, even this third aspect can be viewed as a simple algebraic operation (in linear algebra) with an interest that is independent from probabilistic considerations.

Let's discuss the first aspect. The spectral theorem and its converse, as abstract as they are, tell us that the study of self-adjoint operators is exactly equivalent mathematically to the study of projection valued measures on the measure space $\mathbb{R}$ equipped with the σ-algebra of Borel sets. The phrase 'projection valued measure' has been lumped into 'pvm' and thought of as a single object. Which it is. However, projections are exactly quantum events and so the phrase 'quantum event valued measure' (or 'qevm') is just as reasonable for naming this object. So a qevm is a pvm and *vice versa*. This means that if one admits that self-adjoint operators are a central to the study of quantum theory, one has admitted that qevm's are central too. Even if qevm's are not discussed explicitly, they are still there behind every self-adjoint operator A which has its qevm P_A.

But now that the quantum events $P_A(B)$ are recognized as central to quantum theory (where $A = A^*$ is a quantum observable and $B \subset \mathbb{R}$ is a Borel set), one has to understand what role these self-adjoint operators play in quantum theory. Specifically, we come to the second aspect mentioned earlier. Do these self-adjoint operators $P_A(B)$ correspond to physical observables? Well, we have already seen that they have a physical interpretation! We know that the spectrum of the projection $P_A(B)$ is a subset of $\{0, 1\}$ and so these are the only possible values the corresponding physical observable can take. We have interpreted 0 to mean that the quantum event does not occur and 1 to mean that it does occur.

But there is another argument for accepting that $P_A(B)$ corresponds to a physical observable. We start by discussing a self-adjoint operator A that *does* represent a physical observable. Then there is another version of the spectral theorem (called the *functional calculus*) which says that for any bounded (Borel) function $f : \mathbb{R} \to \mathbb{R}$ there corresponds a self-adjoint, bounded linear operator $f(A) : \mathcal{H} \to \mathcal{H}$ with various nice properties. The operator $f(A)$ is interpreted physically as the observable that consists of this composite of two operations: First, measure A and get a real number a; second, evaluate $f(a) \in \mathbb{R}$. In other words, the physical interpretation of A gives us a physical interpretation of $f(A)$.

Now we take the special case when $f = \chi_B$, the *characteristic function* of the Borel set B. This function is defined for $\lambda \in \mathbb{R}$ by

$$\chi_B(\lambda) := \left\{ \begin{array}{ll} 1 & \text{if } \lambda \in B, \\ 0 & \text{if } \lambda \notin B. \end{array} \right.$$

Then χ_B is a bounded, Borel function. Taking into account the physical interpretation in general for $f(A)$, we see that $\chi_B(A)$ (which is self-adjoint as a consequence of the functional calculus) represents the physical observable that yields the value 1 when A has its observed value in B and has value 0 when A has its observed value in the complement of B, namely in $\mathbb{R} \setminus B$.

At this point it should not surprise the reader to learn that we have $P_A(B) = \chi_B(A)$. This explicitly shows that the self-adjoint operator $P_A(B)$ corresponds to a physical observable. As we have remarked, this physical observable is a type of what is called a *Yes/No experiment*. Such experiments yield exactly two possible values, which can be interpreted as answering a question by Yes or by No. In this case the question is: Did the measurement of A give a value in B?

(Parenthetically, we note that the spectral theorem gives us a qevm P_A associated with a self-adjoint operator A such that $A = \int_{\mathbb{R}} \lambda \, dP_A(\lambda)$. Then we define the functional calculus by $f(A) := \int_{\mathbb{R}} f(\lambda) \, dP_A(\lambda)$. Conversely, a functional calculus for A gives a qevm P_A defined by $P_A(B) := \chi_B(A)$, as noted above. These two operations are inverses to each other. Furthermore, the properties of a qevm translate into those of a functional calculus, and *vice versa*. See a good functional analysis text for all this in appropriate detail.)

Proceeding, we next wish to understand $P_A(B)$ better via its expectation values. This is the third aspect mentioned above. So we take a state ψ and consider the expectation value of the self-adjoint operator $P_A(B)$ in that state. This is colloquially known as using ψ to 'sandwich the observable'. In other words, it is the simple algebraic operation given by forming

$$\langle \psi, P_A(B)\psi \rangle \qquad \text{with } ||\psi|| = 1. \tag{16.8.1}$$

In physics, such an expression is often called a *diagonal matrix entry* of $P_A(B)$. Here one is thinking of a matrix representation of $P_A(B)$ with respect to an orthonormal basis of the Hilbert space and that the state ψ is one element in that orthonormal basis. Indeed, the above real number (16.8.1) does then appear on the diagonal of the associated matrix. So this is actually an operation in linear algebra without an explicit probabilistic interpretation.

Let me indulge in a brief interlude concerning the maps l_ψ in linear algebra defined by

$$l_\psi(T) := \langle \psi, T\psi \rangle,$$

where $T : \mathcal{H} \to \mathcal{H}$ is a linear bounded operator (i.e., $T \in \mathcal{L}(\mathcal{H})$) and $\psi \in \mathcal{H}$ is a state. Then $l_\psi : \mathcal{L}(\mathcal{H}) \to \mathbb{C}$ is a linear map, known as a *(linear) functional* on $\mathcal{L}(\mathcal{H})$. The claim is that knowing the values $l_\psi(T)$ for *all possible* states ψ completely determines T.

Exercise 16.8.1 *Suppose $T_1, T_2 : \mathcal{H} \to \mathcal{H}$ are linear bounded operators. Show that $l_\psi(T_1) = l_\psi(T_2)$ for all states $\psi \in \mathcal{H}$ implies that $T_1 = T_2$.*
Hint: *Use the polarization identity (9.3.8) for complex Hilbert spaces.*

Exercise 16.8.2 *Show there exists a Hilbert space $\mathcal{H}$ as well as operators T_1 and T_2 acting on it such that $l_\psi(T_1) = l_\psi(T_2)$ for all elements ψ in one orthonormal basis of $\mathcal{H}$, but $T_1 \neq T_2$.*
Hint: *Take $\mathcal{H} = \mathbb{C}^2$.*

The point of this brief interlude is that in passing from the study of $P_A(B)$ to the study of all of its diagonal matrix elements (16.8.1) we do not lose information about the operator $P_A(B)$.

As we have already remarked, (16.8.1) is a classical probability function taking B as variable and keeping A and ψ fixed. It is only a question then of giving a name and notation to this classical probability function, something which we have already done in Definition 16.2.4. Recall that notation:

$$P(A \in B \,|\, \psi) = \langle \psi, P_A(B)\psi \rangle.$$

It is important to note that these classical probability measures 'live' on the spectrum $\mathrm{Spec}(A)$ of A meaning that $P(A \in (\mathbb{R} \setminus \mathrm{Spec}(A)) \,|\, \psi) = 0$ for *all* states ψ, since $P_A(\mathbb{R} \setminus \mathrm{Spec}(A)) = 0$. This mathematical fact suggests that $\mathrm{Spec}(A)$ should be given a physical interpretation. And this is what is done. The standard interpretation of $\mathrm{Spec}(A)$ is that it is the set of possible values of measurements of the observable A.

These classical probability functions, known as *spectral measures*, emerge quite naturally from generally accepted standard non-probabilistic aspects of quantum theory. And therefore it behooves us to give them a physical interpretation. Of course, we have already done that. So we have arrived at all of the structures of quantum probability theory by using some generally accepted principles of quantum theory. Excuse me while I dwell on this point. These principles are accepted by just about every physicist, and they do not have any necessary reference to probability theory. Or to put this in yet another way, I am asserting that all of this abstract mathematics of quantum probability theory is a consequence of the commonly accepted physical 'intuition' of quantum theory. The physics and the mathematics are inseparable. The physics leads to the mathematics of self-adjoint operators, which in turn is equivalent to qevm's. The physics leads to the mathematics of states as well. Then the qevm's and the states lead us mathematically to classical probability measures. And having these probability measures (or spectral measures) in hand, how can we not give them some physical interpretation?

Therefore, the intellectual challenge for those who wish to improve or even replace quantum probability is to deal with these comments that place quantum probability as a necessary consequence of other generally accepted quantum theoretical principles that are not at all probabilistic. But for now quantum probability is a feature of quantum theory.

16.9   Spectral Theorem as Diagonalization

It is patently outrageous to say that the spectral theorem as presented earlier is a generalization to the infinite dimensional setting of diagonalization of self-adjoint matrices in the finite dimensional setting. The point of this section is to convince the reader that this is not so nonsensical.

First off, we had better recall what *diagonalization* means in the finite dimensional setting. So we suppose that $A : \mathbb{C}^n \to \mathbb{C}^n$ is a *self-adjoint* (or *Hermitian*) operator, where we assume that $n \geq 1$. This means that $A = A^*$. Notice that the definition (9.3.4) of A^* depends on the existence of an inner product on $\mathbb{C}^n$. For convenience we let this be the standard inner product defined in (9.1.1). Now there is a theorem in elementary linear algebra that says that this operator is *diagonalizable*.

We next describe this theorem without proving it. What this says is that $\mathbb{C}^n$ has some orthonormal basis, say ε_j for $j = 1, \ldots, n$, of eigenvectors of A. This means that there exist scalars λ_j, not necessarily distinct, such that $A\varepsilon_j = \lambda_j \varepsilon_j$ for $j = 1, \ldots, n$. The λ_j's are the eigenvalues of A. Since $A = A^*$, we have that each $\lambda_j \in \mathbb{R}$. The set of all the eigenvalues is called the *spectrum* of A, denoted by $\mathrm{Spec}(A)$. We write this set as

$$\mathrm{Spec}(A) = \{\mu_1, \ldots, \mu_k\} \subset \mathbb{R} \quad \text{where} \quad 1 \leq k \leq n,$$

and the μ_j's are distinct numbers. For $1 \leq j \leq k$ we define the *eigenspace* or *spectral subspace* associated to the eigenvalue μ_j by

$$V_j := \{v \in \mathbb{C}^n \mid Av = \mu_j v\}.$$

Since μ_j is an eigenvalue of A we have that $V_j \neq 0$ for each $j = 1, \ldots, k$. This gives us an orthogonal direct sum decomposition of $\mathbb{C}^n$ as

$$\mathbb{C}^n = V_1 \oplus V_2 \oplus \cdots \oplus V_k, \tag{16.9.1}$$

since eigenspaces for distinct eigenvalues are orthogonal (see Exercise 9.3.13) and the eigenvectors form a basis. Furthermore, A restricted to V_j is a very simple operator, namely, multiplication by the real number μ_j.

We rewrite these facts in terms of some projections. (See Definition 9.3.8.) First, for every $j = 1, \ldots, k$ we have the projection $P_j : \mathbb{C}^n \to \mathbb{C}^n$ whose range is V_j. More explicitly, we take an arbitrary vector $v \in \mathbb{C}^n$ and write it (uniquely!) as $v = v_1 + \cdots + v_j + \cdots + v_k$ with each $v_l \in V_l$. This is just what the direct sum decomposition (16.9.1) gives us. Then we define P_j as $P_j v := v_j$.

Exercise 16.9.1 *Prove for $j = 1, \ldots, k$ that P_j is a projection and $P_j \neq 0$.*

Exercise 16.9.2 *Prove that $P_i P_j = 0$ if $i \neq j$.*
Hint: V_i *is orthogonal to* V_j *for* $i \neq j$.

This is a result that has clear geometrical meaning in terms of the ranges of P_i and P_j. But note that this is very different from the properties of real (or complex) numbers. Here we have $P_i \neq 0$ and $P_j \neq 0$, while their product is zero, $P_i P_j = 0$. One says that P_i and P_j are *zero divisors*.

Now we can rewrite (16.9.1) in terms of these projections as

$$I = P_1 + P_2 + \cdots + P_k. \tag{16.9.2}$$

This equation is called a *resolution of the identity*. There are many, many resolutions of the identity, but this particular one is chosen because A can be written as a pretty formula using it. Actually, it is not so difficult to see that

$$A = \mu_1 P_1 + \mu_2 P_2 + \cdots + \mu_k P_k. \tag{16.9.3}$$

Exercise 16.9.3 *Prove this is correct by restricting A to each V_j.*

We say that the finite sum (16.9.3) is the *spectral resolution of A*. Next, we want to write this finite sum as an integral, but with respect to a pvm instead of a usual measure. To do this we let each point $\mu_j \in \mathbb{R}$ have measure given by the corresponding projection P_j, that is $P(\{\mu_j\}) = P_j$. We say that the points μ_j are *atoms* of the pvm. Also we want $P(B) = 0$, the zero operator, for Borel sets B that do not have any point in common with $\mathrm{Spec}(A)$, that is, if $B \cap \mathrm{Spec}(A) = \emptyset$, the empty set. This is all encoded in the definition of P_A for a general Borel subset B of $\mathbb{R}$ as follows:

$$P_A(B) := \sum_{j \,:\, \mu_j \in B} P_j.$$

The previous equation should be compared with (16.1.1). We recall that a sum over the empty set is defined to be zero, which in this case is the zero operator.

Exercise 16.9.4 *Prove that P_A is a pvm and that $A = \int_{\mathbb{R}} \lambda \, dP_A(\lambda)$.*
Hint: *For the second part show that the integral on the right side is equal to the finite sum (16.9.3).*

Now the full generalization of finite sum (an operation in algebra) to the setting of analysis is integral (an operation *par excellence* in analysis). So when we pass to the infinite dimensional setting, we expect a spectral theorem for self-adjoint operators that has an integral instead of a finite sum. And that is what happens. Of course, in that infinite dimensional setting that integral sometimes reduces to a finite sum and other times to an infinite sum. In those cases, the pvm is supported on *atoms*, that is, those points $\lambda \in \mathbb{R}$ such that $P_A(\{\lambda\}) \neq 0$. But there are self-adjoint operators which have no eigenvalues and hence no atoms. For these the integral form of the spectral theorem is required.

We can also see how the functional calculus works out quite nicely in the finite dimensional setting. We start with the *spectral decomposition* of A:

$$A = \mu_1 P_1 + \mu_2 P_2 + \cdots + \mu_k P_k.$$

Now we evaluate A^2 using simple algebra getting

$$\begin{aligned}
A^2 &= (\mu_1 P_1 + \cdots + \mu_k P_k)(\mu_1 P_1 + \cdots + \mu_k P_k) \\
&= \mu_1^2 P_1^2 + \cdots + \mu_k^2 P_k^2 \\
&= \mu_1^2 P_1 + \cdots + \mu_k^2 P_k.
\end{aligned}$$

In the second equality we used $P_i P_j = 0$ for $i \neq j$, while in the third we used $P_j^2 = P_j$. Iterating this we see that

$$A^m = \mu_1^m P_1 + \cdots + \mu_k^m P_k$$

for all integers $m \geq 2$. But this formula is also true for $m = 0$ (the resolution of the identity (16.9.2)) and for $m = 1$ (the spectral decomposition (16.9.3) of A). Next, for any polynomial $f : \mathbb{R} \to \mathbb{C}$, say $f(x) = c_0 + c_1 x + c_2 x^2 + \cdots + c_d x^d$ with complex coefficients c_j, we see by linearity that

$$f(A) = f(\mu_1) P_1 + \cdots + f(\mu_k) P_k, \qquad (16.9.4)$$

where $f(A) := c_0 I + c_1 A + c_2 A^2 + \cdots + c_d A^d$. Notice that $f(A)$ only depends on the values of the polynomial f on the spectrum $\mathrm{Spec}(A)$. Its values on the set $\mathbb{R} \setminus \mathrm{Spec}(A)$, which has measure 0 according to the pvm P_A, do not enter into the formula (16.9.4).

But what about an arbitrary function $g : \mathrm{Spec}(A) \to \mathbb{C}$? (By the way, such a function g is a Borel function, since the Borel σ-algebra for the discrete topological space $\mathrm{Spec}(A)$ consists of all subsets of $\mathrm{Spec}(A)$. Also, such a function g is also bounded, since $\mathrm{Spec}(A)$ is finite.)

So, the question is: Given such a bounded, Borel function g does there exist a polynomial $f : \mathbb{R} \to \mathbb{C}$ which is equal to g on the finite set

$$\mathrm{Spec}(A) = \{\mu_1, \ldots, \mu_k\}.$$

A problem of this type is called an *interpolation problem*. It is not an *approximation problem*, that is a problem of approximating the function g on the finite set $\text{Spec}(A)$ by a polynomial. In this interpolation problem, the sought-for polynomial must be *exactly equal* to the function g on the finite set $\text{Spec}(A)$. This particular interpolation problem is solved by using the Lagrange polynomials associated with the set $\text{Spec}(A)$. For each $j = 1, \ldots, k$ we define the *Lagrange polynomial*

$$l_j(x) := \frac{\prod_{i \neq j}(x - \mu_i)}{\prod_{i \neq j}(\mu_j - \mu_i)}.$$

Notice that the numerator is a polynomial of degree $k - 1$. Since the μ's are distinct, the denominator is not equal to 0. So l_j is a well defined polynomial of degree $k - 1$. With respect to the set $\text{Spec}(A)$ it acts like a Kronecker delta. Explicitly, $l_j(\mu_j) = 1$ while $l_j(\mu_k) = 0$ for all $j \neq k$. That is, $l_j(\mu_k) = \delta_{j,k}$.

Now it is a straightforward exercise to show that

$$g = g(\mu_1)l_1 + g(\mu_2)l_2 + \cdots + g(\mu_k)l_k =: f$$

as functions on the set $\text{Spec}(A)$. Because the right side is a finite linear combination of polynomials of degree $k - 1$, it follows that g is a polynomial of degree at most $k - 1$. So, the answer is that any arbitrary (bounded, Borel) function $g : \text{Spec}(A) \to \mathbb{C}$ is equal on $\text{Spec}(A)$ to a polynomial f. Thus, we define $g(A) := f(A)$ for such g and its associated f. In short, we have defined the functional calculus of A for any such function g.

Exercise 16.9.5 *Let $\mathcal{F}_{bb}$ denote the space of all bounded, Borel functions $g : \text{Spec}(A) \to \mathbb{C}$, that is the space of all functions $g : \text{Spec}(A) \to \mathbb{C}$. Prove that $\mathcal{F}_{bb}$ is a commutative algebra with identity element. Let $A = A^*$ be a self-adjoint operator acting in $\mathbb{C}^n$. Prove that the map*

$$\mathcal{F}_{bb} \ni f \mapsto f(A) \in \mathcal{L}(\mathbb{C}^n)$$

is linear, preserves product, maps the identity element to the identity element, and satisfies $f^*(A) = (f(A))^*$. Here $\mathcal{L}(\mathbb{C}^n)$ is the algebra of all linear maps $T : \mathbb{C}^n \to \mathbb{C}^n$.

16.10 Two Observables

There is a remarkable difference between classical probability and quantum probability when one considers two (or more) observables. In the case of classical probability there always exists *one* sample space Ω together with its probability measure P, which describes any given probabilistic situation. This is not a theorem but rather just a description of how classical probability works. (More on this in a moment.) And then we define all of the *observables* as real-valued (Borel) functions (also called *random variables*) with domain being that sample space Ω. Suppose $X, Y : \Omega \to \mathbb{R}$ are two such observables, and we wish to consider that X takes a value in the (Borel) set $B \subset \mathbb{R}$ and that Y takes a value in the (Borel) set $C \subset \mathbb{R}$. Of course, that is the event $E := X^{-1}(B) \cap Y^{-1}(C)$ (which is always going to be in the σ-algebra that is the domain of P). So, $P\big(X^{-1}(B) \cap Y^{-1}(C)\big)$ is well defined and tells us the probability of the event E. This is called the *joint probability* that X has a value in B and that Y has a value in C. Let's give it this notation:

$$P(X \in B, Y \in C) := P\big(X^{-1}(B) \cap Y^{-1}(C)\big).$$

The point is that in classical probability this exists for *all* pairs of observables X and Y. And the order is unimportant in the sense that we have

$$P(X \in B, Y \in C) = P(Y \in C, X \in B).$$

Also, the time order of the measurements corresponding to these theoretical observables does not enter the theory at all. But in quantum probability, there simply is no such theoretical, order independent construct that merits the name of joint probability for observables X and Y which do not commute.

However, first here are the promised words about how measure theory guarantees that classical probability models can always be constructed with exactly one sample space. (Those with little interest in measure theory are invited to skip this paragraph.) What happens in practice is that one chooses the theoretical sample space Ω so that one can define the observables (real-valued functions $\Omega \to \mathbb{R}$) of experimental interest. Now suppose that we have found two adequate probability spaces $(\Omega_1, \mathcal{F}_1, P_1)$ and $(\Omega_2, \mathcal{F}_2, P_2)$, each one with its associated observables of interest. You might say that we are stuck with two sample spaces. But no! Because measure theory says that the Cartesian product $\Omega := \Omega_1 \times \Omega_2$ has a unique σ-algebra $\mathcal{F}$ generated by all the product sets $B_1 \times B_2$, where $B_i \in \mathcal{F}_i$ for $i = 1, 2$, and such that the definition $P(B_1 \times B_2) := P_1(B_1) \, P_2(B_2)$ extends uniquely to a probability measure P on the σ-algebra $\mathcal{F}$. Say $X_1 : \Omega_1 \to \mathbb{R}$ was one of the observables of interest for us for the first sample space. Then this observable can be *pulled back* to Ω by the composition

$$\Omega \xrightarrow{\pi_1} \Omega_1 \xrightarrow{X_1} \mathbb{R},$$

where π_1 is the projection map onto the first coordinate, that is we define $\pi_1(\omega_1, \omega_2) := \omega_1$ for all $(\omega_1, \omega_2) \in \Omega$. Intuitively, the composition $X_1 \circ \pi_1$ represents the 'same' experimental observable as X_1. A similar construction works for the observables of interest defined on the second sample space Ω_2. While an expert might complain that this produces independent observables, the answer to that is that dependent observables must be defined in the first place on a common sample space. That's just how dependency works in classical probability. One of Kolmogorov's major achievements was to show how to construct a single sample space Ω on which one can define all the random variables $\{X_t \,|\, t \geq 0\}$ of a *stochastic process*.

Now let's return to the quantum case. Suppose that X, Y are observables for a quantum system, that is, X, Y are self-adjoint operators acting in a Hilbert space $\mathcal{H}$. As above we let B, C be Borel subsets of the real line $\mathbb{R}$. In analogy to the classical case, we want to consider the event that X takes a value in B. This is the quantum event $P_X(B)$, where P_X is the pvm associated to X. Similarly, $P_Y(C)$ is the quantum event that Y takes a value in C.

In general, if E_1 and E_2 are quantum events ($\equiv$ projections), then there is a unique quantum event $E_1 \wedge E_2$, that is the greatest lower bound of E_1 and E_2. We can read $E_1 \wedge E_2$ as 'E_1 and E_2' (a logical connective) or also as 'E_1 meet E_2' (a lattice operation). The easiest way to think about this binary operation is in terms of the closed subspaces $V_1 := \operatorname{Ran} E_1$ and $V_2 := \operatorname{Ran} E_2$, the ranges of E_1 and E_2, respectively. Then the quantum event $E_1 \wedge E_2$ is the projection onto the closed subspace given by their intersection $V_1 \cap V_2$. (See Exercise 16.2.9.) Here are some more exercises about this.

Exercise 16.10.1 *Suppose that the quantum events E_1 and E_2 commute, that is, $E_1 E_2 = E_2 E_1$. Then $E_1 \wedge E_2 = E_1 E_2 = E_2 E_1$.*
In particular, show that $E_1 \wedge E_1 = E_1$.

Exercise 16.10.2 *Suppose that E_1 and E_2 are quantum events such that $E_1 E_2$ is also a quantum event. Prove that E_1 and E_2 commute.*

Exercise 16.10.3 *Find quantum events E_1 and E_2 such that the operator $E_1 E_2$ is not a quantum event, that is, it is not a projection.*

Exercise 16.10.4 *Let $\mathcal{H} = \mathbb{C}^2$ be the Hilbert space with the observables being the spin matrices $X = S_1$ and $Y = S_3$. Prove that each of the quantum events $E_1 := P_X(1/2)$ and $E_2 := P_Y(-1/2)$ is a projection on a 1-dimensional subspace, that these subspaces of $\mathbb{C}^2$ are not equal and so $E_1 \wedge E_2 = 0$. Finally, prove that $E_1 E_2 \neq E_2 E_1$.*
Hint: *Recall the material in Section 16.4. In particular, the spin matrices S_1 and S_3 are defined there.*

The last exercise tells us that the *single* event that S_1 gives the value $1/2$ and that S_3 gives the value $-1/2$ is the event which never occurs, that is, the event 0. This is typical of observables which do not commute. However, we are more interested in the temporal sequence of two events: First S_1 gives the value $1/2$ and then a bit later S_3 gives the value $-1/2$.

Let's see how this works out in general. So we start off with any two quantum observables X and Y of the same quantum system, that is, two self-adjoint operators acting in the same Hilbert space $\mathcal{H}$. Suppose that B and C are Borel subsets of $\mathbb{R}$. We shall consider when X has a measured value in B and Y has a measured value in C. The time order of these two measurements is critically important! So we first suppose that X is measured and that the system starts in the state $\psi \in \mathcal{H}$, that is, $||\psi|| = 1$. Then

$$P(X \in B \,|\, \psi) = \langle \psi, P_X(B)\psi \rangle = ||P_X(B)\psi||^2 \qquad (16.10.1)$$

and, given that this has happened, the state of the system will become

$$\psi_1 := \frac{1}{||P_X(B)\psi||} P_X(B)\psi.$$

This holds by the Collapse Axiom. Now a subsequent measurement is made immediately of the observable Y, but with the system in the new state ψ_1. So we have that

$$P(Y \in C \,|\, \psi_1) = \langle \psi_1, P_Y(C)\psi_1 \rangle$$
$$= \frac{1}{||P_X(B)\psi||^2} \langle P_X(B)\psi, P_Y(C)P_X(B)\psi \rangle. \qquad (16.10.2)$$

Moreover, given that this has happened, then the system will collapse to the new state

$$\psi_2 := \frac{1}{||P_Y(C)\psi_1||} P_Y(C)\psi_1$$
$$= \frac{||P_X(B)\psi||}{||P_Y(C)P_X(B)\psi||} P_Y(C) \left(\frac{1}{||P_X(B)\psi||} P_X(B)\psi \right)$$
$$= \frac{1}{||P_Y(C)P_X(B)\psi||} P_Y(C)P_X(B)\psi.$$

Now we have to consider how to combine the probabilities from (16.10.1) and (16.10.2). My opinion is that by interpreting these two probabilities as theoretical quantities which predict experimental frequencies we are forced to conclude that these probabilities are to be multiplied:

$$P(X \in B; Y \in C \,|\, \psi) = \frac{1}{||P_X(B)\psi||^2} ||P_X(B)\psi||^2 \langle P_X(B)\psi, P_Y(C)P_X(B)\psi \rangle$$
$$= \langle P_X(B)\psi, P_Y(C)P_X(B)\psi \rangle$$
$$= ||P_Y(C)P_X(B)\psi||^2, \qquad (16.10.3)$$

where the left side is to be read as follows: The probability of *first* measuring that $X \in B$ and *second* measuring that $Y \in C$ given that ψ is the initial state of the system. The validity of the formula (16.10.3) does not seem to be derivable from the axioms of quantum theory, and so may have to be considered just as an axiom in its own right. By interchanging X and Y while simultaneously interchanging B and C, we arrive at

$$P(Y \in C; X \in B \,|\, \psi) = \langle P_Y(C)\psi, P_X(B)P_Y(C)\psi \rangle$$
$$= ||P_X(B)P_Y(C)\psi||^2, \qquad (16.10.4)$$

which is the probability of *first* measuring that $Y \in C$ and *second* measuring that $X \in B$ given that ψ is the initial state. If X and Y commute (which is equivalent to saying that the projections $P_X(B)$ and $P_Y(C)$ commute for all Borel sets B, C by functional analysis), then clearly

$$P(X \in B; Y \in C \,|\, \psi) = P(Y \in C; X \in B \,|\, \psi).$$

Furthermore, in this case ψ_2 is an eigenvector with eigenvalue 1 for both $P_X(B)$ and $P_Y(C)$. So further repetitions of measurements of $X \in B$ and $Y \in C$ in any order whatsoever will keep giving the same results and the state repeatedly collapses from ψ_2 to ψ_2. It should be noted that even in this commutative case ψ, ψ_1, and ψ_2 can be 3 distinct states.

Exercise 16.10.5 *Justify the last sentence by constructing an example.*

Of course, at this point it becomes an obligatory exercise to find an example where
$$P(X \in B; Y \in C \,|\, \psi) \neq P(Y \in C; X \in B \,|\, \psi)$$
for some pair of observables X, Y and some pair of Borel sets B, C. This can be seen with the simplest non-trivial Hilbert space for doing quantum theory, namely $\mathcal{H} = \mathbb{C}^2$. In Section 16.4 we saw the spin matrices $X = S_1$ and $Y = S_3$. Using these one has that

$$P_X(B) = \frac{1}{2} \begin{pmatrix} 1 & 1 \\ 1 & 1 \end{pmatrix} \quad \text{and} \quad P_Y(C) = \begin{pmatrix} 0 & 0 \\ 0 & 1 \end{pmatrix} \qquad (16.10.5)$$

where both $B = \{1/2\}$ and $C = \{-1/2\}$ are sets with exactly one element.

Exercise 16.10.6 *Complete this example by verifying (16.10.5) and then continue by computing $P(X \in B; Y \in C \,|\, \psi)$ and $P(Y \in C; X \in B \,|\, \psi)$ for any state $\psi \subset \mathbb{C}^2$. Identify for which states ψ these quantities are not equal.*

This example shows that quantum theory gets rather complicated, but also rather interesting, for observables which do not commute. For example, one might think that for given observables X, Y and a given state ψ the mapping
$$B \times C \mapsto P(X \in B; Y \in C \,|\, \psi)$$
determines a probability measure on $\mathbb{R}^2 = \mathbb{R} \times \mathbb{R}$, where B, C are Borel subsets of $\mathbb{R}$. Of course, this is what happens in classical probability. But not so for quantum probability!

Exercise 16.10.7 *As a challenging exercise, you might try to understand why the last statement is true. Actually, in general neither (16.10.3) nor (16.10.4) defines an additive set function on the rectangular sets $B \times C$.*

However, the *marginals* (to use terminology from classical probability) of (16.10.3) and (16.10.4) do give the classical probability on the other factor. For example,

$$P(X \in \mathbb{R}; Y \in C \,|\, \psi) = \langle P_X(\mathbb{R})\psi, P_Y(C)P_X(\mathbb{R})\psi \rangle = \langle \psi, P_Y(C)\psi \rangle$$
$$= P(Y \in C \,|\, \psi),$$

since $P_X(\mathbb{R}) = I$, the identity map. The other case works out equally quickly.

One moral of this longish story is that for non-commuting observables, the natural definition of their 'joint probability measure' is not a probability measure. But for commuting variables (16.10.3) (which equals (16.10.4)) does define their joint probability measure, which is indeed a probability measure.

Exercise 16.10.8 *For those in the know about measure theory, the very last sentence is a challenge exercise for you.*

These curious properties of measurements of two observables are direct consequences of the Collapse Axiom, which has received much scrutiny and criticism. If this axiom is rejected, it seems that the simplest alternative theory would have that the state does not change when a measurement is made, but the rest of quantum theory is left intact. Continuing with the notation established above, this would mean that starting in the state ψ and measuring X and Y we obtain these probabilities:

$$P(X \in B \,|\, \psi) = ||P_X(B)\psi||^2 \ \text{ and } \ P(Y \in C \,|\, \psi) = ||P_Y(C)\psi||^2. \quad (16.10.6)$$

And the point is that the time order of these measurements would not matter; in either order these would be the probabilities. This already contradicts (16.10.2), which is the probability of getting a value of Y in C *after* first starting in the state ψ and measuring a value of X in B *with collapse*.

But under the hypothesis of no collapse, the probability of getting a value of X in B and of getting a value of Y in C would be the product of the probabilities in (16.10.6):

$$||P_X(B)\psi||^2 \, ||P_Y(C)\psi||^2, \quad (16.10.7)$$

which is actually a joint probability distribution. But this is different, in general, from both (16.10.3) and (16.10.4). Therefore, this alternative to the Collapse Axiom makes a prediction that is at odds with that made by standard quantum theory. And so experiments must be called upon to decide between these two alternatives. I am not sure whether such experiments have been done for many quantum systems. But this has been extensively looked at in the case of spin and, at least for that case, standard quantum theory holds.

Exercise 16.10.9 *Continue with Exercise 16.10.6. Evaluate the expression (16.10.7) and find the conditions under which it is different from the results previously obtained.*

16.11   Notes

The quote from Lewis Carroll is what Alice says in the very first words of Chapter 2 of "Alice in Wonderland." Carroll has Alice realize that her English is not quite right. The beauty in her error is that the meaning comes through anyway, even though it sounds strange. And that is what quantum probability is like. Even more curiouser would it have been if Carroll had lived after quantum theory was developed. (And yes, I realize that my English is not quite right.) Maybe he would have written "Through the Quantum Looking Glass."

My underlying attitude about probability theories of whatever ilk is that they serve to understand the relative frequencies of the results of identically prepared experiments. (The quotation at the beginning of Chapter 8 applies here.) I'm sure that I will be decried in some intellectual circles as hopelessly naïve, quaintly out of date and who knows what other academic shortcoming, if not outright sin. So I am not *dans le vent*, an expression which no doubt is no longer in the wind. Well, gentle reader, if you wish to go down the path of modern philosophical thought on probability theory, I wish you *bon voyage*. It is not a path that interests me. It seems to me that relative frequencies come directly out from experiments and so must be explained. If current probability theories do not serve that end, then they must be replaced by some better way to understand the experimental relative frequencies. Now I don't know what such a theory might be, but I do know what it should be called: a probability theory.

In my terminology *quantum probability* refers to the probability theory that arises in quantum theory, and nothing else. However, it is the first of an infinitude of new probability theories, known generically as *non-commutative probabilities*. The non-commutativity is that of a C^* algebra, a very beautiful theory which is way too advanced for inclusion in this book. In the context of this chapter the relevant C^*-algebra is $\mathcal{L}(\mathcal{H})$, the bounded linear operators mapping a Hilbert space $\mathcal{H}$ to itself. One can think about the Kolmogorov axiomatization of classical probability as analogous to Euclid's axiomatization of geometry. While it took some two millennia after Euclid's highly influential book was written for non-Euclidean geometries to appear in the scientific literature, the seeds of quantum probability theory were already sown before Kolmogorov's seminal work in the 1930s. And it is worthwhile to note that quantum probability and non-commutative probability are very active fields of research as of the moment of writing this. However, most forms of non-commutative probability, such as *free probability*, involve one linear functional that emulates the integral of calculus. In quantum probability, as understood here, there is no such unique linear functional, but rather one such for each quantum state.

There are major differences between classical probability and quantum probability. I think that the most important of these is not at the level of mathematics but rather at the level of physics. Simply put, there is no physics in classical probability; there is no equation for time evolution. It is purely a kinematical model. Of course, it can be part of some more inclusive theory in which something changes in time, say there is a time dependence of the probability measure (i.e., the state), and there is some way to describe that change. However, this is in addition to the basic Kolmogorov theory.

But quantum probability is a part of quantum theory. And that includes the time dependent Schrödinger equation in an essential way as well as the collapse of the state. While we can focus on the particular corner of quantum theory known as quantum probability, we should not forget its fundamental role in its physical context.

It is worth commenting on the analogy of the Collapse Axiom with the idea of *conditional probability* in classical probability theory, as expounded on in any number of fine texts. However, classical physics, despite its name, is not a sub-discipline of classical probability. But there is a sub-discipline of classical physics, known as *classical statistical mechanics*, which uses classical probability as one of its principle tools. So this analogy, as far as I can make out, is just that: an analogy. And not an explanation.

Feynman states in chapter 37 of [12] that the "*only* mystery" of quantum theory is how probabilities are combined. (Emphasis in original.) I would add that an even deeper mystery is why there are probabilities *at all* in quantum theory and why we need complex numbers in order to evaluate them.

The expression 'eye exam' originally arises in the context of a lecture when a student asks the professor a question whose answer is on the blackboard at that very moment. So, the professor says that the answer is an eye exam, thereby challenging the student to see (literally) the answer. Who was the first professor to use this expression? I'm not sure, but it wasn't me, even though I have been known to use it.

Chapter 17

The Heisenberg Picture

> The play's the thing, wherein I'll
> catch the conscience of the King.
> Hamlet in *Hamlet*, William Shakespeare

The *Heisenberg picture* is an equivalent reformulation of the standard quantum theory, which in the present context is referred to as the *Schrödinger picture* and which is what we have seen so far in this book. However, the Heisenberg picture can muddle up the intuitions of quantum theory that you may have acquired from learning the Schrödinger picture. This should be an important lesson to those who think that 'quantum intuition' is easily attainable.

17.1 Kinetics and Dynamics chez Heisenberg

The Heisenberg picture has exactly the same *kinematics*, that is, the same mathematical structures as used in the Schrödinger picture. The Hilbert space, the states and the observables are exactly the same. What is different is the *dynamics*, that is, the time evolution equation. In each picture the (time independent) Hamiltonian H of the system plays a central role in the time evolution equation. In the Schrödinger picture, it is the time dependent Schrödinger equation which tells us how the states change in time. As we know, that equation is

$$i\hbar\frac{d}{dt}\psi(t) = H\psi(t)$$

together with an initial condition $\psi(0) = \varphi$ for some state φ. Then, as we have seen, the solution is $\psi(t) = e^{-itH/\hbar}\varphi$. What was never said explicitly, but is true nonetheless, is that the observables do not change in time in the Schrödinger picture. In the Heisenberg picture we have exactly the opposite

© Springer Nature Switzerland AG 2020
S. B. Sontz, *An Introductory Path to Quantum Theory*,
https://doi.org/10.1007/978-3-030-40767-4_17

situation: the states do not change in time but the observables do! Let A be a self-adjoint operator acting in the Hilbert space. In the following we take $t \in \mathbb{R}$ to be the time. The time evolution of A is given formally by

$$i\hbar \frac{d}{dt} A(t) = [A(t), H] \qquad (17.1.1)$$

together with the initial condition $A(0) = A$. This equation is far from being rigorous. The right side could involve the commutator of two densely defined operators, and consequently could be an operator with a very small domain, possibly the zero subspace. The rigorous time evolution of A is given by

$$A(t) = e^{itH/\hbar} A e^{-itH/\hbar}. \qquad (17.1.2)$$

Exercise 17.1.1 *Prove that (17.1.2) is a rigorous solution of (17.1.1) with the initial condition $A(0) = A$ provided that the dimension of the Hilbert space is finite. In particular, you should understand why (17.1.1) is a rigorous equation in this case.*
Hint: *Recall the definition of the exponential of a matrix in (13.3.8).*

Recall that the *unitary group* $e^{itH/\hbar}$ is well defined for all $t \in \mathbb{R}$, since H is self-adjoint. Because of this we now see that the rigorous path is to take (17.1.2) as the *definition* of $A(t)$, given the initial condition $A(0) = A$.

Exercise 17.1.2 *For readers with a good background in functional analysis, prove the following statements. The rest of you should read and attempt to understand what each statement says.*

- *Let H be the Hamiltonian of a quantum system and define $H(t)$ using (17.1.2). Prove that $H(t) = H$ for all $t \in \mathbb{R}$. (This is the statement in the Heisenberg picture of conservation of energy.)*

- *If the self-adjoint operator A is densely defined in D, a dense subspace of the Hilbert space, show that $A(t)$ is a self-adjoint operator densely defined in $e^{itH/\hbar}(D)$.*

- *The spectrum is time independent, namely $\operatorname{Spec} A(t) = \operatorname{Spec} A$ for all $t \in \mathbb{R}$.*

- *Suppose that E_A is the pvm of the self-adjoint operator A. Show that $B \mapsto e^{itH/\hbar} E_A(B) e^{-itH/\hbar}$ is the pvm of $A(t)$, where B is a Borel subset of the real line. We write this as $E_{A(t)} = e^{itH/\hbar} E_A e^{-itH/\hbar}$.*

The quickest way to see the relation between the Heisenberg picture and the Schrödinger picture is to take a state φ in the Hilbert space and consider

$$\langle \varphi, A(t)\varphi \rangle = \langle \varphi, e^{itH/\hbar} A e^{-itH/\hbar}\varphi \rangle = \langle e^{-itH/\hbar}\varphi, A e^{-itH/\hbar}\varphi \rangle = \langle \psi(t), A\,\psi(t) \rangle,$$

where $\psi(t) = e^{-itH/\hbar}\varphi$ is the solution of the Schrödinger equation.

On each side we have a time dependent diagonal matrix element (or expected value) of an observable in a state. In the Heisenberg picture on the left side, we have a constant state and a time dependent observable. In the Schrödinger picture on the right side, we have a time dependent state and a constant observable. And these two time dependent expressions are equal. With the notation established above, the more general result in terms of quantum probabilities is

$$
\begin{aligned}
P(A(t) \in B \mid \varphi) &= \langle \varphi, E_{A(t)}(B)\, \varphi \rangle \\
&= \langle \varphi, e^{itH/\hbar} E_A(B)\, e^{-itH/\hbar}\, \varphi \rangle \\
&= \langle e^{-itH/\hbar}\, \varphi, E_A(B)\, e^{-itH/\hbar}\, \varphi \rangle \\
&= \langle \psi(t), E_A(B)\, \psi(t) \rangle \\
&= P(A \in B \mid \psi(t)).
\end{aligned}
$$

Again, the Heisenberg picture is on the left side, while the Schrödinger picture in on the right side. But each side gives the same time dependent classical probability measure on the Borel subsets B of $\mathbb{R}$. And this time evolving probability measure is what we verify with experiment. Now the time dependence in the Heisenberg picture resides in the expression $A(t)$ only, while in the Schrödinger picture it resides in the expression $\psi(t)$ only. So the idea that the state of a quantum system carries 'properties' that change in time is completely invalid in the Heisenberg picture, in which all states are stationary. What can change in the Heisenberg picture are the observables, but there are also *stationary observables* which are those that are constant in time. We leave to the reader to verify that stationary observables in the Heisenberg picture are equivalent to stationary states in the Schrödinger picture.

The language used to deal with these facts can be intimidating. The word 'reality' gets thrown around a lot. So does 'quantum weirdness'. Do try to keep thinking in a scientific manner and not be swept away by rhetoric. In fact, the motivation for including this material in an introductory text on quantum theory is to prepare the reader to deal with rhetorical excesses. But also be warned that trying to get one's mind around all this is quite daunting. After all, we are accustomed to thinking of an experimental measurement of a specific physical quantity as the result of using the same measuring device, regardless of the time. Also we are used to thinking of the physical system being measured as changing in time, and this change is reflected in the changing 'state' of the system. But the Heisenberg picture throws out this way of thinking! Colloquially speaking, the Heisenberg picture requires thinking 'outside of the box'.

Notice that in each picture the fundamental mathematical problem is to find an explicit expression for the unitary group $e^{-itH/\hbar}$ for a given self-adjoint operator H acting in a Hilbert space. It is fair to say that the unitary group, and not any particular equation that leads to it, is the dynamics of the system. This means that Schrödinger's razor only applies to the

Schrödinger picture. In this regard a theorem of M. Stone is relevant. That theorem says that any unitary group $U(t) : \mathcal{H} \to \mathcal{H}$ satisfying a mild continuity hypothesis as $t \to 0$ is of the form $U(t) = e^{-itH/\hbar}$ for a unique densely defined, self-adjoint operator H acting in the Hilbert space $\mathcal{H}$. If t has dimensions of time, then H has dimensions of energy.

Notice that in each picture the fundamental physical problem is to find an explicit expression for a densely defined, self-adjoint operator H that is supposed to describe a physical system. This problem is usually solved by *first quantization*, one of the basic items in the tool kit of a quantum physicist.

There any number of other pictures besides those of Schrödinger and of Heisenberg. For example, the *interaction picture* is also used in quantum theory. In these other pictures both the state and the observable are time dependent. However, the probability measure $B \mapsto P(A(t) \in B \mid \psi(t))$ does not depend on the picture, that is, the physics is the same. It is a matter of convenience which picture one uses to study a particular physical system.

17.2   Notes

After Heisenberg published in 1925 his quantum theory, known as *matrix mechanics*, there was a flurry of activity. At long last there seemed to be a consistent way of dealing with atomic and molecular physics. Then in 1926 Schrödinger published his seminal papers with a quantum theory whose foundations were quite different. So, after a generation of searching for the already dubbed *quantum mechanics*, there was an embarrassing abundance of riches. However, very shortly Schrödinger proved that the two theories are equivalent. Our proof is basically the same, but goes to the level of quantum probability. Interestingly enough, Schrödinger never did accept the probabilistic interpretation of the solution of his eponymous equation as being fundamental. He remained a dedicated determinist. Of course, the Schrödinger equation is completely deterministic. It is the interpretation of its solution which is probabilistic.

Chapter 18

Uncertainty (Optional)

> Information is the resolution of uncertainty.
> Claude Shannon

A lot of what is said in this chapter is rejected by someone or other in the scientific community. But it is what I have come to accept as a clear accounting of a topic that touches on both the mathematical and physical interpretations of quantum theory. Unfortunately, it seems to be almost impossible to find an exposition of this topic that does not include some bits of utter nonsense. If you can believe a nearly doubly negative, totally self-referential sentence.

18.1 Moments of a Probability Distribution

Let's consider the probability distribution $\rho(\mathbf{x}) := |\psi(\mathbf{x})|^2$ of a stationary state ψ on $\mathbb{R}^3$ from a mathematical point of view. One such consideration is whether ρ has finite *moments* defined by

$$\mu_k := \int_{\mathbb{R}^3} d\mathbf{x}\, \mathbf{x}^k |\psi(\mathbf{x})|^2,$$

where $\mathbf{x}^k := x_1^{k_1} x_2^{k_2} x_3^{k_3}$ and $k = (k_1, k_2, k_3)$ is a *multi-index* of non-negative integers. This is clearly too many sub-indices, and so we consider the case of a 'wave/particle' on $\mathbb{R}$ instead. Besides, the basic ideas will be present in this simple case. So for $\psi \in L^2(\mathbb{R})$ we consider

$$\mu_k := \int_{\mathbb{R}} dx\, x^k |\psi(x)|^2, \tag{18.1.1}$$

where $k \geq 0$ is an integer. Notice that μ_k depends only on the physical state defined by ψ, that is to say, if we replace ψ by $\lambda \psi$ for some complex

© Springer Nature Switzerland AG 2020
S. B. Sontz, *An Introductory Path to Quantum Theory*,
https://doi.org/10.1007/978-3-030-40767-4_18

number λ with $|\lambda| = 1$, then μ_k remains unchanged provided that the integral converges.

First off, $\mu_0 = 1$, since $|\psi(x)|^2$ is a probability distribution. In general, we can not say much more. If the integral in (18.1.1) converges, then μ_k is a real number. Moreover, $\mu_k \geq 0$ if k is even. All of the remaining moments with $k \geq 1$ could be infinite or undefined. Or they could be finite. But it is common to assume that μ_1 and μ_2 are finite real numbers. And in many interesting examples this is true. But the fascination that these two moments μ_1 and μ_2 have generated in the physics community is all out of proportion. Even knowing all the moments of a probability distribution does not necessarily tell us everything about it. Unfortunately, the meaning of these two particular moments has been exaggerated, if not completely distorted, to the point that these two alone are claimed to 'explain' the difference between classical and quantum theory and that these two moments give us some essential knowledge about 'properties' of the position of the 'particle'.

Clearly, $\mu_2 = 0$ implies $x^2|\psi(x)|^2 = 0$ for (almost) all $x \in \mathbb{R}$, which in turn implies $\psi = 0$ in $L^2(\mathbb{R})$. But as noted above, $\mu_2 \geq 0$. So, we must have $\mu_2 > 0$ if ψ is a state. As for μ_1 we have

$$\mu_1 = \int_{\mathbb{R}} dx\, x\, |\psi(x)|^2 = \int_{\mathbb{R}} dx\, \psi(x)^*\, x\, \psi(x) = \langle \psi(x), x\, \psi(x) \rangle,$$

which is called the *expected value of x in the state* ψ. This is indeed the *average* or *mean* of the probability distribution $\rho(x) = |\psi(x)|^2$ on the real line $\mathbb{R}$ as you can learn from almost any introductory probability book which deals with continuous distributions.

Similarly,

$$\mu_2 = \int_{\mathbb{R}} dx\, x^2\, |\psi(x)|^2 = \int_{\mathbb{R}} dx\, \psi(x)^*\, x^2\, \psi(x) = \langle \psi(x), x^2\, \psi(x) \rangle$$

is called the *expected value of* x^2 *in the state* ψ. However, we are usually more interested in another real number associated with a probability distribution. If μ_1 exists, this is defined by

$$Var = Var(x \,|\, \psi) := \int_{\mathbb{R}} dx\, (x - \mu_1)^2\, |\psi(x)|^2 = \langle \psi(x), (x - \mu_1)^2 \psi(x) \rangle$$

and is called the *variance* (or *dispersion*) of x in the state ψ. Its value indicates how much the probability distribution is 'spread out' around its mean value μ_1. It is sometimes known as the *second central moment* of x in the state ψ. Tinny! In physics especially the variance is often called the *uncertainty*, which misleads one into thinking about lack of knowledge. This is even worse than tinny. It's sophistry!

We note that a well known equality in probability theory also holds here:

$$Var = Var(x \,|\, \psi) = \langle \psi(x), (x - \mu_1)^2 \psi(x) \rangle$$
$$= \langle \psi(x), x^2 \, \psi(x) \rangle - 2\mu_1 \langle \psi(x), x \, \psi(x) \rangle + \mu_1^2 \langle \psi(x), \psi(x) \rangle$$
$$= \mu_2 - 2\mu_1^2 + \mu_1^2$$
$$= \mu_2 - \mu_1^2.$$

So under the assumption that both μ_1 and μ_2 are finite for the state ψ, we conclude that Var, the variance of ψ, is also finite. In this context one often considers as well the *standard deviation* defined by $\sigma := +\sqrt{Var}$. In quantum theory based on the Hilbert space $L^2(\mathbb{R}^3)$ we can say that $\sigma > 0$, while in probability theory we can only say that $\sigma \geq 0$. Just to complicate matters, we often write σ^2 instead of Var. Worse yet, we read σ^2 as the *standard deviation squared*. This otherwise unfortunate expression does have the virtue of showing that in good English grammar sometimes the adjective can follow the noun.

We can use the standard notation $\mathcal{E}$ from probability theory for expected values as well. So we write

$$\mu_1 = \mathcal{E}(x \,|\, \psi), \qquad \mu_2 = \mathcal{E}(x^2 \,|\, \psi)$$

and so forth, where ψ is a state. Actually, for any $f : \mathbb{R} \to \mathbb{R}$ we can write

$$\mathcal{E}(f \,|\, \psi) = \langle \psi(x), f(x)\psi(x) \rangle,$$

which is read as 'the expected value of f in the state ψ'. In this notation, we have that

$$Var(x \,|\, \psi) = \mathcal{E}(x^2 \,|\, \psi) - \mathcal{E}(x \,|\, \psi)^2.$$

Exercise 18.1.1 *Prove that $\sigma > 0$ as claimed above.*

We often omit the state ψ from our notation. And this can mislead the reader. All of these quantities depend on ψ. For example, if we make measurements of x for an ensemble of systems, each of which is identically prepared to be in the state ψ, we will obtain a set of measured values, say $\alpha_1, \alpha_2, \ldots, \alpha_N$. These values will not necessarily cluster closely around one real number as we expect in deterministic classical physics. But from this list we construct the usual *statistical estimator*

$$\overline{\alpha} := \frac{\alpha_1 + \alpha_2 + \cdots + \alpha_N}{N}$$

for the theoretical value μ_1. Similarly, we can construct the usual estimator for the variance. A good introductory statistics book will give you the formula for an *unbiased estimator* of the variance. As is usual in statistical analysis, the conclusions apply to the ensemble rather than to any one thing in that ensemble. It is not correct to think that each individual observation gives μ_1

or even something relatively close to μ_1. The measured values could cluster, for example, at $+5$ and -5 in a symmetric way with almost no values near 0 thereby agreeing with a theoretical probability distribution with peaks at $+5$ and -5 and $\mu_1 = 0$. But then that would most likely give an estimator $\overline{\alpha}$ near $\mu_1 = 0$. Moreover, by taking ever larger ensembles the probability that $\overline{\alpha}$ differs from $\mu_1 = 0$ in absolute value by some given number $\varepsilon > 0$ approaches 0. The previous statement is a theorem in *mathematical statistics*. It is equally incorrect, as this example shows, to think that the state 'has' a numerical value, which experiment ever better approximates through ever more measurements.

The state $\psi \in L^2(\mathbb{R})$ *determines* a *probability distribution* of positions. After all that is what the equation $\rho(\mathbf{x}) = |\psi(\mathbf{x})|^2$ is telling us. It might seem strange to see the words 'determines' and 'probability' in the same sentence (as they do appear above). Get used to it!

Also, given an ensemble of systems in the prepared state ψ, you can not make the variance Var as small as you like. However, by changing the state ψ we can decrease the variance. A given state ψ *determines* a probability distribution, which in turn gives us $Var(x|\psi)$ by using a simple formula. And it could even happen that we are preparing a state for which the variance does not even exist.

Of course, x is an example of a *measurable quantity* which is represented in quantum theory by a self-adjoint operator. Almost all other physically measured quantities are also represented by self-adjoint operators. (Recall that time is a measured quantity for which there is no self-adjoint operator.) Anyway, a measurable quantity associated with a self-adjoint operator will always have a probabilistic interpretation that refers to ensembles of systems prepared in the same state. Any such observable will have an expected value as well as a variance for that state, if the moments μ_1 and μ_2 exist for that probability distribution. Neither of these values can be controlled experimentally once the prepared state has been fixed.

18.2   Incompatible Measurements

In this discussion we have only mentioned experiments which measure *one* quantity. Are there experiments that can measure two or more quantities at the same time? It seems obvious and natural to expect this should always be possible. But it is not! There are some pairs of observables, as already described in Section 16.10, that are compatible in this sense, while there are other pairs which are not. This is radically different from classical mechanics where there is an underlying assumption (supported by experiment) that any pair of observables is compatible. If a car goes roaring down the road, we feel confident that we can measure (at as many moments in time as our current technology allows) its position, velocity, momentum, temperature, kinetic energy, you name it. The only limiting factor is available technology. Or so we think.

But in experiments measuring quantum systems there are incompatible pairs, like it or not. This is a theoretical statement and can, of course, be falsified by experiment. However, it is a consequence of the collapse of the 'wave' function. How so? Well, first note that incompatible pairs of measurements *can* be made in succession, that is, one can be made and then nanoseconds later the other can be made. But the first measurement collapses the 'wave' function and so the second measurement (in general) is not being made on the same initial state as was the first measurement. And then the second measurement also collapses its initial 'wave' function. Clearly, the order of the two measurements can be reversed and, again, there will be two collapses of 'wave' functions. For example, we start with a state ψ and measure A and then measure B, getting the values a and b, respectively. But starting with the same state ψ we can measure B and then measure A, getting the values b' and a'. Given incompatible observables A and B it is perfectly possible that $a \neq a'$ or $b \neq b'$. (And this is an inclusive 'or'. Both may hold.) All this can be discussed in quantum theory, and clear scientific statements can be formulated. However, as shown in Section 16.10 none of this analysis has anything to do with the Heisenberg inequality, which we will present in (18.2.1) below. Nonetheless, this situation of consecutive measurements of incompatible observables is often taken to be the physical interpretation of the Heisenberg uncertainty principle. That is all well and good, provided that this interpretation is not confused with (18.2.1).

Moreover, the theoretical assertion of the collapse condition forbids a simultaneous measurement of a pair of incompatible observables, since which of the two measurements would collapse the 'wave' function? The point here is that these two measurements individually would collapse the same initial 'wave' function in two different ways. Hence a simultaneous measurement of two incompatible observables would force Nature to make the choice of *Buridan's ass*.

The most famous example of such an incompatible pair of operators for $\mathbb{R}$ is the pair of position x and momentum p. (Actually, any pair whatsoever of non-commuting operators will do just as well.) The famous Heisenberg position/momentum uncertainty principle has been encoded as the inequality (due to Kennard, not to Heisenberg)

$$Var(x \,|\, \psi)\, Var(p \,|\, \psi) \geq \hbar^2/4, \qquad (18.2.1)$$

where ψ is a state, that is $||\psi|| = 1$. So we can define the moments for the operator p in the state ψ and then define the variance denoted by $Var(p \,|\, \psi)$, which should be read as the variance of the observable p given the state ψ. Similarly, $Var(x \,|\, \psi)$ denotes the variance of the observable x given the state ψ, as presented above. Notice that the same state ψ appears in these two variances. The mathematical proof of this extremely famous result is quite disappointing. It is just a simple application of the Cauchy-Schwarz inequality. We will return to this in order to give a more self-contained exposition. But for now I would prefer to focus on ideas and calculations.

First though, here is a little interlude on the more common way of writing the Heisenberg uncertainty inequality (18.2.1) in physics. One considers the standard deviation of each quantity instead of its variance. While we used σ before to denote a standard deviation, we now use Δ. By taking positive square roots the Heisenberg uncertainty inequality (18.2.1) is equivalent to

$$\Delta(x \,|\, \psi) \, \Delta(p \,|\, \psi) \geq \hbar/2.$$

Often the state ψ is dropped from the notation. So you will see this version in the literature:

$$\Delta(x) \, \Delta(p) \geq \hbar/2.$$

Next a change of language occurs. Instead of saying 'standard deviation' one says 'uncertainty'. Even though definitions in mathematics can use arbitrary words for the defined structure, this particular word can totally obscure the mathematical meaning. The problem is that the word 'uncertainty' already has a common meaning in English. And that meaning has something to explicitly do with faulty human knowledge. Moreover, that meaning is not equivalent to 'standard deviation'. But the die can not be uncast.

So what does this uncertainty inequality say? Simply that if we estimate, as described above, the variance of x in some state ψ of an ensemble of experiments and we similarly estimate the variance of p of another ensemble of experiments prepared in the *same* state ψ, then the above inequality will be statistically verified. Note that the ensemble for p must be completely disjoint from the ensemble for x, since these are incompatible measurable quantities. Notice that there is no reference in this explanation of knowledge in any way or form about individual elements in the ensembles. We are not speaking of knowledge—or lack of knowledge—about the position or momentum of any observed individual 'wave/particle'. Simply we measure the position for some of these observations (but not the momentum) while we measure the momentum for other observations (but not the position).

If we replace the state ψ by a sequence of states ψ_n with $n \geq 0$ such that $\lim_{n \to \infty} Var(x \,|\, \psi_n) = 0$ (as we can do by taking bump functions ψ_n centered at a given real number), then we have $\lim_{n \to \infty} Var(p \,|\, \psi_n) = \infty$ as a consequence of the Heisenberg uncertainty inequality. Similarly, we can find a different sequence of states ϕ_n such that $\lim_{n \to \infty} Var(p \,|\, \phi_n) = 0$. In that case it follows again by the Heisenberg uncertainty inequality that $\lim_{n \to \infty} Var(x \,|\, \phi_n) = \infty$. So states with measured values of one of these observables tightly distributed around its mean value will be states that have measured values for the other observable widely dispersed around its mean value. The Heisenberg uncertainty inequality also admits the possibility that *both* of the variances are very large. These are not statements about *the* value or the knowledge of *the* value of either of these two observables in a given state, since no state has a single value for either of them, but rather a distribution of values. The Heisenberg uncertainty inequality (18.2.1) says something about the variances of those distributions, no more and no less.

The Heisenberg uncertainty inequality is not about measurement any more or less than any other theoretic prediction properly interpreted. It is not a statement about one type of measurement limiting the information available from another type of measurement, but rather that states with very small variance in x will have large variance in p and similarly that states with very small variance in p will have large variance in x. If one wishes to think of variance as a measure of information (i.e., small variance corresponds to a lot of information), then the increase in information about x (in some state ψ) that one gets by changing that state *can* result in a decrease in information about p for that same change of states. And vice versa. Since we are dealing with an inequality, we only can say that a sufficiently large increase in information about x due to some change of state will *necessarily* result in a decrease in information about p due to the same change of state. Viewed this way the Heisenberg uncertainty inequality is a statement about the change of the information of observables due to a change of state.

Recall that once ψ has been given, the theoretical values of $Var(x \mid \psi)$ and $Var(p \mid \psi)$ are *determined*, at which point the only remaining thing to do is perform the experiment to see whether or not the theory holds. After all, we test *all* theoretic predictions in physics by comparing them with experimental data. And unlike the great Yogi's saying, good experiments beat good theory every time and *not* vice versa.

It is fine if one wishes to summarize this discussion about variances as "the better one knows x the worse one knows p and conversely". But only if this refers, as above, to statements about an ensemble of quantum systems in identical states. Trying to give this discussion an interpretation in terms of any one individual system is not what one can legitimately get from the Heisenberg uncertainty relation (18.2.1).

However, the quantum situation, properly understood, may run counter to your intuition. To see this, the classical one-dimensional system of one massive particle gives an instructive contrast. For such systems a (pure) state is determined by two real numbers, one for its position and another for its momentum. Knowledge about the position, no matter how precise, tells us nothing at all about its momentum even if we perform a large ensemble of identical experiments on identically prepared systems. (And conversely, though we won't get into that.) On the other hand, let's think about an ensemble of identical experiments on identically prepared one-dimensional one-particle quantum systems all of which are in the same quantum state ψ. Then the measured values of the position x for each system gives us information about $Var(x \mid \psi)$ and so, using (18.2.1), at least some, admittedly incomplete, information about $Var(p \mid \psi)$, the variance of p. So, contrary to what you might expect, this can be seen as *better* than the analogous classical case where no information about momentum is obtained.

18.3   Example: Harmonic Oscillator

Let's see how things work out for the ground state of the quantum harmonic oscillator $\psi_0(x) = Ce^{-x^2/2}$ where we now choose $C > 0$ so that $||\psi_0|| = 1$. (See (6.2.2).) Note that

$$\int_{\mathbb{R}} dx \, |e^{-x^2/2}|^2 = \int_{\mathbb{R}} dx \, e^{-x^2} = \pi^{1/2},$$

where the last equality might be familiar to you. If not, you can check it out in the literature. Or better yet, you can try to evaluate it yourself. Anyway, this shows that $\psi_0(x) = \pi^{-1/4}e^{-x^2/2}$ gives a normalized ground state 'wave' function. Then the expected value of position in this state is

$$\mu_1 = \langle \psi_0, x\psi_0 \rangle = \int_{\mathbb{R}} dx \, (\pi^{-1/4}e^{-x^2/2})^* \, x \, \pi^{-1/4}e^{-x^2/2} = \int_{\mathbb{R}} dx \, \pi^{-1/2} \, x \, e^{-x^2} = 0,$$

since the continuous integrand decays to zero sufficiently fast at infinity in order to conclude that it is integrable and the integrand is an odd function. So the measured values of x are predicted to have mean (or average) value $\mu_1 = 0$. But how spread out will they be around that mean value?

$$Var(x \,|\, \psi_0) = \mu_2 - \mu_1^2 = \langle \psi_0, x^2\psi_0 \rangle = \int_{\mathbb{R}} dx \, \pi^{-1/2} \, x^2 \, e^{-x^2} = \frac{1}{2},$$

where the last integral is also well known. In fact for every integer $k \geq 0$, the value of the integral

$$\int_{\mathbb{R}} dx \, x^k \, e^{-x^2}$$

is well known. These are called *Gaussian integrals*.

The fact that $Var(x \,|\, \psi_0) \neq 0$ does *not* imply that the 'wave/particle' is moving when in the ground state ψ_0. The value of $Var(x \,|\, \psi_0)$ simply speaks to the results of position measurements of an ensemble of identically prepared systems. The state ψ_0 is a stationary state. Its time evolution is $\Psi(t) = e^{-itE_0/\hbar}\psi_0$, where E_0 is the ground state energy, a real constant. By our previous discussion $\Psi(t)$ represents the exact same state as ψ_0 does, since $|e^{-itE_0/\hbar}| = 1$. $\Psi(t)$ is a state that does not change with time. That state is all the information we have about the 'wave/particle'. It makes no sense whatsoever to say that it—or something associated to it—is moving. Or changing. Or fluctuating. Or whatever other obscure synonym one might use instead of moving. In this case even Galileo would have to admit that it does not move.

Let's do the same for the momentum observable

$$p = \frac{\hbar}{i} \frac{d}{dx}.$$

The expected value of p in the state ψ_0 is

$$\langle \psi_0, p\psi_0 \rangle = \int_{\mathbb{R}} dx \, \left(\pi^{-1/4}e^{-x^2/2}\right)^* \frac{\hbar}{i} \frac{d}{dx}(\pi^{-1/4}e^{-x^2/2}) = 0,$$

since again the continuous integrand is integrable and is odd. But for the record we note that

$$\frac{d}{dx}(e^{-x^2/2}) = -x\,e^{-x^2/2}$$

and so

$$\frac{d^2}{dx^2}(e^{-x^2/2}) = \frac{d}{dx}(-x\,e^{-x^2/2}) = (-1+x^2)\,e^{-x^2/2}.$$

So the variance of p in the state ψ_0 is

$$Var(p\,|\,\psi_0) = \langle \psi_0, p^2\psi_0 \rangle$$

$$= \int_{\mathbb{R}} dx\,(\pi^{-1/4}e^{-x^2/2})^* \left(\frac{\hbar}{i}\right)^2 \frac{d^2}{dx^2}(\pi^{-1/4}e^{-x^2/2})$$

$$= -\hbar^2\pi^{-1/2}\int_{\mathbb{R}} dx\,(-1+x^2)\,e^{-x^2}$$

$$= -\hbar^2\pi^{-1/2}\left(-\pi^{1/2} + \frac{1}{2}\pi^{1/2}\right)$$

$$= \frac{1}{2}\hbar^2.$$

Again, the fact that $Var(p\,|\,\psi_0) \neq 0$ does *not* imply that the 'wave/particle' or anything else is moving. It is also a statement about an ensemble of momentum measurements. But I repeat myself repeatedly.

Next we compute the left side of the Heisenberg inequality (18.2.1)

$$Var(x\,|\,\psi_0)\,Var(p\,|\,\psi_0) = \frac{1}{2}\left(\frac{1}{2}\hbar^2\right) = \frac{1}{4}\hbar^2,$$

which is consistent with what the Heisenberg inequality says. Actually, for this particular state ψ_0 the inequality becomes an equality. We say that the state ψ_0 *minimizes* the Heisenberg inequality. Equivalently, we say that ψ_0 *saturates* the Heisenberg inequality. Any state that minimizes the Heisenberg inequality is called a *minimal uncertainty state*. This is often taken to be an essential property of the so-called *coherent states*, which form an active area of current research. See the encyclopedic volume [1] for more on this topic.

It is a long, but possibly instructive exercise to compute the left side of the Heisenberg inequality for the rest of the eigenfunctions of the harmonic oscillator. It turns out that for each of them the inequality holds in its strict form, that is to say, the left side is strictly bigger than the right side. Hence none of these *excited states* is a minimal uncertainty state.

Exercise 18.3.1 *So to get a further taste of how this works, compute the left side of the Heisenberg uncertainty inequality for the stationary state that has energy eigenvalue* $(3/2)\hbar\omega$.

So far in this chapter we have been concerned only with a stationary state. What can we say about a solution $\Psi(t, x)$ of the time dependent Schrödinger equation? Well, we simply fix some time t and use $\Psi(t, x)$ instead of $\psi(x)$ in the above discussion. Then all the quantities become time dependent. So the time dependent version of the Heisenberg inequality for the state $\Psi(t, x)$ is simply

$$Var\big(x \,|\, \Psi(t, x)\big) \, Var\big(p \,|\, \Psi(t, x)\big) \geq \frac{1}{4}\hbar^2$$

for every time t. So we have two time varying quantities whose product at any time is bounded below by a universal constant. There are some curious possibilities. One possibility is that $Var\big(x \,|\, \Psi(t, x)\big)$ and $Var\big(p \,|\, \Psi(t, x)\big)$ can oscillate with time between 0 and $+\infty$. In fact, it is even possible that these oscillations bring each of these quantities individually as close to 0 as you like, but not at the same time! This is because when one of them is very, very close to 0 the other must be very, very large indeed. However, there could be times when both are very, very large.

Another fascinating possibility is that, while neither factor is constant in time, nonetheless

$$Var\big(x \,|\, \Psi(t, x)\big) \, Var\big(p \,|\, \Psi(t, x)\big) = \frac{1}{4}\hbar^2$$

holds for every time t. In this case when one of the variances gets close to 0 at some time we know the exact (large) value of the other variance at that same time.

My own opinion is that the importance of the Heisenberg uncertainty inequality has been exaggerated and even, at times, distorted into saying something meaningless at best or false at worst. I do not care to scold my colleagues, whose sincere dedication to doing good science I do not doubt, in more detail on this point. It suffices to note the obvious scientific fact that the Heisenberg uncertainty inequality (18.2.1) is not an equation which allows us to predict the time evolution of a physical system. In fact, it is not even an equation. At best it tells us that certain situations, which we expect to occur according to classical physics, do not occur in nature. And it is a perfectly fine scientific statement since it can be checked by the ultimate scientific test: experiment.

18.4   Proof of the Uncertainty Inequality

In this section we will prove the Heisenberg uncertainty inequality for the position and momentum operators in dimension one. This argument can also be applied to the position and momentum operators associated to the same direction in $\mathbb{R}^n$. There are also generalizations for pairs of operators acting in a Hilbert space. A nice factoid is that for a pair of commuting operators the lower bound of that Heisenberg inequality is zero. But for the quantum (that is, non-commuting) case the lower bound is non-zero.

Before proceeding with that proof I wish to note that, except in very special cases, the energy operator, that is the Hamiltonian H, acting in a Hilbert space $\mathcal{H}$ does not have a conjugate self-adjoint operator that acts in $\mathcal{H}$ and represents the time. (A self-adjoint operator T *conjugate* to H would satisfy $i[H, T] = \hbar I$.) In quantum theory time is a *parameter* and not an *observable*. A way to see physically why this is so is that every measurement is made simultaneously (by definition!) with a time measurement. So the time measurement is compatible with *all* other measurements, and so would have to be represented by a self-adjoint operator that commutes with all operators, including energy operators. And this leads to the time operator being a real multiple of the identity operator. And such an operator has a single point spectrum, that is, only one possible measured value. And time is not like that. You will see an energy-time uncertainty relation written in the literature, but that is not a special case of the argument of this section. In fact, the physical interpretations of such an uncertainty relation, again found in the admittedly enormous literature, are never the interpretation of the Heisenberg uncertainty inequality as given here. To be sure, what I am saying in this paragraph is highly controversial in the physics community. And I have not read all of the relevant literature. Who has?

Now we are going to see how the Heisenberg inequality is a consequence of the Cauchy-Schwarz inequality (9.3.3). Recall that $Var(x|\psi) = \mu_2 - \mu_1^2$ where ψ is a state. Define a new observable y by $y := x - \mu_1$. Then the expected value of y is

$$\mathcal{E}(y \,|\, \psi) = \mathcal{E}\big((x - \mu_1) \,|\, \psi\big) = \mathcal{E}(x \,|\, \psi) - \mu_1 = 0$$

and so

$$\begin{aligned}
Var(y \,|\, \psi) &= \mathcal{E}(y^2 \,|\, \psi) - \mathcal{E}(y \,|\, \psi)^2 \\
&= \mathcal{E}(y^2 \,|\, \psi) \\
&= \mathcal{E}\big((x - \mu_1)^2 \,|\, \psi\big) \\
&= Var(x \,|\, \psi).
\end{aligned}$$

This simply says that by shifting the observable x by the appropriate constant amount the expected value becomes 0 while the variance is unchanged. So without loss of generality we can assume that the original observable x has expected value 0. Similarly, we assume that the expected value of p is 0. These assumptions clean up the algebraic manipulations of the following argument.

Next, since x is a real variable we have

$$Var(x \,|\, \psi) = \langle \psi, x^2 \psi \rangle = \langle x\psi, x\psi \rangle = ||x\psi||^2.$$

A partial integration argument shows that $p = \frac{\hbar}{i}\frac{d}{dx}$ is a symmetric operator, which justifies the second equality here:

$$Var(p \,|\, \psi) = \langle \psi, p^2 \psi \rangle = \langle p\psi, p\psi \rangle = ||p\psi||^2.$$

The inner product and norm used in these equations are those of the Hilbert space $L^2(\mathbb{R})$. Now the Cauchy-Schwarz inequality (9.3.3) and the above identities say that

$$|\langle x\psi, p\psi \rangle|^2 \leq ||x\psi||^2 ||p\psi||^2 = Var(x \mid \psi) \, Var(p \mid \psi). \qquad (18.4.1)$$

So now it is a question of identifying the left side of this inequality. In the next calculation we use the symmetry of the operators x and p as well as the fact that ψ is a state. So we have

$$\begin{aligned}
\langle x\psi, p\psi \rangle &= \langle (px)\psi, \psi \rangle \\
&= \langle (px - xp + xp)\psi, \psi \rangle \\
&= \langle (-i\hbar I + xp)\psi, \psi \rangle \\
&= i\hbar ||\psi||^2 + \langle (xp)\psi, \psi \rangle \\
&= i\hbar + \langle p\psi, x\psi \rangle \\
&= i\hbar + \langle x\psi, p\psi \rangle^*,
\end{aligned}$$

where we also used $px - xp = [p, x] = -i\hbar I$. (See (5.1.4).) This tells us that the imaginary part of $\langle x\psi, p\psi \rangle$ is given by

$$\text{Im}(\langle x\psi, p\psi \rangle) = \frac{1}{2i}(\langle x\psi, p\psi \rangle - \langle x\psi, p\psi \rangle^*) = \frac{\hbar}{2}.$$

This in turn gives us the lower bound

$$|\langle x\psi, p\psi \rangle|^2 \geq |\text{Im}(\langle x\psi, p\psi \rangle)|^2 = \frac{\hbar^2}{4}.$$

Combining this with (18.4.1) we obtain

$$Var(x \mid \psi) \, Var(p \mid \psi) \geq \frac{\hbar^2}{4}.$$

And this is the Heisenberg inequality (18.2.1). This argument has not been completely rigorous. Some consideration must be given to the existence of the integrals that give the inner products. Also some work must be done to show that p is a symmetric operator. This is a question of choosing an adequate domain for the formal operator p. In fact, with a good choice of domain, p is realized as a self-adjoint operator, which is a very special case of a symmetric operator. These considerations are dealt with in many texts, such as [25].

While $Var(x \mid \psi) \geq 0$ and $Var(p \mid \psi) \geq 0$ both follow immediately from the definition of the variance, the Heisenberg inequality actually implies $Var(x \mid \psi) > 0$ and $Var(p \mid \psi) > 0$. Notice that ψ does not depend on time. If ψ is an eigenfunction of the Hamiltonian H with eigenvalue E, then the time evolution of ψ is given by

$$\Psi(t, x) = e^{-itE/\hbar}\psi(x).$$

But the factor $e^{-itE/\hbar}$ lies on the unit circle of $\mathbb{C}$, and it does not depend on x. Hence, $\Psi(t, x)$ represents exactly the same state as ψ does. In other words, the state of the quantum system is *not* changing in time. We already commented on this for the special case of the ground state of the harmonic oscillator. The point here is that this is a general result. So, the inequality $Var(x \mid \psi) > 0$ means that even knowing everything possible according to quantum theory about a system, the measurements of x do not give one unique (within experimental precision) value, but rather a distribution of different values. We already knew this, but now we have a quantitative statement of this fact.

Anyway, this distribution of measured values is often called the *quantum fluctuations* of the observable, which is x in the present case. But the state $\Psi(t, x)$ is not changing or fluctuating in this case. So, we are not supposed to think of the 'wave/particle' as moving around on some trajectory or in some orbit. Similarly, the inequality $Var(p \mid \psi) > 0$ does not mean that the momentum of the 'wave/particle' has a value that is fluctuating around its mean value. Simply put, in quantum theory a 'wave/particle' does not have a single value of position (nor of momentum) that measurement then can give us. Instead identical measurements starting from identical initial states give us a non-trivial distribution of different values of position or of momentum. And that's what probability is all about!

This interpretation of quantum theory is so counter-intuitive to some people that there have been several attempts to 'rescue' quantum theory by introducing so-called *hidden variables*, whose currently unknown values account in a deterministic manner for the distribution of measured values. Unfortunately, emotions often play a role in discussions of such issues. While hidden variables have not been completely ruled out, some ideas along those lines, such as *Bell's inequalities*, have been falsified by experiment. At least, some physicists claim that such experiments have been done and have been correctly interpreted. Nonetheless, the discussion of this and related issues continues to this day in the physics as well as philosophical communities. The purpose of these comments is to present something along the lines of the *Copenhagen interpretation* in order to give context for such discussions.

Exercise 18.4.1 *Suppose $\psi_1, \psi_2 \in L^2(\mathbb{R})$ are states. Prove the following generalization of the Heisenberg inequality:*

$$Var(x \mid \psi_1) \, Var(p \mid \psi_2) \geq \frac{\hbar^2}{4} |\langle \psi_1, \psi_2 \rangle|^2. \qquad (18.4.2)$$

Also show that $0 \leq |\langle \psi_1, \psi_2 \rangle| \leq 1$.

As far as I know the inequality (18.4.2) has no name and is not very much seen in the literature. In the case when ψ_1 and ψ_2 are orthogonal, this gives us only the very weak result

$$Var(x \mid \psi_1) \, Var(p \mid \psi_2) \geq 0$$

even though it is true that

$$Var(x \mid \psi_1) \, Var(p \mid \psi_2) > 0,$$

since both factors on the left are strictly positive, as we have seen.

Exercise 18.4.2 *Let A, B be operators representing physical observables. (You can work under the condition that they are symmetric operators.) Find and prove a generalization of the Heisenberg inequality for these observables.*
Hint: *The lower bound should contain the commutator $[A, B]$.*
N.B. *This exercise is applicable to the spin operators S_1, S_2, S_3 that were introduced in Chapter 14. It is this form of the uncertainty principle which has been most extensively checked in the experiment. The position/momentum inequality (18.2.1) is more difficult to verify experimentally.*

18.5 Notes

The literature on uncertainty relations is enormous. I leave it to the reader to browse through it, though I recommend [24] for a clear discussion. If there is one certainty, it is that no one has read it all.

My point of view is almost heretical, even though it is a straightforward application of probabilistic ideas to a probabilistic statement. My near heresy lies in not taking the uncertainty relation (18.2.1) as a statement about a single experiment. Rather, I prefer to say it is only a meaningful statement when referring to ensembles of identical experiments. Why? Because the correct way to verify theoretical probabilistic statements, as far as I can see, is to compare the predicted probabilities with the experimentally measured relative frequencies by using established statistical methodology. Anything besides that just is not science in my opinion. In any case the experimentally measured relative frequencies have to be explained somehow, and quantum theory does that by predicting a probability measure for any given observable in any given state.

No doubt I will also be criticized for saying the Heisenberg uncertainty principle is an optional topic with almost no importance in quantum theory. But that is my sincere opinion. I included this chapter because the reader most likely has heard about this topic, and so it is worthwhile to explain what it is all about.

Heisenberg did not write a mathematical formulation of the uncertainty principle, and so some scientists dispute whether (18.2.1) actually captures Heisenberg's insights correctly or completely. However, (18.2.1) is commonly accepted as being the Heisenberg Uncertainty Principle (for position and momentum in one dimension), and so I have only commented on it and not on other mathematical statements purporting to represent Heisenberg's ideas better or more fully. For example, in [13] the authors give an alternative interpretation of uncertainty that makes no reference to inequalities with standard deviations.

Of course, many totally rigorous mathematical results have been dubbed as an uncertainty relation or even as a Heisenberg uncertainty relation. That is all fine and dandy. But this is a physics book. And so I want to interpret those mathematical results, if possible, in terms of the measurable properties of physical systems, and most particularly, of quantum systems.

Yogi Berra and his famous sayings, the Yogi-isms, are discussed in the Notes to Chapter 19.

Chapter 19

Speaking of Quantum Theory (Optional)

Couvrir les fautes,
excuser les crimes,
c'est aller a l'abime.
Émile Zola

How we speak about any subject is important because it reflects and influences how we think about it. And it is generally recognized that it is notoriously difficult to think about quantum physics. I think this is due to two factors: the fundamental role played in it by a probability theory with no underlying deterministic theory and the representation of time as a parameter instead of as a self-adjoint operator.

19.1 True vs. False

There is so much nonsense said about quantum physics that it is difficult to know where to even begin. Of course, the self-appointed experts with nary a day in a science course (but often with an overload of New Age rhetoric or cargo cult science at their disposal) are the easiest to confront. Such overwhelming ignorance combined with an amazing level of hubris is too easy a target. So I leave that critique to others. What really concerns me are the people with the supposedly adequate credentials (Ph.D., full professor, or Nobel Prize winner!) who should know better, but somehow end up as DoubleSpeakers. They do the general public a great disservice by creating juicy sound-bites, which are at times misleading and often downright lies. While everyone has the right to express their ideas, no one is above scientific criticism for what they say.

© Springer Nature Switzerland AG 2020
S. B. Sontz, *An Introductory Path to Quantum Theory*,
https://doi.org/10.1007/978-3-030-40767-4_19

19.2   FALSE: Two places at same time

Here is an example. Consider this often repeated FALSE statement:

> An electron can be at two places at the same time. FALSE.

The very first comment has to be that there is no experimental evidence to support this FALSE statement. And I mean absolutely none.

To avoid technical details, we discuss this in the simpler one-dimensional case first. We will get back to the three-dimensional case in an exercise. Suppose that Q represents the position observable, that is, for all $x \in \mathbb{R}$

$$(Q\psi)(x) = x\psi(x).$$

Then Q is a self-adjoint operator when we take ψ in the dense linear subspace of $L^2(\mathbb{R})$ defined by

$$D := \{\psi \in L^2(\mathbb{R}) \mid x\psi(x) \in L^2(\mathbb{R})\}.$$

Exercise 19.2.1 *The quantum event that Q has a value in a Borel subset B of $\mathbb{R}$ is $E_Q(B)$, which turns out to be the projection operator that maps $\psi \in L^2(\mathbb{R})$ to*

$$E_Q(B)\psi = \chi_B \psi.$$

Here E_Q denotes the pvm associated with the self-adjoint operator Q, and χ_B is the characteristic function of B. (Recall that $\chi_B(x) = 1$ if $x \in B$ and otherwise $\chi_B(x) = 0$.)

  Therefore, the range of $E_Q(B)$ consists of those functions that are zero (almost everywhere, to be completely correct) on $\mathbb{R} \setminus B$.

Recall that if E_1 and E_2 are quantum events with ranges V_1 and V_2, respectively, then the quantum event that both events occur at the same time is the projection operator (denoted by $E_1 \wedge E_2$) associated with the closed subspace $V_1 \cap V_2$. (See Exercise 16.2.9.)

Exercise 19.2.2 *Let B_1 and B_2 be disjoint Borel subsets of $\mathbb{R}$. Let χ_{B_1}, resp. χ_{B_2}, denote the characteristic function of B_1, resp. B_2. Prove that $\operatorname{Ran} E_Q(B_1) \cap \operatorname{Ran} E_Q(B_2) = 0$, the zero subspace. Then the corresponding quantum event is the zero operator.*

  In other words $E_Q(B_1) \wedge E_Q(B_2) = 0$ is the quantum event that never occurs no matter what state the system is in. Colloquially, an electron (or any other 'particle') is never in two disjoint regions at the same time.

  In any state ψ, the probability that the value of Q is both in B_1 and B_2 is

$$P(Q \in B_1 \cap B_2 \mid \psi) = P(P_Q(\emptyset) \mid \psi) = \langle \psi, 0\, \psi \rangle = 0.$$

So how do so many experts fall into this linguistic trap? Well, for two given disjoint non-empty open intervals I_1 and I_2 of the real line, there exist many states ψ such that

$$0 < P(Q \in I_1 \,|\, \psi) < 1 \quad \text{and} \quad 0 < P(Q \in I_2 \,|\, \psi) < 1. \tag{19.2.1}$$

But the statement that these two probabilities are non-zero does *not* say that the electron (or whatever) is at one and the same time both in I_1 and in I_2. Rather, these are probability statements that can be tested against measured frequencies in an ensemble of identical experiments of occurrences of values in I_1 and I_2 given that the system is in state ψ. So what is the reformulation of the FALSE statement so that it becomes true? It is that there exist states ψ that satisfy (19.2.1). Nontheless, (19.2.1) does not say that the 'particle' is simultaneously in I_1 and in I_2. Besides, if an electron was simultaneously in both I_1 and I_2, the *conservation of electric charge* would be violated! Another wild consequence of this FALSE statement would be that the electron can be in 3 places at the same time. And at 4 places at the same time. And so on *ad absurdum*.

Moreover, given such disjoint intervals I_1 and I_2, there exist many states ϕ_1 such that

$$P(Q \in I_1 \,|\, \phi_1) = 1 \quad \text{and} \quad P(Q \in I_2 \,|\, \phi_1) = 0, \tag{19.2.2}$$

and also there exist many states ϕ_2 such that

$$P(Q \subset I_1 \,|\, \phi_2) = 0 \quad \text{and} \quad P(Q \in I_2 \,|\, \psi_2) = 1. \tag{19.2.3}$$

Exercise 19.2.3 *Find states ψ, ϕ_1, ϕ_2 that satisfy the conditions (19.2.1), (19.2.2) and (19.2.3), respectively. In the process to doing this, you should understand why in each case there are many such states.*

If you think that this is *déjà vu* all over again, you are right. This is a reprise of the discussion about the spin matrices. As noted there, the same argument applies to any observable which can have at least two values. The fact that the Hilbert space for the the spin matrices is 2-dimensional while the Hilbert space for Q is infinite dimensional is not pertinent. What matters is that all of these observables have two or more possible values. And again, the counter-intuitive aspect of quantum probability is that even complete knowledge of the state of a system does not always tell us with probability 1 what the value of a particular measurement will be.

The discussion above was presented in order to show how the general theory of quantum probability, including pvm's and quantum events, can deal with misinterpretations of quantum theory. In this particular case there is an other way to correctly understand what quantum theory says.

Exercise 19.2.4 *Understand why the above statement is FALSE by considering the integral of the probability density function over the appropriate sets. Show that this argument easily generalizes to $\mathbb{R}^3$.*

19.3   FALSE: Any state can be determined

There are many other erroneous statements that come from respectable sources. It is not worthwhile examining all of them, though I believe that most are the result of not understanding quantum probability. Here is just one more that is rarely explicitly stated, but can be a hidden assumption that leads to incorrect conclusions. Consider this FALSE statement:

It is possible with one single measurement to determine the state of any arbitrarily given quantum system. FALSE.

Well, if the Hilbert space of the system has dimension 1, this statement is sort of true, because then there is only one possible state! But that is not what I mean by a *quantum system*, which must have a Hilbert space of at least dimension 2. Sorry about giving this criterion for being a quantum system afterwards! The FALSE statement above contradicts quantum theory, so if someone can actually find a measurement technique that does measure the state of any arbitrary quantum system, then quantum theory will have to be modified or replaced.

Why is this statement false? In general, with a given quantum system in some initial state ψ_i, a measurement (self-adjoint operator) of it will give a real number and collapse the system into a final state ψ_f. But the only thing we learn about the initial state ψ_i is that the *probability* $|\langle \psi_i, \psi_f \rangle|^2 \neq 0$. This is very far from measuring ψ_i, since it only excludes ψ_i from lying in the subspace orthogonal to ψ_f.

Of course, given an *ensemble* of quantum systems *all in the same state*, we can gain some information about what that state is by performing a corresponding *sequence* of single measurements. But the previous sentence, which is true, is not to be confused with the above FALSE statement. On the other hand, using the collapse of the state as a tool, one can prepare a system so that it is in a known state *after* a measurement. The moral is that measurement is a probabilistic, not deterministic, process in quantum theory, and so its result does not give sufficient information to determine the given, prior state of the system. (Parenthetically, we note that quantum systems do carry information, but of a new type known as *quantum information*, what else?) Therefore, this error is also due to a flawed understanding of quantum probability.

Exercise 19.3.1 *While reading the previous paragraph, you most likely were thinking about measurements that yield more than one possible value. Now understand why that paragraph is valid also for the case of a measurement which always yields one and the same value.*

19.4   Expected Values—One Last Time

It is often stated by persons who should know better that quantum theory does not give information on individual events but *only* predicts expected (or average) values. While it is true that individual events typically can not be predicted, this misconception underestimates what quantum theory does, which is to predict the classical probability distribution of any observable in a fixed, given state. (Already stated in 1932 in [32].) This is more information that just the first moment (also called expected value or average) of that probability measure. To put it even more bluntly, if the observed values do not coincide statistically with the predicted probability distribution, then the theory is in error although the observed average turns out to be the predicted expected value. See the comments on Axiom 5 in Chapter 21 for more details.

19.5   Tunneling

A third example is given by *tunneling*. This quantum property of matter can be seen in many models, though the first was using *Fowler-Nordheim (FN) tunneling* to understand electron field emission. (Parenthetically, we note that this was also an early use of Fermi-Dirac statistics.) The idea is to consider a potential energy that has a valley or *well* surrounded by an enclosing ridge region (called the *barrier*) which then gives way to outlying lowlands. Here is such a potential energy in dimension one. So for $x \in \mathbb{R}$ we let

$$
V(r) = \begin{cases} 0 & \text{for } |x| < A, \\ B & \text{for } A \leq |x| \leq A + 1, \\ B/4 & \text{for } |x| > A + 1. \end{cases}
$$

Here A and B are strictly positive real numbers. The region $|x| < A$ is the well, the barrier region is $A \leq |x| \leq A + 1$, and the outlaying lowlands is $|x| > A + 1$. On the other hand B is referred to as the *height* of the barrier. This potential energy is a 'toy model' for the combined strong and electromagnetic energies felt by an alpha particle in and near a heavy nucleus.

Of course, B has dimensions of energy. One then considers the quantum Hamiltonian for this potential energy:

$$
H = -\frac{\hbar^2}{2m}\frac{d^2}{dx^2} + V(x).
$$

Now it is an exercise to solve the time independent Schrödinger's equation $H\psi = E\psi$. Say we do this for $E = B/2$, half the height of the barrier. One solution ψ has a probability density $\rho(x) = |\psi(x)|^2$ that is concentrated in the well $|x| < A$, continues to be positive in the barrier region $A \leq |x| \leq A + 1$, and also is positive (though very small) for $|x| > A + 1$. This solution is called a *tunneling solution*. Call it what you will, what does it mean?

Some highly qualified physicists will say that it means that a 'particle' placed in the well can tunnel through the barrier region and appear in the outlying region. They say that the particle tunnels out of the well. But that understanding is based on an *initial condition* in time and a notion that the 'particle' is moving in time. However, the solution ψ does not depend on time; it is a stationary state. Also the time independent Schrödinger's equation never has an initial condition in time. There is no 'placing' the particle in the well to start with. And there is no motion, since neither $\psi(x)$ nor $\rho(x)$ depends on time.

Considering a classical particle moving in this same potential V is usually presented as a contrast to the quantum system. The only fact from classical mechanics that you need to know in order to understand this paragraph is that the total energy (defined as kinetic energy plus potential energy) of a classical particle is constant as it moves along a trajectory satisfying Newton's equation of motion. In the classical case one does impose initial conditions. Say we place the particle in the well region (zero potential energy) and give it initial kinetic energy $E = B/2$. So the total energy of the particle is $B/2$. Then its trajectory remains in the well for all time, since its kinetic energy would have to become strictly negative in the barrier region. But the formula $(1/2)mv^2$ for the kinetic energy of a body with mass $m > 0$ shows that it can not be negative. (It turns out $m < 0$ does not occur in nature.) In this classical case, we simply can not place the particle in the barrier region with total energy $B/2$, since that again would imply negative kinetic energy, an impossibility. Finally, we can place the particle in the outlying region with total energy $B/2$, half of which will be potential energy and the other half kinetic energy. In this final case, again by conservation of total energy, the particle will never enter the barrier region. Consequently, it can not get beyond the barrier and into the well region.

What do we learn from the classical situation? First and foremost we see that the time independent Schrödinger's equation provides a solution that is dramatically different from any classical particle solution. Furthermore, that quantum solution does not display any interference effect which is taken as being the indelible fingerprint of a wave. So tunneling is a purely quantum phenomenon in which 'wave/particle duality' is not found.

If we wish to discuss time evolution in quantum theory, we start with the time dependent Schrödinger's equation. In that case we do start with an initial state ϕ_0, and we must find $\psi_t = \exp(-itH/\hbar)\phi_0$. To get non-trivial time evolution we must pick ϕ_0 to be different from a stationary state. For example, we now can pick ϕ_0 to be strictly positive in a small interval $|x| \leq \epsilon << B$ and zero in the complement of that small interval. Then the solution $\psi_t(x)$ and its probability density $\rho_t(x) = |\psi_t(x)|^2$ will depend non-trivially on the time t. One can find these functions numerically and produce videos of them using computer programs. These videos can be very impressive, especially since they show some sort of change with the passage of time, that is, some sort of motion. But motion of what?

The change of a time dependent probability density is not the motion of a particle nor of a wave. In classical physics the motion of a particle in space tells how its non-zero mass moves along a continuous trajectory, and the motion of a wave tells how its non-zero energy propagates continuously in many spatial directions. Neither of these is the motion of a probability density in space. In order to compare a computer generated time dependent probability density function with experiment, one has to measure position in an ensemble of identically prepared experiments. And, as expected, the measurement of the position will collapse the wave function. So that must be taken into consideration, too. As far as I am aware, there is nothing in the classical physics of a massive particle that corresponds to this collapse. So speaking sensibly about tunneling solutions certainly is possible, but it must be done within the framework of quantum theory. Using language from classical physics can be misleading or, even worse, wrong.

19.6 Superposition

Any unit vector $\psi \in \mathcal{H}$, a Hilbert space, can be expanded in any orthonormal basis $\{\phi_\alpha \mid \alpha \in A\}$ as $\psi = \sum_\alpha c_\alpha \phi_\alpha$ for unique complex numbers c_α satisfying $\sum_\alpha |c_\alpha|^2 = ||\psi||^2 = 1$. This is simply a statement about the geometry of Hilbert space. In quantum physics, we interpret any unit vector as a possible state. If we have a quantum system in the state ψ, then it has infinitely many different such expansions if $\dim \mathcal{H} \geq 2$. This geometric fact is called *superposition* and is not a mystery requiring further considerations. The state of the quantum system is ψ, no matter what (if any) expansion we use. If now a measurement is made that collapses the quantum system to one of the states ψ_α, then that is also not a mystery requiring further rumination, but rather an instance of the Measurement Condition. The new final state is simply ϕ_α and, in general, no longer ψ. Of course, we can expand ϕ_α itself in an infinite number of distinct orthonormal bases, and so it has the same basic property as any other state, such as ψ itself, has.

However, in many expositions, especially for non-technical audiences, some statement is made that the initial state ψ is fundamentally different in some way (being 'coherent') from the final state ϕ_α (being 'decoherent'). Sometimes emphasis is placed in the value of the measurement that induced the collapse to ϕ_α, as if that gave ϕ_α some extra virtue. But what if that value is lost even though the measurement occurred? Or what if ψ had been produced by the collapse associated with a measurement? The only possible distinction between the states ψ and ϕ_α is that we might know no value of any measurement for ψ while we know the value of one measurement for ϕ_α. That is what the Measurement Condition gives us, and no more than that. If there is any mystery here, then it resides in the Measurement Condition itself. The point of this brief section is that there is no essential difference between one state of a quantum system and another. Beware of attempts to say the contrary!

19.7   Quantum Fluctuations

> Mieux vaut comprendre peu
> que comprendre mal.
> Anatol France

As a final example consider *quantum fluctuations*. This expression is often used to increase intuition by supposing that a time independent state of a quantum system is undergoing changes as time goes by. Put this way, this is a contradiction and thus thwarts my intuition. What about your intuition? One is again uncomfortably confronting the probabilistic interpretation of quantum theory. Quantum theory tells us that the measurements of a time independent quantum state can give various different values or, in common parlance, these measured values *fluctuate*. But the underlying state remains time independent; there is nothing changing (or fluctuating) about that state itself. So beware of explanations invoking quantum fluctuations. They may be wrong. But they may be right. It depends on what exactly is being said.

19.8   Notes

American readers will recognize the expression "*déjà vu* all over again" as a Yogi-ism. These sayings are named for the Hall of Fame baseball player, coach, and manager Yogi Berra. Whether he actually said this is summed up in another famous Yogi-ism: "I really didn't say everything I said." This puts Yogi (or somebody else) in the same ballpark with Bertrand Russell.

The other Yogi-ism referred to earlier: "Good hitting beats good pitching every time. And *vice versa*."

As the reader may be aware by now, I take exception to explanations which could unkindly be called sophistry, that is, the use of some jargon without any clear idea of what is being said. Such rhetorical tricks have been known since antiquity. In quantum theory, they include cavalier appeals to complementarity, duality, and uncertainty among others. But many of my physics colleagues justify statements which are objectively false by claiming that such statements help them (subjectively) understand quantum theory. I can hardly imagine how. My own (subjective) reaction is that for me such statements only add confusion to an already very complicated topic. But we all, as scientists, should only rely on objectively (that is, experimentally obtained) verifiable criteria for accepting or rejecting any statement. This seems to me to be at the heart of the famous dictum of E. Rutherford, Nobel Laureate in chemistry, that all science should have an explanation understandable by a barmaid. The psychological, or possibly even philosophical, analysis required to understanding why many people accept objectively false statements as true is far beyond my area of expertise. I leave it to others to ponder this important question.

Chapter 20

Complementarity (Optional)

> There is something fascinating about science.
> One gets such wholesale returns of conjecture
> out of such a trifling investment of fact.
>
> Mark Twain

20.1 Some Ideas and Comments

Complementarity is another one of those phrases that intimidates beginners
without adding much to quantum theory. It is easily confused with duality,
because neither of these expressions is clearly defined in most expositions.
But whatever might be meant is mostly irrelevant to the everyday business
of quantum theory, which has to do with Schrödinger's equation and its
solutions.

One crude attempt is to define two observables A and B as complementary
if they do not commute. But in that case one can simply say they are
non-commuting observables and be done with it. Usually, the position operator
Q and the momentum operator P in a one-dimensional system are taken to
be the standard case of complementary observables. Of course, a motivating
example is far from being a general concept.

Complementarity is attributed to N. Bohr and is regarded by many to
be his crowning contribution to quantum theory. One common element is
that it refers to a pair of 'properties' of a quantum system. However, it is
interpreted variously as a principle of physics or of philosophy of science. Some
maintain that it is a manifestation of the Heisenberg uncertainty principle
while others say not. When it is exemplified by physical observables, these are
always non-commuting. But it is usually maintained that non-commutativity
is not the essential property of complementarity. Sometimes complementarity
is described by a series of examples without a clear statement about what

© Springer Nature Switzerland AG 2020
S. B. Sontz, *An Introductory Path to Quantum Theory*,
https://doi.org/10.1007/978-3-030-40767-4_20

is common to all these examples. In fairness, sometimes a common property is given, and this has to do with properties of measurement or knowledge. The assertion is that the measurement (resp., knowledge) of one of the pair of properties impacts the measurement (resp., knowledge) of the other property. However, this assertion (if formulated as a property of observables) is simply a consequence of collapse for non-commuting observables.

I find this mixture of epistemology with duality to be quite troubling. The collapse of the state clearly implies that measuring just one observable will not be enough to know what the initial state was. But measuring two non-commuting observables is not better when they are done on the same system consecutively, where we invoke the collapse criterion which precludes their simultaneous measurement. Consequently we are led to consider an ensemble of measurements on identically prepared systems, which throws us into the jaws of quantum probability.

There is some cultural preference for thinking that duality is a primary property that unifies nature and that multiplicity is to be shunned. Strange to think that duality gives rise to unity, no? But such thinking sometimes seeps into science!

Often complementarity is also exemplified by the so-called 'wave/particle duality' for which explicit observables do not exist. (I mean that there is no wave observable nor is there a particle observable, where 'observable' means self-adjoint operator.) Since the basic constituents of matter are not waves (that is, quantum 'particles' never spread out in all spatial directions) and are not point particles following classical trajectories, the concept of 'wave/particle duality' is not clearly defined, except to the extent that certain systems display *some* wave properties as well as *some* particle properties. The idea is that different (i.e., complementary) measurements reveal this dual 'wave/particle' structure. But the underlying concept of 'wave/particle duality'—and how it can be checked experimentally in general—is typically not a concern. If this 'duality' is a fundamental principle of *all* quantum systems, then it must be described in *full* generality.

Moreover, in quantum physics there are neither waves nor particles. (See Chapter 3.) This renders the whole idea of 'wave/particle duality' suspect. In this context the famous *two-slit experiment* is usually appealed to as an important example. But one example does not suffice. And a few examples do not suffice. What both the two-slit experiment and the associated *one-slit experiment* show is that the basic constituents of matter are neither waves nor particles. See Feynman's explanation of this in chapter 37 of [12] in 1963. Another curious aspect of the literature is the almost total lack of a discussion of the three-slit experiment, which leads to confused thinking about a 'particle' being in 3 places at the same time. Or even worse the four-slit experiment! Going down this particular rabbit hole leads one to the infinite-slit experiment, wherein the screen is gone having been replaced with an infinite number of infinitely thin slits. Then the 'particle' is at infinitely many places at the same time. This particular way of incorrect thinking about quantum theory has already been analyzed in Chapter 19.

But there is a duality in quantum theory that everyone knows about, but is never (as far as I know) emphasized in introductory, or even advanced, texts. This is the *conjugate duality* of a Hilbert space with itself. By the Riesz representation theorem 9.3.2, the space of bounded linear functionals on a Hilbert space $\mathcal{H}$ is anti-unitarily identified with $\mathcal{H}$ canonically. I have no idea whether this has any physical interpretation, though it is the fact behind Dirac notation.

Perhaps some of the ideas of complementarity aid in the comprehension of quantum theory for some people. But for me complementarity is at best a side issue. As mentioned earlier the central business of quantum theory is understanding Schrödinger's equation.

20.2 Notes

This short chapter does not come close to doing justice to the enormous literature on complementarity. Nor is it intended to. Since the importance of this topic is so highly stressed by so many, I felt obliged to comment on it. But my comments are highly controversial. An overwhelming multitude of experts of all sorts take strong objection to what I say here. I let them speak for themselves. The reader will have no trouble finding them.

Chapter 21

Axioms (Optional)

> I had been told that Euclid
> proved things and was much
> disappointed that he started
> with axioms.
> Bertrand Russell

In this chapter we collect axioms for quantum theory in one place and make some comments about them. However, this is very much a work in progress. In fact, this is one of the reasons that quantum theory remains so perplexing. And that is why the axiomatization is so important. Of course, on a day to day basis most scientists can take the appropriate Schrödinger equation as their starting point and work effectively with its quantum theory in order to understand new quantum systems and applications of quantum theory. But the desire to establish first principles should be acknowledged for its value in providing a clearer understanding, which some day might even lead to another, better theory. In some sense this is what is hoped for in the desired unification of quantum theory with the theory of general relativity, which I will comment on a bit more in Chapter 22.

I feel that this aspect of quantum theory should be presented to beginners because in the first place it is accessible to them and in the second place because it is good to learn early on that there is still a tentative element in quantum theory. The discussion about axioms for quantum theory is ongoing, opinions are divided, and the eventual outcome is, well, uncertain.

21.1 A List of Axioms

Axiom 1: (Kinematics) Every isolated system in nature has an associated Hilbert space $\mathcal{H}$ together with a collection of relevant self-adjoint (though not

© Springer Nature Switzerland AG 2020
S. B. Sontz, *An Introductory Path to Quantum Theory*,
https://doi.org/10.1007/978-3-030-40767-4_21

necessarily bounded) operators acting in that Hilbert space. Each of these operators represents in the quantum theory a physical observable for that system. At least two of these operators do not commute. Either the Hilbert space $\mathcal{H}$ or at least one of these operators must depend explicitly on Planck's constant $\hbar$.

Axiom 2: (States) Every system is described completely at every moment in time by its *state*, which is defined, up to a multiplicative complex constant of absolute value 1, as a unit vector in the Hilbert space, that is, by a vector $\psi \in \mathcal{H}$ with $||\psi|| = 1$.

Axiom 3: (Statistics) Multi-particle systems have a Hilbert space that is a tensor product of Hilbert spaces, each of which corresponds to a distinct type of boson or distinct type of fermion. The number and types of the bosons and fermions depends on the system being considered and this in turn determines the number and types of the Hilbert spaces in the tensor product.

Axiom 4: (Dynamics) The time evolution of a quantum system is given by Schrödinger's equation

$$i\hbar \frac{\partial \psi}{\partial t} = H\psi,$$

where H, the *Hamiltonian* of the system, is a self-adjoint operator acting in $\mathcal{H}$ and that characterizes the system, together with an initial condition on the solution at time t_0. The solution of this equation is a function $\psi : \mathbb{R} \to \mathcal{H}$. This initial condition has the form $\psi(t_0) = \varphi$, where $\varphi \in \mathcal{H}$ is a state of the system and $t_0 \in \mathbb{R}$ represents the initial time.

Axiom 5: (Measurement and Collapse Axiom) Suppose an experiment is done on a system for measuring the physical quantity corresponding to a self-adjoint operator T acting in $\mathcal{H}$. Let $\mathrm{Spec}\,(T)$ denote the spectrum of T. Then the set of all the values obtained by measurements of the observable corresponding to T is exactly the set $\mathrm{Spec}\,(T)$. (By the way it is a theorem that $\mathrm{Spec}\,(T)$ is a non-empty, closed subset of $\mathbb{R}$.)

We let P_T be the unique quantum event (or projection) valued measure associated with T by the spectral theorem. Let $B \subset \mathrm{Spec}\,(T)$ be a Borel set. Then the probability that the measurement of the observable corresponding to T has its value in B, given that the system is in the state ϕ at the time of the measurement, is given by *Born's rule*

$$\mathrm{Prob}(T \in B \,|\, \phi) = \langle \phi, P_T(B)\, \phi \rangle.$$

Given that the measurement did indeed yield a value in B, then $P_T(B)\phi \neq 0$ must hold and the system has also *collapsed* (i.e., changed) into the state determined by the unit vector

$$\frac{1}{||P_T(B)\,\phi||} P_T(B)\, \phi.$$

Axiom 6: ($\hbar$ **Dependence of Non-commutativity**) For every pair A, B of bounded non-commuting, self-adjoint operators representing observables (whose existence is posited in Axiom 1) their (ring) commutator

$$f_{A,B}(\hbar) := [A, B] = AB - BA$$

must be a function of Planck's constant $\hbar > 0$ for which this limit exists: $\lim_{\hbar \to 0^+} f_{A,B}(\hbar) = 0$.

For a pair of unbounded non-commuting self-adjoint operators A, B which represent observables, this axiom must be rephrased as follows: The (group) commutator

$$g_{A,B,t}(\hbar) := e^{itA} e^{itB} e^{-itA} e^{-itB}$$

for each $0 \neq t \in \mathbb{R}$ must be a function of Planck's constant $\hbar > 0$ for which this limit exists: $\lim_{\hbar \to 0^+} g_{A,B,t}(\hbar) = I$.

The topology in which these limits are taken is discussed below.

Axiom 7: (**Planck's Law**) Suppose that a system has states ψ_1 and ψ_2 with respective energies E_1 and E_2. Suppose also that $E_2 > E_1$. If the system changes from state ψ_2 to state ψ_1 with the emission of a photon, then that photon has angular frequency $\omega = (E_2 - E_1)/\hbar$. Furthermore, if the system is in state ψ_1 and changes to state ψ_2 with the absorption of a photon, then that photon has angular frequency $\omega = (E_2 - E_1)/\hbar$.

Axiom 8: (**Quantization**) Given a physical system with a description in classical theory, there is a way to construct the corresponding quantum theory of the system from that classical theory. This construction is called *quantization* and must satisfy all of the previous axioms.

21.2 A Few Comments

Axiom 1 is satisfied by all systems on the atomic and smaller spacial scales. I call these *quantum systems*. The best guess is that every physical system is a quantum system, and thus there is no need to define quantum systems. However, it seems pointless to try to describe many larger systems by using quantum theory when they are already so well described by classical theory. It is not even clear how to define *classical system*. One uses experience when deciding which description to use with a given system. The description that works better wins, and that's that. Moreover, this axiom does not tell us how to find the Hilbert space and self-adjoint operators that are appropriate for a given system. And theorists sometimes do play around with distinct formulations, such as one based on a *configuration space* or on a *phase space*, for example. But these concepts come from classical physics and so do not seem to be conceptually important in quantum theory, though they can be useful theoretical techniques.

While Axiom 1 requires at least one pair of non-commuting observables, it does not require at least one non-trivial pair of commuting (that is, classically related) observables. However, this extra condition is always satisfied in practice and so might merit inclusion in this Axiom.

Axiom 1 requires quantum theory to have two characteristic properties: a Hilbert space and Planck's constant. It is rather difficult for me to imagine a quantum theory without these.

Axiom 2 is incomplete. The states described there are actually called the *pure states*. There are other states as well; these are the *mixed states*, which are important in *statistical quantum mechanics*. The definition and properties of mixed states can be found manywhere in the literature. But I have decided not to discuss them in this book.

In some theories not all unit vectors represent states. In other words there are extra *super-selection rules* which also pose a puzzle for this axiom. These rules exclude some unit vectors from being states for certain quantum systems. For example, in a spin 1/2 system we not only have the states ↑ of spin up and ↓ of spin down as eigenstates of the observable S_3, but we also have convex combinations of these such as

$$\frac{1}{\sqrt{2}}(\uparrow + \downarrow). \tag{21.2.1}$$

And this is a perfectly feasible state for this system. However, letting ↑ represent a proton and letting ↓ represent a neutron as is done in the theory of *isospin*, the state (21.2.1) is not a valid state. A reason for justifying this is that electric charge is a conserved quantity, and so it does not make sense to have a state that has charge +1 with probability 1/2 and has charge 0 with probability 1/2. But these super-selection rules are in conflict with Schrödinger's razor since they are *a posteriori* conditions that do not arise from a differential equation.

Axiom 3 to date looks correct. But the division of particles into different types (beyond just bosons and fermions) is not completely understood. This is the domain of the *Standard Model* of Particle Physics. While this model has been wonderfully successful, including the prediction of the *Higgs boson* as now seen in experiment, many experts seek something more fundamental behind it. We don't know how this will pan out. Also, to understand Axiom 3 you have to devote the time and effort to learn about tensor products.

Axiom 4 is rock solid to date. Of course, this is the Schrödinger picture. But having good axioms in one picture, it is an easy matter to translate them into any other picture. This axiom is also called the *unitary condition* or simply *unitarity*. The point is that the Hamiltonian H being self-adjoint determines the unitary group $e^{-itH/\hbar}$ acting on the Hilbert space $\mathcal{H}$ for every time $t \in \mathbb{R}$ (by the spectral theorem of functional analysis) and that unitary group then solves the initial value problem (also called the Cauchy problem) for the Schrödinger equation, namely the solution is $\psi(t) = e^{-itH/\hbar}\,\varphi$ where φ is the initial condition for $\psi(t)$ at $t = 0$. So, unitarity is a consequence of Schrödinger's razor. Some prefer to put unitarity itself as an axiom.

The success of quantum theory is principally due to the multitude of Hamiltonians that clever physicists have come up with in order to describe a corresponding multitude of quantum systems. This is the power of Axiom 4. This is also the origin of the folklore that one should not be caught doing quantum theory without having a Hamiltonian in play. The other axioms serve to support Axiom 4.

Also, Axiom 4 is the only axiom which involves complex numbers directly through the presence of $i = \sqrt{-1}$ in it. But Planck's constant $\hbar$ appears in this axiom and in others too. However, a mysterious aspect is that time appears in this equation as a parameter and time does not appear in quantum theory as a genuine observable, that is, a physical quantity represented by a self-adjoint operator. Of course, the Schrödinger equation is a *non-relativistic equation* and so this may be the best we can get without going relativistic.

It is worth noting that many of the Hamiltonians that do arise in physics have the form of a *Schrödinger operator*, which in its simplest form is

$$H = -\frac{1}{2m}\Delta + V,$$

where $m > 0$ is the mass of a 'particle', Δ is the Laplacian operator acting in $\mathcal{H} = L^2(\mathbb{R}^3)$, and $V : \mathbb{R}^3 \to \mathbb{R}$ determines a *multiplication operator*, also acting in $\mathcal{H}$. This leads to a research area in *mathematical physics* concerning this class of operators. Often, the potential energy term V studied defies physical intuition, and so one gets results that are mainly mathematical in nature. Also, one can change the dimension of the Euclidean space from 3 to other values, study complex valued or time dependent potentials, consider m to be dependent on $\mathbf{x} \in \mathbb{R}^3$, and so forth. However, it happens that even these cases can sometimes shed some light on more physically reasonable situations.

Axiom 5 continues to be the most controversial of the axioms. This is its complete formulation. In Chapter 10 we saw a special case of Axiom 5, and the general case was presented in Chapter 16. Before getting into that, let's recall from Exercise 16.2.11 that

$$\mu_{T|\phi}(B) := \langle \phi, P_T(B)\,\phi \rangle$$

for fixed self-adjoint T and state ϕ is a classical probability measure defined for Borel sets $B \subset \mathbb{R}$. So it makes sense to relate this probability measure to experimentally determined frequencies. Of course, it is an experimental issue to see if that relation is true. Being a probability measure, $\mu_{T|\phi}$ can have moments. For example, its first moment m_1 (also called the *expected value* of T or the *average* of T) is given by another version of Born's rule:

$$m_1 := \int_{\mathbb{R}} \lambda\,\mu_{T|\phi}(\mathrm{d}\lambda) = \int_{\mathbb{R}} \lambda\,\langle \phi, P_T(\mathrm{d}\lambda)\,\phi \rangle = \langle \phi, \int_{\mathbb{R}} \lambda\,P_T(\mathrm{d}\lambda)\,\phi \rangle = \langle \phi, T\phi \rangle,$$

$$(21.2.2)$$

provided that ϕ is in the domain of definition of T. (The way to understand (21.2.2) is to read it from left to right, though the rigorous proof of it reads from right to left.) Some texts emphasize the importance of this formula for the first moment to the extent that the probability measure $\mu_{T|\phi}$ is never mentioned and only the expression $\langle \phi, T\phi \rangle$ appears. But here we understand that behind this version of Born's rule is a well defined classical probability measure, namely $\mu_{T|\phi}$, whose first moment enters the theory but does not replace the role played by the probability measure $\mu_{T|\phi}$ itself. So, Axiom 5 is the most fundamental Born's rule, while (21.2.2) is a consequence of it.

Axiom 5 is where probability enters quantum theory and that stirs up a lot of concerns, doubts, and dissents. The jury is out on this axiom, even though it has strong experimental support. There are even some who advocate that this axiom should be entirely eliminated from quantum theory, though it should then be replaced with something else to deal with the (apparent) probabilistic aspect of quantum systems. Note that what I call collapse is also called *quantum jump*, *reduction* of the state or the *projection postulate* of von Neumann. See Isham's book [16] for a thoughtful, detailed discussion on probability, collapse, and many other important topics in quantum theory.

Axiom 5 is also where measurement enters quantum theory. This is so conceptually different from how measurement is handled in classical theory that it leads to what is called the *measurement problem* in quantum theory. This is actually a collection of problems related to the nature and importance of the measuring process. In the Copenhagen interpretation, measurement was characterized as being an interaction of a 'classical' system (producing measured values in a way consistent with classical physics) with a quantum system. This approach is still advocated to this day, though many objections have been made to it. The simplest objection is that all systems are basically quantum in nature, and so one must address the problem of deciding which of these systems are to be deemed 'classical' and what makes them different.

To fully understand what Axiom 5 says one has to fully understand the definition of the spectrum of an operator, which was given in the optional Section 9.5. The reader who skipped over that might wish to read it now.

I won't give my opinion of Axiom 5, since that doesn't count. Sorry, dear reader, your opinion also does not count. This has to be settled in terms of an objectively verifiable scientific analysis that the physics community can come to accept solely on its merits for explaining experiments. For now, the axiom just sits there looking at us. But we do use it! For example, experiments are designed to put a system into a specific initial state by using the deterministic collapse condition. This is known as *preparation* of the state. The experiment then continues, for example, by studying how the system in this known initial state interacts with other systems.

Axiom 5 applies to situations where the state is time dependent both before and after the measurement. In this case ϕ refers to the state of the system at the moment of the measurement.

One of the earliest physics insights in quantum theory is that for any state $\phi \in L^2(\mathbb{R}^3)$, its absolute value squared $|\phi|^2$ is the probability density for the location measured in $\mathbb{R}^3$ of the quantum system in that state. This basic aspect of quantum theory seems not to be found in the axioms. But it is a consequence of Axiom 5 and an explicit understanding of what the observable is for location. To simplify the exposition we first discuss the one-dimensional case. So we consider the Hilbert space $L^2(\mathbb{R})$ and the position operator Q, where $Q\psi(x) = x\psi(x)$ for $\psi \in L^2(\mathbb{R})$ and $x \in \mathbb{R}^3$. The fact we need from the functional analysis of the unbounded, self-adjoint operator Q is that its pvm $P_Q(B)$ is the projection operator given as multiplication by χ_B, the characteristic function of the Borel set $B \subset \mathbb{R}$. (See Exercise 19.2.1.) Then for a given state $\phi \in L^2(\mathbb{R})$ we have the probability that a measurement of Q lies in B is

$$P(Q \in B \mid \phi) = \langle \phi, P_Q(B)\phi \rangle_{L^2(\mathbb{R})} = \langle \phi, \chi_B\phi \rangle_{L^2(\mathbb{R})} = \int_{\mathbb{R}} dx\, \phi^* \chi_B \phi = \int_B dx\, |\phi|^2,$$

showing that the integral of the far right side has the standard probabilistic interpretation. To generalize this to $L^2(\mathbb{R}^3)$, we consider the three *commuting* operators Q_1, Q_2, Q_3 associated with the three Cartesian coordinates x_1, x_2, x_3 of $\mathbb{R}^3$. Therefore the pvm's of these three unbounded, self-adjoint operators commute among themselves, and so we get a joint probability distribution for the three of them. By a slight modification of the result in Section 16.10 for commuting operators, we see for a state $\phi \in L^2(\mathbb{R}^3)$ that

$$P(Q_1 \in B_1, Q_2 \in B_2, Q_3 \in B_3 \mid \phi) = ||P_{Q_1}(B_1)P_{Q_2}(B_2)P_{Q_3}(B_3)\phi||^2$$

$$= ||\chi_{B_1}(x_1)\chi_{B_2}(x_2)\chi_{B_3}(x_3)\phi(x_1, x_2, x_3)||^2 = \int_{B_1 \times B_2 \times B_3} dx_1 dx_2 dx_3\, |\phi(x_1, x_2, x_3)|^2,$$

where B_1, B_2, B_3 are Borel subsets of $\mathbb{R}$. The left side can also be interpreted as $P((Q_1, Q_2, Q_3) \in B_1 \times B_2 \times B_3)$. With an appropriate magic touch of measure theory, we can replace the product Borel set $B_1 \times B_2 \times B_3$ with a general Borel subset B of $\mathbb{R}^3$ to get $P((Q_1, Q_2, Q_3) \in B) = \int_B |\phi|^2$. And this is the result sought for.

Axioms 4 and 5 (and Axiom 7 implicitly) are the only ones which refer to how quantum systems change in time, although these changes are quite different. Axiom 4 concerns a deterministic, continuous time evolution of the state of a quantum system, while Axiom 5 describes a probabilistic, instantaneous time evolution. And it is a serious problem with quantum theory that there are two distinct ways to describe the time evolution of a physical system. As far as I am aware every other theory in physics entails exactly one way for describing time evolution. This anomalous situation in quantum theory leads one to suspect that this axiomatization of quantum theory is not completely correct. Or, to put it explicitly, it seems there should be one unified way for describing time evolution in quantum theory. Of course, Axiom 4 refers to the time evolution of an isolated quantum system for which unitarity holds, while Axiom 5 refers to a quantum system that is being measured by another, interacting system in which case unitarity does not always hold for the measured quantum system.

The statement of Axiom 6 is trivial if A and B commute. That is why this axiom is only about non-commuting observables. Dirac is famous for saying on many occasions that the key property of quantum theory is that the observables do not commute. Axiom 1 incorporates that observation. This axiom strengthens this by saying how they do not commute as Planck's constant $\hbar$ (taken to be a parameter) approaches 0. This particular property of quantum theory has certainly been recognized before. Anyway, it seems to me to be an essential ingredient in quantum theory that is not included in the other axioms. Moreover, it amplifies on the role, already mentioned in Axiom 1, that $\hbar$ plays in quantum theory. Consequently, this property deserves its own axiom, something which I have never seen in other axiomatic presentations of quantum theory. This axiom can be overlooked in some specific examples since one tends to use units in physics for which $\hbar = 1$. The limit in this Axiom is with respect to some topology on operators. Since we often deal with densely defined, unbounded operators, we do not want to use the operator norm. The exact topology as well as domain considerations are technical issues which can be left on a side for now, though they must be dealt with in a complete exposition.

It must be emphasized that Axiom 6 is about quantum theory and nothing else. It is not to be confused with quantization schemes that have some relation involving Planck's constant between the Poisson bracket in classical mechanics and the commutator in quantum mechanics. It seems that Dirac was the first to discuss such a quantization scheme, which unfortunately did not work out as expected. Nonetheless, Dirac's ideas have survived in modified form in *deformation quantization*. As will be proposed later on *any* quantization scheme should satisfy Axioms 1 to 7.

Axiom 7 is the historical starting point of quantum theory in Planck's article of 1900, but for whatever reason is rarely considered a basic axiom. Yet it is independent of the other axioms. Notice that it is not asserted that the transition from the initial state to the final state *must* occur under these circumstances, because in some systems there are other restrictions which prohibit the transition. Also, this axiom should be included, since it is essential in the interpretation of the experimentally determined spectra of light. In a more comprehensive set of axioms it could be subsumed into an axiom on the *conservation of energy* or, even better, into an axiom on the time evolution of quantum systems that includes matter and light. Of course, there are quantum theories which include time dependent interactions between matter and light, but what I am speaking of seems not yet to exist, namely a complete axiomatic approach to that sort of non-relativistic quantum theory.

Axiom 8 is not about quantum theory *per se*, but rather a relation of it with classical theory. Maybe it would be better to think about quantum theory strictly in its own terms without reference to the historically prior classical theory, in which case quantization is simply the process of finding the correct quantum theory for each quantum system. However, most physicists use this axiom to start off their analysis of a system with quantum theory. The quantization is often done so quickly that it is not even given explicit recognition. A physicist will simply write down an expression and declare that it is the correct *quantum Hamiltonian* for the quantum theory of the system being considered. What this amounts to usually is just the canonical quantization of a specific function in classical theory; that function is called the *classical Hamiltonian*, and it plays an important role in classical theory, though that does not interest us now.

The exact quantization to produce quantum theory is deliberately not specified in Axiom 8. However, it is (universally?) accepted by physicists that the position and momentum observables of classical physics must be quantized so that the canonical commutation relations (5.1.4) hold. Perhaps this should be included in Axiom 8.

Typically, the most difficult part of quantization consists in *quantizing* the functions that represent observables in a classical theory (for example, functions on a classical *phase space*) to produce the self-adjoint operators that represent the *same* observables in the quantum theory. Mathematically speaking, there are many distinct ways to quantize a classical theory. Clearly, nature does not recognize all these possibilities. Somehow, only one of them is physically correct. So this is a problematic aspect of this axiom, so much such that some experts reject it completely as being irrelevant to quantum theory. But by doing so they make it more difficult, though not impossible, to implement Axiom 1. In any case Axiom 8 is more of a wish than anything else, since it does not specify at all the quantization to be used.

The saying "quantization is operators *instead of* functions" is based on these ideas. It turns out that a form of the spectral theorem, called the *functional calculus*, almost gives us a quantization. We suppose that T is a bounded, self-adjoint operator acting on a Hilbert space $\mathcal{H} \neq 0$. We let $\text{Spec}(T)$ (a closed, non-empty subset of $\mathbb{R}$) denote the spectrum of T, and let $\mathcal{A}_T$ denote the set of bounded Borel functions $f : \text{Spec}(T) \to \mathbb{C}$. Then $\mathcal{A}_T$ is a complex vector space that also is equipped with a multiplication operation, namely point-wise multiplication of functions, under which the constant function 1 is the identity element. One says that $\mathcal{A}$ is an *algebra with identity*. An *algebra map* is a linear map between algebras that preserves the multiplication. Given this setup, here is a relevant version of the spectral theorem.

Theorem 21.2.1 (Spectral Theorem: Functional Calculus Version)
There is an algebra map $\phi_T : \mathcal{A}_T \to \mathcal{L}(\mathcal{H})$ that satisfies $\phi_T(1) = I$, the identity operator, and $\phi_T(\text{id}) = T$ if id is the identity function $\text{id}(\lambda) = \lambda$ for all $\lambda \in \text{Spec}(T)$. (The function id is bounded, since $\text{Spec}(T)$ is bounded.)

Remarks: There is also a version of this theorem for unbounded, self-adjoint operators, which is an important situation in quantum physics. We often use the notation $f(T) := \phi_T(f)$. So, we get the operator $f(T)$ *instead of* the function f, which looks something like a quantization. For example, if $f(x) = a_0 + a_1 x + \cdots + a_n x^n$, a polynomial, then $f(T) = a_0 I + a_1 T + \cdots + a_n T^n$.

Exercise 21.2.1 *Prove the last sentence.*

Theorem 21.2.1 does not look much like the Spectral Theorem 16.2.1, but the proof shows the close connection. The idea is to let P_T denote the pvm associated with T by that version of the Spectral Theorem and then define $f(T) := \int_{\mathbb{R}} f(\lambda) \, dP_T(\lambda)$. Next one shows that this defines an algebra map. Conversely, given that Theorem 21.2.1 holds, one defines $P_T(B) := \chi_B(T)$ for every Borel subset B of $\mathbb{R}$, where χ_B is the characteristic function of B. Then one shows that this defines a pvm P_T that satisfies $T := \int_{\mathbb{R}} \lambda \, dP_T(\lambda)$.

The image of the map ϕ_T is a *commutative* sub-algebra of $\mathcal{L}(\mathcal{H})$, since its domain $\mathcal{A}_T$ is a *commutative* algebra. Explicitly, for all $f, g \in \mathcal{A}_T$ we have

$$\phi_T(f)\phi_T(g) = \phi_T(fg) = \phi_T(gf) = \phi_T(g)\phi_T(f),$$

where we used $fg = gf$ and that ϕ_T preserves multiplication. Consequently, the self-adjoint elements in the image of ϕ_T commute among themselves, and so Axiom 1 is not satisfied. So this powerful theorem in functional analysis does not give us a quantization. Also, note that Planck's constant $\hbar$ does not appear in the functional calculus.

21.3 Notes

Many physicists discount the importance of an axiomatic foundation for quantum theory. Even some mathematicians have the same opinion. Of course, one can always push proposed axioms to a side and proceed as one wishes, starting with some sort of intuition rather than with a complete mathematical model. But the Schrödinger equation can not be pushed aside! Ultimately, I think that the axiomatic method will prevail as it has in other areas of physics. D. Hilbert thought that the axiomatization of physics was important enough to be included as his 6th problem for the 20th century, although he was not thinking about quantum theory whose first inkling was given by Planck in the last hours of the 19th century. Hilbert's 6th problem is controversial with some claiming that it is not clearly stated and others saying that it is essentially unsolvable. For a sympathetic, though dated, defense of the axiomatic approach, see [36]. How will this all turn out? Time will tell.

However, here is a discouraging word from B. Russell, the logician *par excellence*. Of course, it bemoans that human belief tends to work exactly backwards with respect to the rules of logic. Russell said:

"We tend to believe the premises because we can see that their consequences are true, instead of believing the consequences because we know the premises to be true."

The point here is that in logic the truth of the premise p and of the implication $p \Rightarrow q$, where p and q are propositions, tells us that the consequence q is true. However, even though the implications $p \Rightarrow q_1, \ldots, p \Rightarrow q_n$ are true and the consequences $q_1, \ldots, q_n$ are also true, this does not mean that the premise p is itself necessarily true. Logically, it could be false.

The quote at the start of the chapter is from Russell's autobiography. It can be interpreted as telling us that a logician is made and not born as such.

As a final note, let's remark that the axiomatic approach to quantum field theory (QFT) has had some success, although so far not as much as originally expected. However, various axiomatizations of QFT have been given, and it is not clear which of them, if any, is the correct one for explaining the relevant experimental physics. This ambiguity may be in part behind why the Millennial Prize for a rigorous gauge theory is so difficult.

Chapter 22

And Gravity?

> ... such stuff as dreams are made on.
> Prospero in *The Tempest*, William Shakespeare

In the first quarter of the 20th century, physics was advanced two times as it has never been since. First came Einstein's theories of relativity, both special relativity and general relativity. The former supplanted Newton's absolutist (and quite intuitive to many to this day) world picture of space and time, and the latter replaced Newton's instantaneous action at a distance theory of gravity with a gravitational theory based on a unified curved spacetime that mediates gravitational interactions at the large, though finite, speed of light rather than instantaneously. Still this remained within the realm of classical physics in that the framework is deterministic and motion along trajectories makes sense. Second was quantum theory, which brought into question such classical ideas as deterministic interpretations as well as the intuitive idea of motion itself. By any stretch of the imagination these both were revolutionary developments in scientific thought.

However, there is to date no theory which incorporates in a unified way the established gravitational and quantum theories, at least in some appropriate sense. This hoped for but still non-existent unification is often called *quantum gravity*. It seems safe to assert that physicists quite generally expect that some sort of unification exists. And the consensus is that to achieve this unification, quantum theory will not be changed while gravitation theory will be changed. This consensus could well be wrong. We simply do not know. It could be that there are more pieces of the puzzle, such as dark matter, that must be included into a unified theory. But again, we do not know.

Quantum theory presents many unsolved problems. But among them this unification is one of the biggest challenges in contemporary physics.

© Springer Nature Switzerland AG 2020
S. B. Sontz, *An Introductory Path to Quantum Theory*,
https://doi.org/10.1007/978-3-030-40767-4_22

Appendix A

Measure Theory: A Crash Course

> . . . ils se tenaient aussi loin que
> possible de la troublante réalité
> et ne s'en occupaient pas plus
> que l'algébriste de l'existence
> des quantités qu'il mesure.
> André Gide, L'immoraliste

To successfully use measure theory requires a lot less knowledge than the experts would lead you to believe. In particular, mathematics students are very vulnerable to feeling insecure about this topic. But think of it this way: What do you really need to know about the real numbers $\mathbb{R}$ in order to use them? The rationals $\mathbb{Q}$ form a sub-field, actually the smallest in terms of the partial order given by set inclusion. Then there are zillions of sub-fields $\mathbb{F}$ lying between: $\mathbb{Q} \subset \mathbb{F} \subset \mathbb{R}$. Do you need to know about this in order to do physics? No! Also there are zillions of types of irrational numbers, that is, elements in $\mathbb{R}$ that are not in $\mathbb{Q}$. Again, is this relevant for physics? I think not. Of course, if you want to be an expert in the theory of real numbers, then details like these become important. Much the same applies to measure theory. If you want to be an expert in measure theory, then go for it. But the goal of this Appendix is to present in about 14 pages the most essential details needed for doing physics. Maybe only half of what is here will actually be needed in practice. And from time to time maybe a bit more. My policy is that proofs are totally optional, while definitions are quite helpful.

© Springer Nature Switzerland AG 2020
S. B. Sontz, *An Introductory Path to Quantum Theory*,
https://doi.org/10.1007/978-3-030-40767-4_A

A.1   Measures

A *measure space* is a triple $(\Omega, \mathcal{F}, \mu)$ satisfying various properties. First, Ω is any non-empty set whatsoever. Next $\mathcal{F}$, called a σ-*algebra*, is a collection of subsets of Ω such that $\emptyset, \Omega \in \mathcal{F}$, $\Omega \setminus B \in \mathcal{F}$ for all $B \in \mathcal{F}$, and $\cup_{j \in J} B_j \in \mathcal{F}$ whenever each $B_j \in \mathcal{F}$ and J is a finite or countably infinite set. The sets in $\mathcal{F}$ are called *measurable sets*. Lastly, $\mu : \mathcal{F} \to [0, \infty]$, which is called a *measure*, satisfies $\mu(\emptyset) = 0$ and

$$\mu(\cup_{j \in J} B_j) = \sum_{j \subset J} \mu(B_j) \tag{A.1.1}$$

for every finite or countably infinite family $B_j \in \mathcal{F}$ of *pair-wise disjoint* sets, which means that $B_k \cap B_l = \emptyset$, whenever $k \neq l$. The condition (A.1.1) is called σ-*additivity*. However, in the special case when J is finite, this is called *finite additivity*, which in itself is a weaker property than σ-additivity. If the sum on the right side of (A.1.1) is an infinite sum, it is taken to be equal to ∞ if it diverges. Also, if any term on the right is ∞, then the sum is ∞. Look out! If the right side of (A.1.1) evaluates to ∞, then the left side also must be equal to ∞.

If $A \subset B$ are measurable sets, then just by finite additivity we have that $\mu(A) \leq \mu(B)$, that is to say, μ is *monotone*. A set $B \in \mathcal{F}$ is called a *set of measure zero* if $\mu(B) = 0$.

If $\mu(\Omega) = 1$, then we say that μ is a *probability measure* and that Ω is a *probability measure space*. And that opens the door to the wonderful world of classical probability theory. More generally, if $\mu(\Omega) < \infty$, we say that μ is a *finite measure*. If $\Omega = \cup_{j \in \mathbb{N}} A_j$ with each $A_j \in \mathcal{F}$ and $\mu(A_j) < \infty$ (but the family A_j need not be disjoint), we say that μ is σ-*finite*.

The arithmetic of the set $[0, \infty] := \mathbb{R}^+ \cup \{\infty\}$ should be clarified before continuing with measure theory *per se*. Here $\mathbb{R}^+ := \{r \in \mathbb{R} \, | \, r \geq 0\}$ has the usual arithmetic operations of sum and product as well as linear order. The only novelty concerns the element $\infty \notin \mathbb{R}^+$. The order is extended so that $r < \infty$ for all $r \in \mathbb{R}^+$. As for sums we define $a + \infty = \infty + a = \infty$ for all $a \in [0, \infty]$. The rules for products are $a \cdot \infty = \infty \cdot a = \infty$ for all $a > 0$ and $0 \cdot \infty = \infty \cdot 0 = 0$. The very last rule may appear strange at first sight, but it is standard in mathematical analysis.

There is also a standard topology on $[0, \infty]$. The basic neighborhoods of $a \in \mathbb{R}^+$ are given by $(a - \varepsilon, a + \varepsilon)$ for $a > 0$ and $[0, \varepsilon)$ for all $\varepsilon > 0$, while ∞ has basic neighborhoods given by $(a, \infty]$ where $a \in \mathbb{R}^+$. With this topology every monotone increasing sequence in $[0, \infty]$ converges to a unique limit in $[0, \infty]$ and that limit is ∞ if and only if the sequence is not bounded above by some $a \in \mathbb{R}^+$.

There are many examples of measure spaces. For the most basic example we take $\Omega = \mathbb{R}^n$, $\mathcal{F} = \mathcal{B}(\mathbb{R}^n)$ to be the smallest σ-algebra that contains all the open subsets of $\mathbb{R}^n$, and μ to be the *Lebesgue measure*, denoted as μ_L. This measure is simply the appropriate concept of volume in n-dimensional space. The characteristic property of μ_L is that its value on open rectangular sets is given by

$$\mu_L\big((a_1, b_1) \times (a_2, b_2) \times \cdots \times (a_n, b_n)\big) = (b_1 - a_1)(b_2 - a_2) \cdots (b_n - a_n),$$

where $a_1 < b_1, \ldots, a_n < b_n$ are real numbers. This simply says the volume of an n-dimensional box is the product of the lengths of its sides. The sets in the *Borel algebra* $\mathcal{B}(\mathbb{R}^n)$ are called *Borel sets*. The exact definition of μ_L depends on the author, but always boils down to the same thing. So, don't worry about it. The proof that μ_L is σ-additive is long and tedious. Also, not to be worried about. Two basic facts: μ_L is not a finite measure, but it is σ-finite. You might need to know these two regularity properties of μ_L:

$$\mu_L(B) = \sup\{\mu_L(K) \mid K \subset B \text{ and } K \text{ is compact}\}, \tag{A.1.2}$$
$$\mu_L(B) = \inf\{\mu_L(U) \mid B \subset U \text{ and } U \text{ is open.}\} \tag{A.1.3}$$

The supremum and infimum are taken with respect to the linear order of $[0, \infty]$. We say that (A.1.2) is *interior regularity* and (A.1.3) is *exterior regularity*. Note that every compact set is measurable, so $\mu_L(K)$ is defined. Since there are open rectangular sets with arbitrarily large Lebesgue measure, we see that $\mu_L(\mathbb{R}^n) = \infty$. Also, by exterior regularity the Lebesgue measure of any one-point set is zero. By σ-additivity it follows that the Lebesgue measure of any finite or countably infinite set is zero. (However, there are examples, such as the famous *Cantor set*, which are uncountably infinite and yet have Lebesgue measure zero.) By monotonicity, if $A \subset \mathbb{R}^n$ is a bounded set, then $\mu_L(A) < \infty$. But the converse is false. For example, $(0, 1/2) \cup (1, 1 + 1/4) \cup (2, 2 + 1/8) \cup \cdots$, an open subset of $\mathbb{R}$, has Lebesgue measure $1/2 + 1/4 + 1/8 + \cdots = 1$ by σ-additivity, but is unbounded.

It is not too difficult to see that for any Borel set B and vector $v \in \mathbb{R}^n$ the translated set $B + v := \{b + v \mid b \in B\}$ is Borel and that $\mu_L(B + v) = \mu_L(B)$. We say that μ_L is *translation invariant*. It is much trickier to show that $T(B) := \{T(b) \mid b \in B\}$ is Borel if B is Borel and $T : \mathbb{R}^n \to \mathbb{R}^n$ is linear. In that case, $\mu_L(T(B)) = |\det T| \mu_L(B)$, where det denotes the determinant. In particular, if T is orthogonal, then $\mu_L(T(B)) = \mu_L(B)$. Since rotations R are orthogonal linear maps, μ_L is *rotation invariant*, that is $\mu_L(R(B)) = \mu_L(B)$.

Unlike Lebesgue measure which has these properties of invariance with respect to translations and rotations, other measures μ on $\mathcal{B}(\mathbb{R}^n)$ may have the property that $\mu(U) = 0$ for some non-empty open sets U, but not for others. One defines the *support* of μ, denoted as $\mathrm{supp}\,\mu$, as the set of points $p \in \mathbb{R}^n$ such that every open set U containing p has $\mu(U) > 0$. If μ is σ-finite, then $\mathrm{supp}\,\mu$ is the smallest closed set whose complement has μ measure 0.

The next step in measure theory is to specify the appropriate functions. For a general measure space $(\Omega, \mathcal{F}, \mu)$ we say that $f : \Omega \to \mathbb{C}$ is *measurable* if $f^{-1}(B) := \{\omega \in \Omega \mid f(\omega) \in B\} \in \mathcal{F}$ for every Borel subset B of $\mathbb{C}$. (We use the Borel subsets that come from the identification of $\mathbb{C}$ with $\mathbb{R}^2$.) Sometimes, we want to use $[0, \infty]$ as the co-domain and special consideration is made for the extra element ∞ in this set. Usually, one requires that $f^{-1}(\infty)$ is measurable and that, as before, $f^{-1}(B) \in \mathcal{F}$ for all Borel subsets of $\mathbb{R}$ that are also subsets of $[0, \infty)$. We use the same definition if $\mathbb{C}$ is replaced by $\mathbb{R}$ or $\mathbb{R}^n$. The simplest example is a *characteristic function* $\chi_A : \Omega \to \mathbb{C}$ for $A \subset \Omega$ defined by $\chi_A(\omega) = 1$ if $\omega \in A$ and $\chi_A(\omega) = 0$ otherwise. Then χ_A is measurable if and only if $A \in \mathcal{F}$.

The set of measurable functions is closed under the four basic arithmetical operations. This means that if $f, g : \Omega \to \mathbb{C}$ are measurable, then so are $f + g$, $f - g$, and fg on the domain Ω. Also f/g is measurable on the domain $\Omega \setminus g^{-1}(0)$, which is easily seen to be a measurable set. Measurable function are also closed under countably infinite limiting operations. For example, if $f_n : \Omega \to \mathbb{C}$ is measurable for every integer $n \geq 0$, then $\lim_{n \to \infty} f_n$ is measurable, provided that the limit (defined point-wise) exists. Similarly, if $f_n : \Omega \to \mathbb{R}$ or $f_n : \Omega \to [0, \infty]$ are sequences of measurable functions, then $\sup_n f_n$ and $\inf_n f_n$ are measurable. At some point one learns about two more limiting type operations: $\limsup$ and $\liminf$. These also preserve measurability when applied to sequences of measurable functions. More on this later.

A.2   Integrals

As far as we are concerned, measure theory exists in order to define integrals that are better behaved than the Riemann integral. As we will see, this new theory has nice properties when limits are taken. You could already see this coming in the discussion of measurable functions. To give an idea of what we will be doing, the integrable functions will be a special type of measurable function. We do not integrate any other sort of function, nor do we need to. The standard procedure for defining the integral is to do this for various types of functions. At each step we get a nice theory, but only at the final step do we get the general theory. And we will still have the basic properties of the Riemann integral such as linearity and non-negative functions have non-negative integrals. But we will have more properties, and these are the big theorems of measure theory.

Let $(\Omega, \mathcal{F}, \mu)$ be a measure space. We define a *simple function* $f : \Omega \to \mathbb{R}$ to be a measurable function whose image consists of a finite subset of $\mathbb{R}$, say $\{r_1, \ldots, r_k\}$ with exactly k distinct values, and $A_j := f^{-1}(r_j)$ has finite measure, namely $\mu(A_j) < \infty$ for $j = 1, \ldots, k$. (Note that $k \geq 1$, since Ω is non-empty.) Then we have $f = r_1 \chi_{A_1} + \cdots + r_k \chi_{A_k}$. Then we define its integral by

$$\int_\Omega f = \int_\Omega f(\omega) \, d\mu(\omega) := r_1 \mu(A_1) + \cdots + r_k \mu(A_k).$$

In particular, if $f = \chi_A$ with $\mu(A) < \infty$, then $\int_\Omega f = \mu(A)$. The set of all simple functions is a vector space under the usual, point-wise defined sum of functions and multiplication by scalars, which in this context are real numbers. Then we have the linearity of the integral, namely

$$\int_\Omega (af + bg) = a \int_\Omega f + b \int_\Omega g$$

for f, g simple functions and $a, b \in \mathbb{R}$. This seeming triviality is a bit tricky to prove. Also, the integral is *positivity preserving*, which means that if f is a simple function that satisfies $f \geq 0$ point-wise, then $\int_\Omega f \geq 0$.

The next step is to consider measurable non-negative functions $f \geq 0$. We can even admit functions $f : \Omega \to [0, \infty]$. This excludes simple functions with negative values, but it is nonetheless at step forward. So here is a crucial fact. For every $f \geq 0$ there exists an increasing sequence of simple functions f_n such that $\lim_{n \to \infty} f_n = f$ point-wise. So we have an increasing sequence of real numbers $\int_\Omega f_n$ (previously defined), which either has an upper bound, and hence a limit in $\mathbb{R}$, or does not have an upper bound, and hence no limit. In both cases we define

$$\int_\Omega f := \sup_n \int_\Omega f_n.$$

In the first case this gives a real number, and we say that f is *integrable*. In the second case the integral is equal to ∞, that is to say, the integral has a value. However, in this second case we say that f is *not integrable*. Now a second crucial fact is needed. It turns out that this definition of $\int_\Omega f$ does not depend on the particular choice of the increasing sequence of simple functions that converges to f. Moreover, this new integral is linear *provided* that the scalars are non-negative. It also preserves positivity.

The next step is to consider measurable functions $f : \Omega \to \mathbb{R}$. Then we define $f_+(\omega) := \max(0, f(\omega))$ and $f_-(\omega) := \max(0, -f(\omega))$ for $\omega \in \Omega$. Then we clearly have $f_+ \geq 0$ and $f_- \geq 0$. Also, $f = f_+ - f_-$. And, of course, we just can not get out of the class of measurable functions so easily. In fact, it turns out that both the *positive part* f_+ and and the *negative part* f_- are measurable non-negative functions. Therefore, the integrals $\int_\Omega f_+$ and $\int_\Omega f_-$ have values in $[0, \infty]$. We then define

$$\int_\Omega f := \int_\Omega f_+ - \int_\Omega f_- \tag{A.2.1}$$

exactly when the arithmetic difference on the right is defined. To make sense of this we must define the difference of a pair of elements $a, b \in [0, \infty]$, which will give a result in $[-\infty, \infty]$. If $a, b \in (0, \infty)$, then $a - b \in \mathbb{R}$ is the usual difference of real numbers. Two new cases are $\infty - a := \infty$ and $a - \infty := -\infty$ for $a \neq \infty$. But the only case when this difference is not defined is when $a = b = \infty$, that is $\infty - \infty$ is not defined. This is the standard convention in analysis. So, when both integrals on the right of (A.2.1) have the value ∞, the difference is not defined and hence $\int_\Omega f$ is also not defined. When $\int_\Omega f$ is defined it can have values in general in $[-\infty, \infty]$. If the integral is defined and has its value in $\mathbb{R} = (-\infty, \infty)$, then we say that f is *integrable*. Otherwise we say that f is *not integrable*. So f is integrable if and only if both $\int_\Omega f_+$ and $\int_\Omega f_-$ are real numbers.

Watch out! There are two cases when f is not integrable. In one case the integral has value either $-\infty$ or ∞. In the other case, the integral does not have a numeric value. This integral is linear and positivity preserving when one deals with integrable functions. These properties can be extended to functions whose integral has a value $\pm\infty$, but the reader is advised to proceed with due caution.

Next, we consider a measurable $f : \Omega \to \mathbb{C}$. We write $f = \operatorname{Re}(f) + i \operatorname{Im}(f)$ and define $\int_\Omega f := \int_\Omega \operatorname{Re}(f) + i \int_\Omega \operatorname{Im}(f)$, provided that both of the integrals on the right side are finite real numbers. Consequently, $\int_\Omega f \in \mathbb{C}$.

Since measure theory is just a way to get a decent theory of integration (for us!), we have to remark that these integrals are more stable than the Riemann integral with respect to changes in the function f being integrated, which is called the *integrand*. We first remark that an integrable function f that is non-zero only on a set of measure zero has integral zero: $\int_\Omega f = 0$ if $\Omega \setminus f^{-1}(0)$ has measure zero. Next, we consider integrable functions f and g such that $N := \{x \in \Omega \mid f(x) \neq g(x)\}$ is a set of measure zero. Then, $f - g$ is an integrable function which is non-zero only on N. Therefore, $\int_\Omega (f - g) = 0$, and so $\int_\Omega f = \int_\Omega g$.

Putting this into words, we say that integrable functions which differ only on a set of measure zero have the same integral. Even more colloquially, we say that the values of a function can be changed on any set of measure zero without any change in its integral. This even applies to functions that are not integrable in the sense that a non-integrable function modified on a set of measure zero will again give us a non-integrable function. For example the function $f : (-1, 1) \to \mathbb{R}$ defined by $f(x) = 1/x$ for $x \neq 0$ and $f(0)$ equal to any real number whatsoever (or even ∞) is not integrable with respect to Lebesgue measure. The point here is that the value of f on the singleton set $\{0\}$ of Lebesgue measure zero is irrelevant. The singular behavior of f very near but not equal to 0 is what makes f non-integrable. However, one often says informally (and incorrectly) that it is the singularity *at* 0 that makes f non-integrable.

So far we have considered what are called *positive measures* μ, that is, $\mu(B) \in [0, \infty]$ for all $B \in \mathcal{F}$. But this theory can be done almost identically for *signed measures* for which $\mu(B) \in \mathbb{R}$ or for *complex measures* for which $\mu(B) \in \mathbb{C}$. First, for a signed measure μ by the Jordan Decomposition Theorem there is a canonical way of writing $\mu = \mu_+ - \mu_-$, where both μ_+ and μ_- are positive measures. Next one defines $\int_\Omega f \, d\mu := \int_\Omega f \, d\mu_+ - \int_\Omega f \, d\mu_-$ with the usual caveats for avoiding the case $\infty - \infty$. Second, for complex measures μ one writes $\mu = \mu_r + i\mu_i$, where both μ_r and μ_i are signed measures. Next one defines $\int_\Omega f \, d\mu := \int_\Omega f \, d\mu_r + i \int_\Omega f \, d\mu_i$. The properties of these measures and their integrals then follow immediately. The resulting integral will have values in $\mathbb{R}$ or $\mathbb{C}$, respectively. And it will be linear, but not positivity preserving.

The spectral theorem tells us to consider projection valued measures E, which have values in the vector space $\mathcal{L}(\mathcal{H})$. Here there are no infinite values to worry about, and the value of an integral will be an element in $\mathcal{L}(\mathcal{H})$. This starts by defining $\int_{\Omega} \chi_B \, dE := E(B)$, where χ_B is the characteristic function of the Borel set B. Next, for simple $f = r_1 \chi_{A_1} + \cdots + r_k \chi_{A_k}$, where each χ_j is the characteristic function of $f^{-1}(r_j)$, we define its integral by

$$\int_{\Omega} f(\omega) \, dE(\omega) = \int_{\Omega} f(\omega) \, dE(\omega) := r_1 E(A_1) + \cdots + r_k E(A_k) \in \mathcal{L}(\mathcal{H}).$$

So far, this mimics the theory for positive measures. Again for measurable $f \geq 0$ we use the fact that $f = \lim_{n \to \infty} f_n$ point-wise, where each f_n is simple and $0 \leq f_n \leq f_{n+1}$ in order to define

$$\int_{\Omega} f(\omega) \, dE(\omega) := \lim_{n \to \infty} \int_{\Omega} f_n(\omega) \, dE(\omega). \tag{A.2.2}$$

But there is a technical detail here that is more than a mere technicality! And that detail is which topology on the vector space $\mathcal{L}(\mathcal{H})$ should be used for evaluating the limit in (A.2.2). And there are a plethora of distinct topologies when $\dim \mathcal{L}(\mathcal{H}) = \infty$, although there is only one reasonable topology in case of $\dim \mathcal{L}(\mathcal{H}) < \infty$. It turns out that the 'correct' topology is not that induced by the norm, but rather that for which a sequence T_n of operators converges to an operator T if and only if $\lim_{n \to \infty} \| T_n \psi - T \psi \|_{\mathcal{H}} = 0$ for all $\psi \in \mathcal{H}$. This is called the *strong operator topology*. Its exact, rather technical, definition is best left to functional analysis texts. But the crucial point is that the limit in (A.2.2) does not depend on the sequence f_n we have chosen.

We next define the integral with respect to the pvm E for functions f with values in $\mathbb{R}$ by writing $f = f_+ - f_-$. Finally for functions f with values in $\mathbb{C}$ we write $f = \mathrm{Re}(f) + i \, \mathrm{Im}(f)$. In both cases we proceed analogously to the case when the measure is positive.

Despite the clarity of this step-by-step procedure for defining integrals, we almost never use these definitions for evaluating integrals, but rather for developing tools for doing that. The first such tool is absolute integrability.

A.3 Absolute Integrability

I can not overestimate the importance of the two theorems in this section.

Theorem A.3.1 *The (measurable) function $f : \Omega \to \mathbb{R}$ is integrable if and only if $|f| : \Omega \to [0, \infty)$ is integrable. If either condition is true (and hence both conditions are true), then we say that f is absolutely integrable. Also, in this case we have $| \int_{\Omega} f | \leq \int_{\Omega} |f|$. This is called the* triangle inequality *for the integral.*

Remark: We are taking here the absolute value of real numbers. The import of this result is that the integral of $|f|$ always has a value in $[0, \infty]$ that we can try to calculate. If that value is not ∞, then not only is $|f|$ integrable but f itself is also integrable at which point (and not before) it makes sense to speak of the real number $\int_\Omega f$. In particular, notice that we can apply the triangle inequality only after establishing that f (or equivalently, $|f|$) is itself integrable.

Proof: We start with the simple observation that $|f| = f_+ + f_-$. We first suppose that f is integrable, which implies that both $\int_\Omega f_+$ and $\int_\Omega f_-$ are real numbers. Then by the additivity of the integral for non-negative functions it follows that

$$\int_\Omega |f| = \int_\Omega f_+ + \int_\Omega f_- < \infty, \qquad (A.3.1)$$

which says that $|f|$ is integrable.

Conversely, if $|f|$ is integrable we have that (A.3.1) is true, and so each of the two integrals on the right side is a real number. Consequently, f is integrable.

Next, we suppose that at least one of the functions f and $|f|$ is integrable. (And hence both are integrable.) Then by the triangle inequality for real numbers and positivity preservation we have

$$\left| \int_\Omega f \right| = \left| \int_\Omega f_+ - \int_\Omega f_- \right| \le \left| \int_\Omega f_+ \right| + \left| \int_\Omega f_- \right| = \int_\Omega f_+ + \int_\Omega f_- = \int_\Omega |f|. \quad \blacksquare$$

Next we turn to the case of measurable $f : \Omega \to \mathbb{C}$.

Theorem A.3.2 *The (measurable) function $f : \Omega \to \mathbb{C}$ is integrable if and only if $|f| : \Omega \to [0, \infty)$ is integrable. If either condition is true (and hence both conditions are true), then $|\int_\Omega f| \le \int_\Omega |f|$ which is also called the* triangle inequality *for the integral.*

Remark: In this theorem we are considering the absolute value of complex numbers. As in the previous theorem $\int_\Omega |f|$ has a value in $[0, \infty]$. However, in this case $\int_\Omega f$ is a complex number *provided* that f is integrable. We omit the proof, which may not be overly important. But the reader's understanding of what is being said is critically important.

Most often, we establish the absolute integrability of a function in order to establish the equivalent property of its integrability, since it is easier to work with non-negative functions. At this point enter three theorems that help us do that.

A.4   The Big Three

We are now ready to take on the big theorems of measure theory. Their proofs are not very complicated, but not really relevant to someone who just wants to use them. However, it is vital that such a casual user should understand what they say and so be able to use them correctly.

Theorem A.4.1 *(Monotone Convergence) Let $f_n : \Omega \to \mathbb{R}$ be a monotone increasing sequence of non-negative, integrable functions, i.e., $0 \leq f_n \leq f_{n+1}$ for all integers $n \geq 0$. Then*

$$\int_\Omega \lim_{n\to\infty} f_n = \lim_{n\to\infty} \int_\Omega f_n \in [0, \infty]. \qquad (A.4.1)$$

By positivity preservation we have $0 \leq \int_\Omega f_n \leq \int_\Omega f_{n+1}$, that is, this is an increasing sequence of real numbers, and, as such, always has a limit in $[0, \infty]$. If we define $f := \lim_{n\to\infty} f_n$, then $f : \Omega \to [0, \infty]$ is measurable, but need not be integrable. However, $\int_\Omega f$ has a value in $[0, \infty]$ and (A.4.1) holds as an equality between elements in the set $[0, \infty]$. Of course, f is integrable exactly in the case when $\int_\Omega f = \lim_{n\to\infty} \int_\Omega f_n$ has a value in $[0, \infty)$.

Theorem A.4.2 *(Dominated Convergence) Let $f_n : \Omega \to \mathbb{C}$ be a sequence of measurable functions for $n \geq 0$ for which the point-wise limiting function $f := \lim_{n\to\infty} f_n$ exists. Suppose that there exists some integrable function $g : \Omega \to [0, \infty]$ such that $|f_n| \leq g$ for all $n \geq 0$. Then each f_n is integrable and $f = \lim_{n\to\infty} f_n$ is integrable. Moreover,*

$$\int_\Omega \lim_{n\to\infty} f_n = \lim_{n\to\infty} \int_\Omega f_n \in \mathbb{C}. \qquad (A.4.2)$$

Remark: The non-negative function g is called the *dominating function*. It is assumed to be *integrable*, since counter-examples show that otherwise the conclusion of this theorem can be false. The hypothesis that g is integrable implies that the (measurable) set $g^{-1}(\infty)$ has measures zero, that is g assumes the value ∞ only on a set of measure zero. Therefore $f_n^{-1}(\infty)$ also has measure zero for every n. As noted earlier the values of a function on a set of measure zero in its domain have no effect on the value of its integral.

Theorem A.4.3 *(Fatou's Lemma) Let $f_n : \Omega \to [0, \infty]$ be a sequence of non-negative, measurable functions for $n \geq 0$. Then*

$$\int_\Omega \liminf_n f_n \leq \liminf_n \int_\Omega f_n \in [0, \infty]. \qquad (A.4.3)$$

The three previous theorems allow the interchange of a limit with an integral. Fatou's Lemma has the most general hypothesis, but its conclusion is the weakest, namely an inequality, while the first two theorems give us an equality. The downside of Fatou's Lemma is that you have to know what $\liminf$ means. The upside of Fatou's Lemma is that the $\liminf$ exists for *any* sequence of elements in $[0, \infty]$. The idea is that one can define $\liminf$ of a given sequence in $[0, \infty]$ by looking at all of its sub-sequences which *do* have a limit in $[0, \infty]$ and then taking the smallest of all of those limits to be the $\liminf$ of the given sequence. That's quite a mouthful. Just as a quick example, $\liminf (-1)^n = -1$ even though $\lim (-1)^n$ does not exist. Also, if the given sequence does have a limit, then its $\liminf$ is precisely that limit.

So, $\liminf$ is a generalization of limit. If the $\liminf$'s on both sides of (A.4.3) are actually $\lim$'s in an application, then this theorem is easier to use.

The standard definition, which does realize the idea given above, is

$$\liminf_n a_n := \sup_{l \geq 0} \left(\inf_{n \geq l} a_n \right),$$

where $a_n \in [0, \infty]$ is a sequence for $n \geq 0$. This definition is a veritable bag of worms and takes a lot of time to get a hold on. In particular, you have to come to terms with the *infimum* (inf) and *supremum* (sup) of a sequence of real numbers. But sometimes Fatou's Lemma is the only one of these three big theorems that gets the job done. But if you are just starting out learning this material, then you can ignore it for a while, though eventually at your own risk.

Exercise A.4.1 *Suppose that* $a = \lim_{n \to \infty} a_n$ *exists, where* $a_n \in [0, \infty]$ *for each integer* $n \geq 0$ *and* $a \in [0, \infty]$. *Prove that* $\liminf_n a_n = a$.

A.5   Counting Measure and Infinite Series

Besides Lebesgue measure, another quite different and useful measure is *counting measure* μ_C. For this we take Ω to be any non-empty set and $\mathcal{F}$ to be the σ-algebra of all subsets of Ω. We then define $\mu_C : \mathcal{F} \to [0, \infty]$ by $\mu_C(A) := \operatorname{card}(A)$, the number of elements in A, if $A \subset \Omega$ is a finite subset. Otherwise, we define $\mu_C(A) := \infty$. It is a bit of a chore to prove this is σ-additive. But it is. In this case every function $f : \Omega \to \mathbb{C}$ is measurable. Counting measure μ_C on Ω is σ-finite if and only if the cardinality of Ω is finite or countably infinite.

The most famous example of counting measure is when we take $\Omega = \mathbb{N}$, the set of non-negative integers $n \geq 0$. Then a function $f : \mathbb{N} \to \mathbb{C}$ is usually not written using the notation $f(n)$ but rather f_n or even more commonly as a_n in order to indicate that what we are talking about is a sequence of complex numbers. What is $\int_{\mathbb{N}} a_n \, d\mu_C(n)$? And what does it mean to say that a_n is integrable? To address these questions we start with the characteristic function χ_j of the one-point subset $\{j\} \subset \mathbb{N}$ for some $j \geq 0$. By definition $\mu_C(\{j\}) = 1$ and so $\int_{\mathbb{N}} \chi_j \, d\mu_C = 1$, thinking of χ_j as the sequence that has all elements equal to 0, except the jth element which is 1.

Next, suppose that a is a simple function, that is, a sequence which has only finitely many non-zero values a_j which occur only for j in a subset of finite μ_C measure. So a has non-zero values only on some finite subset of $\mathbb{N}$, which we can take without loss of generality to be $I_n := \{j \mid 0 \leq j \leq n\}$ for some integer $n \geq 0$. Then, $a = \sum_{j=0}^{n} a_j \chi_j$ (think about it!) and so by linearity of the integral of simple functions we get

$$\int_{\mathbb{N}} a \, d\mu_C = \int_{\mathbb{N}} \sum_{j=0}^{n} a_j \chi_j \, d\mu_C = \sum_{j=0}^{n} a_j \int_{\mathbb{N}} \chi_j \, d\mu_C = \sum_{j=0}^{n} a_j = \sum_{j=0}^{\infty} a_j,$$

which is the sum of the non-zero values of a. As we already knew from the general theory, all simple functions are integrable.

We move on to the next step in the definition of integral by considering a function $a : \mathbb{N} \to [0, \infty)$. Since a having the value ∞ at some point implies that a has the value ∞ on a set of positive measure, we exclude ∞ from the co-domain here. Now we are going to construct a sequence of non-negative, increasing, simple functions s_n $(0 \le s_n \le s_{n+1})$ and with $\lim_n s_n = a$. Here the limit is taken point-wise, namely $\lim_n s_n(k) = a(k)$ for every $k \ge 0$. Any such sequence will do, but thinking a bit about the previous paragraph, we are motivated to define the *partial sums* $s_n := \sum_{j=0}^{n} a_j \chi_j$. And this does satisfy the required conditions as the reader can check. So, by definition of the integral in this case and using the result from the previous paragraph, we obtain

$$\int_{\mathbb{N}} a \, d\mu_C = \lim_{n \to \infty} s_n = \lim_{n \to \infty} \sum_{j=0}^{n} a_j \in [0, \infty].$$

Moreover, a is integrable if and only if this limit exists in $[0, \infty)$. But this is the standard definition for convergence of an infinite series. Consequently, a non-negative function $a : \mathbb{N} \to [0, \infty)$ is integrable if and only if $\sum_{j=0}^{\infty} a_j$, the corresponding infinite series, is convergent. And if either of these conditions holds (and hence both hold), then the integral is equal to the infinite series.

The next case is when $a : \mathbb{N} \to \mathbb{R}$, that is, we have a sequence a_j of real numbers. Using the general theory we have that a is integrable if and only if $|a|$ is integrable. And by the previous result $|a|$ is integrable if and only if $\sum_{j=0}^{\infty} |a_j|$ is convergent, that is, if and only if the series $\sum_{j=0}^{\infty} a_j$ is absolutely convergent. We get the same result when $a : \mathbb{N} \to \mathbb{C}$, namely that a is integrable if and only if the series $\sum_{j=0}^{\infty} a_j$ is absolutely convergent. And this is the general case which covers all the prior special cases. In short, with respect to counting measure on $\mathbb{N}$ we have that a function a is integrable if and only if the corresponding series $\sum_{j=0}^{\infty} a_j$ is absolutely convergent.

However, in the theory of infinite series there are also some series which are only *conditionally convergent*, that is, the series is convergent but not absolutely convergent. For a conditionally convergent series the limit of the partial sums, $\lim_{n \to \infty} s_n$ exists, but for real-valued a neither a_+ nor a_- is integrable. These conditionally convergent infinite series are analogous to *improper integrals*, which is a topic we do not wish to discuss.

A.6 Fubini's Theorem

The final big theorem is about two σ-finite measure spaces $(\Omega_1, \mathcal{F}_1, \mu_1)$ and $(\Omega_2, \mathcal{F}_2, \mu_2)$. Then we can construct on the Cartesian product $\Omega = \Omega_1 \times \Omega_2$ a σ-algebra $\mathcal{F}$ and a measure μ. First, $\mathcal{F}$ is defined to be the smallest σ-algebra containing all of the 'rectangular' sets $B_1 \times B_2$, where $B_k \in \mathcal{F}_k$ for $k = 1, 2$. (The σ-algebra $\mathcal{F}$ in general contains an enormous variety of sets that are not rectangular. An example is the Borel σ-algebra on $\mathbb{R}^2 = \mathbb{R} \times \mathbb{R}$, which is the smallest σ-algebra containing the product sets $B_1 \times B_2$, where B_1, B_2 are Borel subsets of $\mathbb{R}$.) Then μ is the (unique!) measure on $\mathcal{F}$ with the property that the measure of a rectangular set is as expected:

$$\mu(B_1 \times B_2) = \mu_1(B_1)\mu_2(B_2). \tag{A.6.1}$$

We skip over the numerous, boring pages needed to justify all this rigorously. However, we do pause to note that the product on the right side of (A.6.1) is that of elements in $[0, \infty]$ as previously defined. So $0 \cdot \infty = \infty \cdot 0 = 0$. And now we cut to the chase.

Theorem A.6.1 (Fubini's Theorem) *Let $(\Omega_1, \mathcal{F}_1, \mu_2)$ and $(\Omega_2, \mathcal{F}_2, \mu_1)$ be σ-finite measure spaces. Let $f : \Omega_1 \times \Omega_2 \to \mathbb{C}$ be measurable with respect to the σ-algebra $\mathcal{F}$. Then f is integrable with respect to μ if and only if at least one of these iterated integrals is finite:*

$$\int_{\Omega_2} \Big[\int_{\Omega_1} f(\omega_1, \omega_2) \, d\mu_1(\omega_1) \Big] d\mu_2(\omega_2), \ \int_{\Omega_1} \Big[\int_{\Omega_2} f(\omega_1, \omega_2) \, d\mu_2(\omega_2) \Big] d\mu_1(\omega_1).$$
$$\tag{A.6.2}$$

In that case both of these iterated integrals are equal to the integral

$$\int_{\Omega_1 \times \Omega_2} f(\omega_1, \omega_2) \, d\mu(\omega_1, \omega_2).$$

Let $g : \Omega_1 \times \Omega_2 \to [0, \infty]$ be measurable with respect to $\mathcal{F}$. Then

$$\int_{\Omega_1 \times \Omega_2} g(\omega_1, \omega_2) \, d\mu(\omega_1, \omega_2) = \int_{\Omega_2} \Big[\int_{\Omega_1} g(\omega_1, \omega_2) \, d\mu_1(\omega_1) \Big] d\mu_2(\omega_2) \tag{A.6.3}$$

in the sense that one side is ∞ if and only if the other side is ∞.

Here is a standard way of applying Fubini's theorem. We start off with a measurable function $f : \Omega_1 \times \Omega_2 \to \mathbb{C}$. We want to know if it is integrable. So we first consider $g = |f| : \Omega_1 \times \Omega_2 \to [0, \infty)$. We use the second part of the theorem for the function g by computing or *estimating* the iterated integral on the right of (A.6.3) (or the other iterated integral, using the symmetry in the first part). To estimate a quantity means to find another quantity (preferably finite) that is bigger or equal to the quantity of interest. This always gives a value in $[0, \infty]$ for the estimate of the iterated integral of g. If that value actually lies in $[0, \infty)$, then we know that the integral for g on the left is finite, and it satisfies the same estimate. This says that $g = |f|$ is integrable with respect to μ, which immediately implies that f is also integrable with respect to μ. Going back to the first part of Fubini's theorem, we see that the integral of f can be calculated or estimated by using either of the (equal) iterated integrals of f in (A.6.2).

A.7 $L^1(\Omega)$

Let $(\Omega, \mathcal{F}, \mu)$ be a measure space. We define the associated space of integrable functions by $\mathcal{L}^1(\Omega, \mathcal{F}, \mu) := \{f : \Omega \to \mathbb{C} \,|\, f \text{ is integrable}\}$. One often denotes this space as $\mathcal{L}^1(\Omega, \mu)$ or even as $\mathcal{L}^1(\Omega)$. It is a vector space over $\mathbb{C}$ and has a semi-norm defined by

$$||f|| := \int_\Omega |f| \, d\mu \in \mathbb{R}^+.$$

In general, a *semi-norm* on a complex vector space V is a function $V \to \mathbb{R}^+$, denoted as $v \mapsto ||v||$, that satisfies these properties:

- (Triangle Inequality) $||v + w|| \le ||v|| + ||w||$ for all $v, w \in V$.

- (Homogeneity) $||\lambda v|| = |\lambda| \, ||v||$ for all $\lambda \in \mathbb{C}$ and all $v \in V$.

It follows from the second property by taking $\lambda = 0$ that $||0|| = 0$. If we have a semi-norm that also satisfies

- $||v|| = 0$ implies that $v = 0$,

then we say that $|| \cdot ||$ is a *norm*. In general, there are non-empty measure zero subsets of Ω and so there are non-zero functions $f : \Omega \to \mathbb{C}$ satisfying $||f|| = 0$. (Take N as such a non-empty measure zero subset and $f = \chi_N$.)

The point is that one needs a norm $|| \ ||$ in order to have the associated function, defined by $d(v, w) := ||v - w||$, be the *distance function* of a *metric space*. The defining properties of a distance function $d : X \times X \to \mathbb{R}^+$, where X is any set whatsoever, are

- (Positivity) $d(x, y) \ge 0$ for all $x, y \in X$.

- (Symmetry) $d(x, y) = d(y, x)$ for all $x, y \in X$.

- (Triangle Inequality) $d(x, z) \le d(x, y) + d(y, z)$ for all $x, y, z \in X$.

- $d(x, y) = 0$ if and only if $x = y$.

One says that the pair (X, d) is a *metric space*. When it is obvious what the metric being used is, we simply say that X is a metric space.

In order to get a norm, and hence a metric, we introduce the following equivalence relation $\cong$ on $\mathcal{L}^1$:

$$f \cong g \text{ if and only if } f = g \text{ almost everywhere.}$$

We say that a property holds *almost everywhere* if there exists a measure zero subset N of Ω such that the property holds on $\Omega \setminus N$. We then define $L^1(\Omega, \mu)$ to be the set of equivalence classes of $\mathcal{L}^1(\Omega, \mu)$ under this equivalence relation. So the elements of $L^1(\Omega, \mu)$ are equivalence classes of functions that differ only on measure zero sets. For $f \in \mathcal{L}^1(\Omega, \mu)$ we denote its equivalence class by $[f] \in L^1(\Omega, \mu)$. It turns out that one can add these equivalence classes and multiply them by complex numbers as follows:

$$[f] + [g] := [f + g] \quad \text{and} \quad \lambda[f] := [\lambda f].$$

And in this way $L^1(\Omega, \mu)$ becomes a complex vector space. Also, we define $\|[f]\| := \|f\|$, where the right side uses the semi-norm of $\mathcal{L}^1(\Omega, \mu)$. Then this is a norm on $L^1(\Omega, \mu)$. So, $L^1(\Omega, \mu)$ becomes a metric space with the associated distance function given by $d([f], [g]) = \|[f] - [g]\|$.

Now metric spaces have a convenient necessary property for a sequence to be convergent. But, inevitably, even more definitions!

Definition A.7.1 *Let $x_n \in X$ for $n \in \mathbb{N}$ be a sequence in a metric space (X, d). Then we say that this sequence is* convergent *with* limit $x \in X$ *if for every $\varepsilon > 0$ there exists $N \geq 0$ such that $d(x_n, x) < \varepsilon$ for all $n \geq N$.*

We say that $x_n \in X$ is a Cauchy *sequence if for every $\varepsilon > 0$ there exists $N \geq 0$ such that $d(x_n, x_m) < \varepsilon$ for all $n, m \geq N$.*

Every convergent sequence in a metric space is a Cauchy sequence, but not always conversely. This leads to the next definition.

Definition A.7.2 *Let X be a metric space. Then we say that X is* complete *if every Cauchy sequence in X is convergent.*

The standard example is $\mathbb{R}$ with the distance $d(r, s) := |r - s|$ for $r, s \in \mathbb{R}$. This is a complete metric space. On the other hand $\mathbb{Q}$, the set of rational numbers with the distance defined by the same formula (but now only for $r, s \in \mathbb{Q}$) is not a complete metric space. The big theorem here is the following.

Theorem A.7.1 *(Riesz-Fischer) $L^1(\Omega, \mu)$ is a complete metric space.*

A variant of the discussion in this section, including the technicality of the equivalence relation for functions equal almost everywhere, works just fine for $L^2(\Omega, \mu)$. So another version of this big theorem is as follows.

Theorem A.7.2 *(Riesz-Fischer) $L^2(\Omega, \mu)$ is a complete metric space.*

A.8 Concluding Remarks

Before ending this crash course, let us emphasize again the importance of neglecting what happens with functions on sets of measure zero, since this plays a large role in applications. This allows us to change as we wish any measurable function on any measure zero set without changing the results. We remark that it is also used for *properties*. If $P(x)$ is a proposition about the points $x \in \Omega$, a measure space, then we say that $P(x)$ is true *almost everywhere* if there is a set N of measure zero such that $P(\omega)$ is true for all $\omega \in \Omega \setminus N$. In the case of probability measure spaces one says that the property holds *almost surely* or with *probability* 1.

There is a lot more to measure theory, but this should be more than enough to keep you afloat for a while, or maybe forever.

A.9 Notes

J.E. Littlewood's famous three principles of real analysis, that were intended to show the accessibility of measure theory, are at an even more advanced level than these brief notes. They involve results such as Lusin's theorem and Egorov's theorem. Other advanced topics include Vitali's Lemma, the Radon-Nikodym theorem, the Riesz representation theorem (for measures) and all L^p analysis, including Hölder's inequality and interpolation theory. The reader might wish to study some of these topics if the need arises.

Improper integrals are not what I would call an advanced topic, and they do come up in physics. For example, the *Cauchy principle value* is an improper integral. I did not include them for the sake of brevity. I hope that they are accessible to the reader with the tools presented here.

The actual evaluation of definite integrals is an art as well as a science unto itself that relies on the Fundamental Theorem of Calculus, the theory of residues in complex analysis, and a whole lot of ingenuity. And it's a lot of fun! For example, see [7], [14], [19], and [23].

Bibliography

[1] Ali, S. T., Antoine, J.-P., & Gazeau, J.-P. (2013). *Coherent states, wavelets and their generalizations* (2nd ed.). Berlin: Springer.

[2] Araki, H., & Ezawa, H. (2004). *Topics in the theory of Schrödinger operators.* Singapore: World Scientific.

[3] Ashcroft, N. W., & Mermin, N. D. (1976). *Solid state physics.* San Diego: Harcourt College Publishers.

[4] Bengtsson, I., & Życzkowski, K. (2006). *The geometry of quantum states.* Cambridge: Cambridge University Press.

[5] Berezin, F. A., & Shubin, M. A. (1991). *The Schrödinger equation, mathematics and its applications (Soviet series)* (Vol. 66). Dordrecht: Kluwer Academic Publishers.

[6] Birman, M. S., & Solomjak, M. Z. (1987). *Spectral theory of self-adjoint operators in Hilbert space.* Dordrecht: D. Reidel.

[7] Boros, G., & Moll, V. (2004). *Irresistible integrals.* Cambridge: Cambridge University Press.

[8] Busch, P., et al. (2016). *Quantum measurement.* Berlin: Springer.

[9] Courant, R., & Hilbert, D. (1953). *Methods of mathematical physics* (Vol. 1). Woburn: Interscience.

[10] Cycon, H. L., et al. (1987). *Schrödinger operators.* Berlin: Springer.

[11] Dell'Antonio, G. (2015). *Lectures on the mathematics of quantum mechanics I.* Paris: Atlantis Press.

[12] Feynman, R. P., Leighton, R. B., & Sands, M. (1963). *The Feynman lectures on physics* (Vol. 1). Boston: Addison-Wesley.

[13] Feynman, R. P., & Hibbs, A. R. (1965). *Quantum mechanics and path integrals.* New York: McGraw-Hill.

© Springer Nature Switzerland AG 2020

S. B. Sontz, *An Introductory Path to Quantum Theory,*
https://doi.org/10.1007/978-3-030-40767-4_A

[14] Gogolin, A. (2014). *Lectures on complex integration.* Berlin: Springer.

[15] Hislop, P. D., & Sigal, I. M. (1996). *Introduction to spectral theory with applications to Schrödinger operators.* Berlin: Springer.

[16] Isham, C. J. (1995). *Lectures on quantum theory.* London: Imperial College Press.

[17] Kato, T. (1966). *Perturbation theory for linear operators.* Berlin: Springer.

[18] Kolmogorov, A. N. (1933). *Grundbegriffe der Wahrscheinlichkeits-rechnung.* Berlin: Springer. English translation: *Foundations of the theory of probability* (2nd ed.). White River Junction: Chelsea Publishing Co. 1956.

[19] Krylov, V. I. (1962). *Approximate calculations of integrals.* New York: Macmillan. (Reprint edition, New York: Dover (2006)).

[20] Manoukian, E. B. (2006). *Quantum theory, a wide spectrum.* Berlin: Springer.

[21] Mermin, N. D. (2007). *Quantum computer science.* Cambridge: Cambridge University Press.

[22] Miller, W, Jr. (1972). *Symmetry groups and their applications.* Cambridge: Academic.

[23] Nahin, P. J. (2015). *Inside interesting integrals.* Berlin. Springer.

[24] Peres, A. (1993). *Quantum theory: Concepts and methods.* Dordrecht: Kluwer Academic Publishers.

[25] Prugovecki, E. (1981). *Quantum mechanics in Hilbert space* (2nd ed.). Cambridge: Academic.

[26] Schmüdgen, K. (2012). *Unbounded self-adjoint operators on Hilbert space.* Berlin: Springer.

[27] Schrödinger, E. (1926). Quantisierung als Eigenwertproblem (Erste Mitteilung). *Annals of Physics, 79,* 361–376.

[28] Sinha, K. B., Srivastava, S. (2017). *Theory of semigroups and applications.* Berlin: Hindustan Book Agency and Springer.

[29] Sternberg, S. (1994). *Group theory and physics.* Cambridge: Cambridge University Press.

[30] Teschl, G. (2014). *Mathematical methods in quantum mechanics, with applications to Schrödinger operators* (2nd ed.). Providence: American Mathematical Society.

[31] Varadarajan, V. S. (1984). *Lie groups, lie algebras, and their representations*. Graduate texts in mathematics (Vol. 102). Berlin: Springer.

[32] von Neumann, J. (1932). *Mathematische Grundlagen der Quantenmechanik*. Berlin: Springer; English translation: *Mathematical foundations of quantum mechanics*. Princeton: Princeton University Press (1955).

[33] von Neumann, J., & Wigner, E. P. (1929). Über merkwürdige diskrete eigenwerte. *Zeitschrift für Physik, 30*, 465–467.

[34] Weidmann, J. (1980). *Linear operators in Hilbert space*. Berlin: Springer.

[35] Weyl, H. (1928). *Gruppentheorie und Quantenmechanik*. Berlin: Springer; English translation: *Theory of groups and quantum mechanics*. London: Methuen and Co. (1931). Reprint: New York: Dover (1950).

[36] Wightman, A. S. (1976). Hilbert's sixth problem: Mathematical treatment of the axioms of physics. In: Proceedings of Symposia in Pure Mathematics. Part I, Mathematical developments arising from Hilbert problems (Vol. XXVIII, pp. 147–240). Providence: American Mathematical Society.

[37] Wigner, E. P. (1939). On unitary representations of the inhomogeneous Lorentz group. *Annals of Mathematics (2), 40*(1), 149–204.

[38] Woit, P. (2017). *Quantum theory, groups and representations*. Berlin: Springer.

Index

© Springer Nature Switzerland AG 2020
S. B. Sontz, *An Introductory Path to Quantum Theory*,
https://doi.org/10.1007/978-3-030-40767-4

Printed in the United States
by Baker & Taylor Publisher Services